THE BRITISH GENTRY, THE SOUTHERN PLANTER, AND THE NORTHERN FAMILY FARMER

THE BRITISH GENTRY THE SOUTHERN PLANTER AND THE NORTHERN FAMILY FARMER

AGRICULTURE AND SECTIONAL ANTAGONISM IN NORTH AMERICA

JAMES L. HUSTON

Louisiana State University Press

Baton Rouge

Publication of this book is made possible in part by support from the Department of History at Oklahoma State University.

Published by Louisiana State University Press

Manufactured in the United States of America
First printing

DESIGNER: Michelle A. Neustrom
TYPEFACE: Cassia
PRINTER AND BINDER: Maple Press

Library of Congress Cataloging-in-Publication Data are available at the Library of Congress.

ISBN 978-0-8071-5918-7 (cloth: alk. paper) — ISBN 978-0-8071-5919-4 (pdf) — ISBN 978-0-8071-5920-0 (epub) — ISBN 978-0-8071-5921-7 (mobi)

The paper in this book meets the guidelines for permanence and durability of the Committee on Production Guidelines for Book Longevity of the Council on Library Resources. ♾

This book is dedicated to my wife, Kathy, and her love of cats

CONTENTS

PART III

TABLES

PREFACE

Most studies of the southern plantation have dwelt almost exclusively on slavery, slave culture, and interactions between plantation masters, overseers, and the enslaved. But when one places northern and southern agriculture in a comparative framework, the size of farm is just as illustrative of dramatic contrasts as slave ownership. Thus northern and southern agricultural evolution can be viewed at the forefront of the sectional controversy. Material and ideological friction was inherent between the family farm and the plantation. To make the differences in northern and southern agriculture comprehensible, and to see the pattern over time, I will depict the evolution of English farming over four centuries. English agriculture can serve as a standard to judge the development of American agriculture. This comparison allows far more emphasis on landownership than have most American historians.

The sectional controversy was fueled by the material incompatibility between the family farm of the North and the great plantation of the South. Northern owner-occupiers of farms—the term now used to describe farmers who owned and worked their own farms—wanted land to be carved into chunks of 80 to 120 acres for future family members; the plantation owners wanted land available to them in thousands of acres. The two agricultural regimes thus clashed over the distribution of land; it was a battle over a natural resource. Slavery entered this struggle by being the backbone of the plantation system. For the plantation to reach its immense size, the planter, like the English gentry, had to have access to cheap labor. The British obtained that labor from a surplus population; the southern planter got it by inventing African slavery. But without cheap labor, both the large British estate and the great southern plantation could not sustain their size.

The question of which agricultural regime was to determine land distribution in the territories by itself generated the realignment of the 1850s, or at the least was the predominant factor. The Kansas-Nebraska Act played the key role. When the threat appeared that the plantation system might invade the land claimed by the family farm system, the family farm system retaliated politically by creating an anti-extension-of-slavery party to stop it. I demonstrate this action by dividing the northern congressional districts into agricultural districts, industrial districts, and those districts balanced between the two, to show that the surge in Republican power was distinctly rural and agricultural, not urban and industrial. When southern leaders, not necessarily the planters only, realized that future expansion was to be dedicated to family farmers and not to plantation slaveholders, they departed the Union and created an independent nation.

Three themes supplement this general argument. The first is the rise and application of what has become known as the "free labor ideology." As of late, this ideology, composed of the ideas of working for a wage, saving, then obtaining property and eventually rising to an independent livelihood, has been interpreted as a contemporary rhetorical device to justify factory wage labor (industrial capitalism) by making wage labor more ethically acceptable than slave labor.[1] That interpretation is seriously misguided. The free labor ideology, which actually started in England during the English Civil War and the rise of Puritanism, was in the United States an agricultural ideology that accurately described how sons of farmers eventually obtained farms of their own. Moreover, writers specifically mentioned that rising by wage labor to independent ownership of property occurred in rural areas, not urban ones. Most tellingly, I apply the free labor ideology of Americans to the system that made the most sense: to British agricultural labor, not to factory labor. The elements of the free labor ideology come out with striking force and emphatically did not justify subsistence wages but distinctly posited a wage rate that made saving possible and opened a route to property ownership. Even in the 1850s many free labor proponents recognized that, while the set of ideas worked well in the northern countryside, it operated strangely in the urban, industrial environment; in fact, free labor advocates had to make revisions about the idea of social mobility to fit the urban landscape.

The second theme is something of a reversal of a longstanding viewpoint about the North and South: to wit, the patriarchal South versus the capitalist North. For historians of the early American republic, the march of the northern states to a capitalist marketplace has been the dominant theme of research for perhaps three decades.[2] However, the depiction is correct only for northern cities, not the northern rural countryside. And the northern urban areas were not at odds with the plantation. They were not clashing over any specific resource or economic endeavor. So long as the plantation remained a hinterland and as long as the plantation complex required financial and transportation services from the core (that is, northern urban areas), there was no necessary material conflict between the urban North and the plantation South. If the plantation states had developed their own service and manufacturing sectors, then economic collisions would have surfaced.

The grand interpretive scheme of the industrial North versus the rural South took its modern form in the works of the late Marxist historian Eugene Genovese.[3] Genovese found the structural difference between the North and South in the relationship of the laborer to the owner of property. In the plantation, ownership of another person led to the fermentation of the ideal of patriarchy; people were still exploited, and a few lived in luxury off the labors of others, but, because slaves were the property of the master, the relationship inculcated a sense of obligation to take care of the owned people. In the North, free labor emerged as the key institution in which the owners of capital acknowledged no obligation whatsoever to their employees. The owners of capital were thus free to exploit their workers to the fullest extent possible; the workers got the lowest wage that human existence allowed (subsistence wages); and the workers' fate was left to themselves—that constituted the reality of freedom for the worker under capitalism. In the Genovese presentation, the owners of capital were aggressive, growing, and disgruntled with the lackadaisical South. Here was the source of North-South tensions; it was all summarized in the laborers' relationship to property.

In my search through the account books, diaries, journals, and statistics of northern farms and southern plantations, I have concluded that for antebellum America the reverse of the Genovese proposition was the reality. The definition used by Genovese (and now a host of others) to describe capitalism—the set of relations that came out of the laborers' position—was inadequate. If capitalism were looked upon as a system of continual accumulation by individuals, of reinvestment to obtain more property and income even when the mountain of ownership served no practical purpose of consumption, then one encountered the standard stereotype of a capitalist. This is not to say that the labor definition should be tossed aside but simply that it is not comprehensive. When one uses this definition of constant and eternal accumulation without end, one surmises that the northern family farmer failed the test of capitalism. The northern family farmer, from the Revolution to World War II, sought competency and independence, not accumulation. This facet of northern agriculturists resides in the statistics of the northern farm in the states, except California and Oregon, that formed the Union in 1860.[4] Northern farmers used the market to achieve nonmarket goals—to obtain the income necessary to acquire competence—but they did not extend themselves much beyond that. This conclusion goes against the grain of a considerable body of research.[5]

And if one applied the definition of constant reinvestment for continued accumulation beyond rational necessity, one ends up seeing the plantation as a capitalist enterprise, regardless of patriarchy to the slaves (whether it was actually there or not remains a bone of contention). Planters continually sought to add more land and slaves to their enterprise, to grow larger and larger. Often historians have attributed to the great planters the aristocratic attitude of noblesse oblige to their charges, comparable to the ideals upheld by the English aristocracy and gentry. However, it is a standard conclusion now that the English aristocracy and gentry by 1600 were more capitalist than medieval in their outlook.

I propose that the core ideological difference between North and South was neither the free labor ideology nor the patriarchy of the plantation. The clash was between the self-mastery of the family farm and mastery-over-others of the plantation. What drove northerners to fear the plantation was that planters were accustomed to giving orders to others and having them obeyed. Northern farmers were most concerned with mastering themselves and the chores of the farm rather than controlling other people. Out of the plantation came a celebration of hierarchy, inequality, and a fear of the masses; out of the family farm came an individualism that lauded social mobility, equality, and democracy. The ideological contest between North and South was not capitalism versus premodern patriarchy; it was over the intellectual and psychological derivations that spun off from the structures of the family farm and the great plantation.

The third argument is that agriculture was the chief characteristic of the North. Time and again, the phrase "industrial North versus agrarian South" is trotted out. I find the phrase guilty on two counts. First, the states that composed the North in 1860 (excepting California and Oregon) were overwhelmingly rural. A simple accounting shows that some 72 percent of northerners lived in rural areas. Only Rhode Island and Massachusetts were less than 50 percent rural and had moved to a modern urban-industrial condition. Even New York and Pennsylvania, the economic powerhouses of future industrialism, were two-thirds rural in 1860. Slightly less than one-half of the northern male working population in 1860 was directly engaged in agriculture, and if one included the auxiliary occupations that farmers required (mechanics, teamsters, merchants), probably over 70 percent of the population was directly tied to agricultural activity. This agrarian aspect of the North was clearly visible in congressional representation; over 70 percent of northern congressional districts had four to five times as many farms in their districts as manufacturing employees. The dominant aspect of the North in 1860 was agricultural, not industrial.[6]

The second error of the generalization of the "industrial North versus agrarian South" is that it misrepresents the South. The South's major characteristic was not agrarian, it was the plantation. The plantation dominated the South because it dominated landownership; it engrossed so much land that the "yeomen" of the South were being pushed out. Thus the sectional conflict was not between northern industry and southern agrarianism, but between two agricultural regimes; it was the small-farm North versus the plantation South. By redefining the sectional collision in this fashion, two features immediately stand out. First, industrialization was irrelevant to the sectional conflict. Second, the American Civil War did not anticipate the future modern America but was mired in an agrarian past.

The amount of secondary literature on these topics is so immense that using the typical full citation in the notes per chapter produced an ungainly mess. Therefore, I used short-title forms in the notes for secondary works, leaving the full citation only for the bibliography. For primary material, I included date of publication as that has significance for the topic under consideration. It must as well be admitted that space limitations played a major role in which historians I have cited and which ones I had to eliminate. To those I have omitted, I apologize and offer as an excuse the emphasis on brevity that academic publication now prizes.

In the completion of this work I have debts to acknowledge. Most belong to the kindness, efficiency, and courtesy of the historical societies I visited: the Bentley Library at the University of Michigan, the Wisconsin Historical Society, the University of Illinois, the Abraham Lincoln Presidential Library, the Indiana Historical Society, the Indiana State Library, the Ohio Historical Society, the Minnesota Historical Society, the Western Reserve Historical Society, the University of Virginia, the Historical Society of Virginia, and, online, the University of North Carolina. To a large degree I give my gratitude to Oklahoma State University, its College of Arts and Sciences, and its Department of History for granting me the funds to do this research and the time to write the manuscript.

In putting together the interpretations of this book, I have had the assistance of talented, perspicacious, and critically adept friends. My greatest debt is to Paul Paskoff, who read this book in its long version, offered great encouragement, and by his interlineations reminded me that much of my deathless prose belonged in a coffin, not on a printed page. Likewise, Frank Towers assisted me mightily in fortifying my interpretations and loaned me his expertise on the plantation South. I received a generous review of the manuscript from an anonymous reader, and I express my gratitude for her/his advice on the areas that required sharpening. Because English history in my life of publication has been terra incognito, I required some assurance that my narrative and analysis had merit; that reassurance, with hesitation over some points, came from Kristen Burkholder, whose reading of my chapters on England was enormously appreciated. Jeff Cooper of my department gave me warnings on colonial history, especially the old Cavaliers-versus-Roundheads interpretation, and his admonitions prompted me to be more precise in my explanation of cultural differences between the colonial North and South. Parts of this manuscript were read by Jane Flaherty and Kathleen Hilliard, and I am grateful for their criticisms. Dan Feller and Heather Cox Richardson gave me encouragement to finish the task.

The publication of a manuscript inevitably comes with a number of tussles, but I have enjoyed a comparatively smooth journey through editing, revision, and copyediting, for which I earnestly thank Rand Dotson, Lee Sioles, Catherine Kadair, and the rest of the Louisiana State University publishing staff, as well as freelance copyeditor Stan Ivester. I greatly appreciate their efforts. As always, the indefatigable voice urging me on was my wife, Kathy, who besides being the one who lifts the sagging spirit has also been the best editor one can have, and, if I let her, would be the best publicity agent a person could find.

ABBREVIATIONS

(used in tables, notes, and bibliography)

AgH	*Agricultural History.*
AgHEW6	Joan Thirsk and G. E. Mingay, eds., *The Agrarian History of England and Wales* 6: *1750–1850*. Cambridge, U.K.: Cambridge University Press, 1989.
AgHEW7pt1	Joan Thirsk and E. J. T. Collins, eds., *The Agrarian History of England and Wales* 7, part 1: *1850–1914*. Cambridge, U.K.: Cambridge University Press, 2000.
AHR	*American Historical Review.*
ALLib	Abraham Lincoln Presidential Library, Springfield, Ill.
CWH	*Civil War History.*
EEH	*Explorations in Economic History.*
EHR	*Economic History Review.*
GBCensus1851	Great Britain, Registrar-General, *The Census of Great Britain in 1851*. London: Longman, Brown, Green, and Longmans, 1854.
HStatUS76	U.S. Bureau of the Census, Dept. of Commerce. *Historical Statistics of the United States: Colonial Times to 1970*. Washington, D.C.: Government Printing Office, 1975.
IndHL	Indiana Historical Library, Indianapolis.
JAH	*Journal of American History.*
JCWE	*Journal of the Civil War Era.*
JEH	*Journal of Economic History.*
JER	*Journal of the Early Republic.*
JSH	*Journal of Southern History.*
MnnHS	Minnesota Historical Society, St. Paul.
MVHR	*Mississippi Valley Historical Review.*
OHS	Ohio Historical Society, Columbus.
UIll	Lincoln and Illinois Room, University of Illinois, Champaign-Urbana.
UMich	Bentley Historical Library, University of Michigan, Ann Arbor.
UMich2	William L. Clements Library, University of Michigan, Ann Arbor.

UNC	Southern Historical Collection, Wilson Library, University of North Carolina, Chapel Hill.
UVa	Albert and Shirley Small Special Collections Library, University of Virginia, Charlottesville.
VHS	Virginia Historical Society, Richmond.
VMHB	*Virginia Magazine of History and Biography.*
WesResHS	Western Reserve Historical Society, Cleveland, Ohio.
WisHS	Wisconsin Historical Society, Madison, Wis.
WMQ	*William and Mary Quarterly.*

PUBLISHERS

CamUP	Cambridge, U.K.: Cambridge University Press.
CorUP	Ithaca, N.Y.: Cornell University Press.
DukeUP	Durham, N.C.: Duke University Press.
HUP	Cambridge, Mass.: Harvard University Press.
IaUP	Ames: Iowa State University Press.
IndUP	Bloomington: Indiana University Press.
JHUP	Baltimore: Johns Hopkins University Press.
KSUP	Kent, Ohio: Kent State University Press.
LSUP	Baton Rouge: Louisiana State University Press.
OUP	Oxford, U.K.: Oxford University Press.
PrinceUP	Princeton, N.J.: Princeton University Press.
UAlaP	University: University of Alabama Press.
UChiP	Chicago: University of Chicago Press.
UGaP	Athens: University of Georgia Press.
UIllP	Urbana: University of Illinois Press.
UMassP	Amherst: University of Massachusetts Press.
UMoP	Columbia: University of Missouri Press.
UNCP	Chapel Hill: University of North Carolina Press.
UPKan	Lawrence: University Press of Kansas.
UPKy	Lexington: University Press of Kentucky.
UPennP	Philadelphia: University of Pennsylvania Press.
UTnP	Knoxville: University of Tennessee Press.
UPVa	Charlottesville: University Press of Virginia.
YUP	New Haven, Conn.: Yale University Press.

PART I

The English and British Agricultural Standard, 1300–1880

English agricultural history is vitally important to understanding North American agricultural practices. Of course, the English transmitted to their colonies not only preferences for crops and production techniques, but also institutional procedures (such as property rights, setting of boundaries, riparian rights) and a disposition toward commercial exchange. Beyond those features, however, were three qualities of British agriculture that played a crucial role in American agriculture. One was the physical environment: foliage, temperature, climate, soil variety, and geographical size. Cultural expectations and evaluations centered on geographical features much more than one might expect. The second was the landownership pattern that the Americans inherited. At the time of American colonization in the beginning of the seventeenth century, England confronted a decision about the disposition of its lands, whether in great estates or in smaller yeoman parcels; that English decision transmitted itself in a crude way to the colonists to adopt either the plantation or the family farm. Third, England set a standard by which to evaluate American agricultural practices because in the final analysis the two areas took decidedly different paths in regard to husbandry. Those paths included not only the subjects of animals, crop production, use of labor and capital, and the adoption of machinery, but also much broader questions of social welfare, the role of women, the meaning of work and labor, and finally the matter of social and political egalitarianism.

1

The English Civil War and the Birth of the Free Labor Ideology

English historiography for the mid-seventeenth century grapples with two mighty subjects: the transition from feudalism to capitalism that occurred between 1350 and 1650 and the origins of the English Civil War, 1640–60. Almost everyone accepts that by 1500 England had undergone a substantial economic change and had become a market society; the question that has raised a century of scholarly furor is how and if the civil war and the economic transformation were intertwined, and if the major actors in the civil war represented the forces that arose from the emergence of capitalism. The debates have famously involved the topics of the rise of the gentry, the origins of capitalist farming, and the appearance of a bourgeois class accompanied by wage laborers. A host of writers resolutely say these large-scale economic alterations had no influence on the English conflict between Cavaliers and Roundheads. Rather, the civil war was due to, depending upon which historian one reads, Puritanism, a religious hostility toward Catholicism, the political ineptitude of Charles I, a division between court and country, the problems of bringing together three kingdoms—England, Scotland, and Ireland—under one rule, the political upheavals generated by the Little Ice Age that resulted in worldwide famine conditions, and the activities of determined minorities.[1]

These subjects and the war of interpretations are not central to the concerns of this work, but within the English transition to capitalism and the outbreak of the English Civil War are slices of material relevant to the formation of American husbandry. The free labor ideology so prominent in the antebellum North among Free Soilers, abolitionists, and Republicans had its roots in English social and political developments between 1550 and 1660. The genesis came from the aristocracy and gentry, who pronounced a new rationale for their existence that celebrated the leisured life while simultaneously deprecating manual labor and virtually all labor in general.

In modern historiography this is known as the code of gentility. That set of aristocratic values generated a stark opposition from those who found labor perhaps debilitating when taken to extremes but necessary and worthy of respect—from the English "middle"—craftsmen, yeomen, and Puritans. It took verbal form in the doctrine of calling, a doctrine that had the seeds of a total rejection of the aristocratic ideal of leisure. The emphasis on the primacy of labor and its universal application to all men (and eventually women) erupted in the Leveller movement in the late 1640s. If one conceives of the cult of gentility as the aristocratic path that had as its economic foundation great landlords using tenant farmers who hired wage-earners, and the embryonic free labor ideology as the path of yeoman farmers whose productive activity was based on a multitude of small farms run by family labor, then the choice the people of England faced in 1640–60 becomes clear: a country of a few landed magnates or a nation of family farmers.[2] The Levellers lost, the gentry and aristocrats won, and the path taken was luxury for a privileged few and debasement for the multitude. Not until the 1920s would the decision of 1640–60 be undone.

English Change, 1066–1550

The England that would fall into chaos in the mid-seventeenth century was the product of major agricultural changes and dramatic swings in population numbers. After the Norman conquest, England practiced a brand of European feudalism in which the population dwelt in villages organized around manors. The manor was essentially an agricultural unit in which various types of peasants engaged in husbandry. The lord of the manor was a knight or aristocrat, and was responsible for keeping the peace and protecting the realm. His was a military function. Local rulers dominated as the crown was weak for centuries until Henry VII established the Tudor monarchy in 1485 by defeating Richard III at the battle of Bosworth.[3]

The great force that reshaped England's agriculture was disease. The plague came to England in 1348, 1361, and 1369. Population totals fell from roughly 5.0 million in 1350 to 2.1 million by 1430. Suddenly the manor lords lacked the labor to till their holdings (the demesne) and, to acquire income, the lords sold some of their holdings to prosperous and successful peasants. This, according to Richard Allen, was the birth of the yeoman farmer—the independent family farmer who owned his own land (50 to 150 acres), and farmed it himself with his children and one or two live-in servants as agricultural laborers. Then came a reshuffling of the categories of landholders. Henry VIII broke away from the Catholic Church, created the Anglican Church, and substituted the King for the Pope as head of the church. In doing so, he seized the land held by the Catholic Church, some 20 to 30 percent of the realm. By a series of extravagant adventures, Henry VIII had to sell much of this seized land.[4] The end result was a striking change in the pattern of landownership. The aristocracy maintained over the years 1450 to 1700 an ownership share of 15 to 23 percent, but the gentry—those landowners who were not noble but who owned over 300 acres (the lesser gentry) and sometimes more than 10,000 acres (the high gentry)—increased their share from 25 to nearly 50 percent. The yeomen also grew in ownership from 20 percent to a range of 25 to 33 percent. This new pattern of landownership and the potential it held for the direction of society was a subtle question surrounding the political upheavals in England after 1640.[5]

Population dynamics then again beset the country. The numbers expanded again until by 1600 nearly 4.3 million dwelt in the land, and by 1650 5.2 million. England and Wales only encompassed slightly less than 60,000 square miles and had at the time brought perhaps 21 million acres under cultivation. The population density in 1600 was over 70 persons per square mile (in 1066 it had been 21 to 27 persons per square mile).[6] This excess of people flooded over a land with an economy unprepared for the increase, resulting in pauperism, unemployment, vagabonds, and a high incidence of poverty. Coupled with the distresses accompanying a rising tide of human population was a determination on the part of the aristocracy and gentry to earn more money by ending the strip farming of the manor as well as the commons area open to all living in the manor, and to consolidate landholdings into larger farms—the procedure of enclosure, prominent in the years 1450 to 1525 until discouraged by the Tudor monarchs. Ultimately the aristocrats and knights were losing their military functions and becoming more capitalist in behavior—they were seeking more income to sustain a luxurious lifestyle. To manage the farms, the magnates contracted with tenant farmers—the leases varied considerably at this time—while the tenants hired wage laborers to do the actual work of farming. At first the wage-earners were small owners of land, from cottagers (2 to 3 acres) to peasants (15 to 30 acres), who still had access to the commons of the manor (land used by all for wood and grazing) and so were not completely dependent on income from wages. That condition would severely erode over the next one-and-one-half centuries.[7]

Beyond the alteration in agriculture, two other economic developments were noteworthy. The first was the awesome growth of London, which by 1600 housed some 200,000 inhabitants. London acted as a market that stimulated commercial agriculture. Its other effects were to expand the number of merchants and to increase cloth manufacturing. Manufacturing at the time, however, was artisanal, nonmechanical, and conducted in small shops. It was an important element in the economy but small compared to agriculture; England at this time was 80 percent rural.[8]

Not insignificantly, political change came to England in the form of a more powerful kingly government in London. One of the achievements of the Tudor monarchs was the suppression of baronial misbehavior and the extension of the law of the realm into the counties and parishes (church geographical divisions). To do so, a tentative agreement was made between the aristocracy, or the landed magnates in the counties, and the monarch. The magnates (in this capacity called "lord lieutenants") would nominate individuals to be appointed justices of the peace, the primary arm of county government, and then obtain approval from the Privy Council acting for the monarch. This was a delicate balance between the landed aristocracy wanting dominion over their small realms and the monarch desiring a uniform national code of behavior and administration, but the Tudors made it work.[9] The Stuarts did not, and the worst of them, Charles I, was beheaded for it.

The Aristocracy and Gentry, 1550–1640

Aristocracy had its orders and numbers. The high aristocracy was composed of those who were anointed by the king and who sat in the House of Lords. The order was, highest to lowest, duke, marquess (and marchioness), earl, viscount, and baron. Their numbers were never very great, being about 97 in 1400, reduced by the War of the Roses to perhaps fewer than 30, then recovering to around 60 under Elizabeth I, but swelling under James I and Charles I to nearly 160.[10] The other peers were the religious leaders of the church, largely the bishops, who sat in the House of Lords and who numbered as many as 30. Beneath the aristocracy were the knights of the shire, those individuals who owned land and owed the king military service. Not nobles, they may have comprised 200 families in the 1520s and 500 by 1600. Finally came landholders who claimed the title of esquire or "Mr." and who owned land from 300 to 10,000 acres. They obtained their title from the College of Heraldry by providing evidence of landownership and family continuity. Landowners have been lumped into the greater and lesser gentry, and they numbered between 3,500 esquires and 15,000 gentlemen in 1690. Their numbers fluctuated considerably. The last group of landholders was the yeomen, who could be equal to the lesser gentry in finances and landholding, but they had no title and were not considered gentlemen because they performed physical labor.[11]

The transition away from feudalism for the English upper class involves the contraction of military duties for the peers and the knights and the search for a new justification of nobility. Feudalism was a political and social system geared to providing a military presence to protect against invasion. The Tudors changed the military function of the nobility. They extended the king's law, arranged for internal police powers to be exercised by the county landholders, and under Elizabeth I refrained from conflict with France and Europe. It is perhaps symbolic that the last castle and moat built for defense against invaders was constructed in 1511–12.[12] In time, the military function was taken over by the Royal Army and Navy; the aristocracy participated by having their sons serve as the officer corps. By the 1550s it was clear that the military-protection aspect of feudalism had totally collapsed. Along with it collapsed the rationale for the existence of the aristocracy and the knights of the realm. They found a new rationale in landownership, conspicuous consumption, and a gentleman's life.

Thesis: The Cult of Gentility

The seventeenth-century aristocracy and gentry had a number of traits inherited from their medieval past which continued to operate in their lives. They were the ones who made land the most prized possession in England, both for economic and status reasons. They also fervently believed in family lineage and intended to maintain their own family's preeminence by the laws of primogeniture and entail, although entailing an estate for centuries was disallowed in the Tudor courts. Overall, the aristocracy and gentry prized stability and belittled social change; they had no concern for the sufferings of the poor and were more focused on avoiding and suppressing peasant riots. They had, in J. V. Beckett's phrase, a "habit of authority" and wanted to impress on commoners their duty to obey their superiors.[13]

The peers from the twelfth to the sixteenth century were never known for living a monastic life of simplicity. They had always been the great landowners and the wealthy of the land. Yet their activities had not been concentrated in economic undertakings; their role had been military and political. With the withering away of their military function, their relation to the economy changed. They had to assume responsibility for the operation of their estates. They had to insure the collection of rents from tenants to maintain their standard of living; to do so they hired supervisors, called bailiffs and stewards. The peers, in short, had to become more aware of the market, they had to become partial capitalists. Few historians of the English aristocracy argue that the peers and gentry became market-driven folk, but most admit that the peers assumed a new attitude toward wealth accumulation.[14]

The key to the position in which the aristocracy (and gentry) found themselves after 1550 is that landed wealth was almost the sole criterion that separated them from the generality of the English people. To display their superiority in a manner to remind others continuously of their elevated status, they had to spend lavishly, entertain lavishly, dress lavishly. And to do all this, the peers required a vast income. By the ownership of hundreds or even thousands of acres, a peer or one of the great gentry could find tenants to oversee agricultural operations and to obtain the necessary labor from villagers. The gentry and the aristocracy became rentiers; they did not labor, and their income, their rent, was derived from the labor of others.[15] The gentry and aristocracy were supervisors of supervisors, overseeing their bailiffs and stewards while sometimes actively engaging in managing the tenants themselves. However, the situation posed a dilemma. Now that they were shorn of their military duties, how could the gentry and aristocracy justify their position at the top of the economic hierarchy when they did so little? The answer came in the form of the code of gentility, a class ideology.[16]

In its broadest application, the gentility standard simply explained how a person could become a refined individual worthy of the title of "gentleman." In its seventeenth-century manifestation, the code of gentility had a stark economic aspect that clearly gave it class connotations. A gentleman did no manual labor. The famous definition came from Sir Thomas Smith in the late sixteenth century: "he is a gentleman who is so commonly taken and reputed . . . whosoever studieth in the universities, who professeth the liberal sciences, and to be short, who can live idly and without manual labour and will bear the port, charge, and countenance of a gentleman."[17] This ideal made its appearance either simultaneously with the rise of Puritanism or before it. The economics contained within the code are clear enough. They exclude from the start 80 percent of the English population by stipulating that manual labor disqualified one from being a gentleman. Thus all craftsmen, all farm laborers, all yeomen, and most tenant farmers could not be gentlemen because they worked with their hands. The next step excluded about 15 percent more: anyone who engaged in trade, no matter how lucrative, could not be a gentleman. Edward Chamberlyne in 1609 wrote that businessmen were always inferior to gentlemen: "Tradesmen in All Ages and Nations have been reputed ignoble," and that, if a gentleman were to engage in trade, he "thereby loses his Gentility for ever."[18] Gentlemen had a fortune that enabled them to live without working at all; their lives were devoted to leisure, to pursue the art of living.[19]

This definition of a superior group within a society was tailor-made to justify the landholding aggrandizement that the aristocracy and gentry had accomplished. All that needed to be done was to explain the uses of leisure time and why society should rejoice that it had a leisured class. This part of the code of gentility altered over time. At first the primary use of leisure was that it afforded a group the opportunity to offer service to the government. The leisured should run the government because they possessed the intelligence necessary to formulate good laws in Parliament and had the time to examine political issues; the majority existed "onlie to be ruled, and not to rule other[s]." And a corollary to their political power was to insure order reigned in the counties and to suppress the riots and challenges to authority which seemed so prolific. In the eighteenth century, the code of gentility would focus more on deportment, dress, and manners. As the code moved into the nineteenth century, gentlemen and gentlewomen would be expected to patronize the arts and offer society a model for the rest of society to emulate.[20]

How the peers and the high gentry behaved, however, opened a large gap between the theory of gentility and its performance. The aristocracy demanded and obtained deference; inferiors had to acknowledge in words and behaviors the superiority of the peers by bowing, by averting eye contact, by taking off headgear, and by using appropriate language to show the aristocracy's superiority. The peers also reinforced their status by being the only ones allowed to carry weapons openly on their persons. In the sixteenth and seventeenth centuries, but not by the nineteenth, deference was often enforced by physical violence.[21]

Most visible to commoners was the conspicuous consumption that aristocrats engaged in. Instead of using leisure time to cultivate more virtue, the aristocracy used it to enjoy every bodily excess imaginable and to engage in overt hedonism, sensuality, and sport. One of the first activities they engaged in was building mansions in the countryside where they could act out being little kings. They exhibited their magnificence by the number of liveried servants in their hire. Although written about the peers in the nineteenth century, this description probably was accurate for the sixteenth and seventeenth centuries: the gentleman "rides, hunts and pursues various kinds of sport, as do also his sons, while his wife enters local society and his daughters learn music and painting."[22] The great show—the location for real conspicuous consumption, to show off the material well-being of the magnate—took place in London. Besides country estates, the elite purchased homes in the city. There entertainment was done lavishly. London was the site of Parliament, and the London "Court" became famous for its display of fashion, flow of alcohol, swearing, and outlandish displays of wealth.[23]

This display of wealth has been interpreted by historians as a means of cowing the majority of the English people into admitting the superiority of the landed gentry and aristocrats. The magnificence and splendor of the trappings of wealth and a leisured life, coupled with some shows of paternalism, was supposed to extract deference from commoners and a willingness to be ruled by the upper class.[24] Commoners probably did accept rule by the gentry and peers, but whether willingly or not remains a question. Instead of being awed or perhaps even inspired by such displays of wealth, many might have just as easily been outraged and incensed. The civil war gave those with contrary ideas an opportunity to voice their disgust at the ideology of wealth without labor.

Antithesis: The Doctrine of Calling

A host of issues and concerns assailed the people of England, and it should be distinctly understood that the population was not obsessed with the single question of the justness of the code of gentility. They, or some elements among them, had apprehensions about the direction of kingly government, the religious establishment and the policies of Archbishop Laud, enclosures, vagabonds, and foreign affairs. The question about the fairness of the code of gentility was tucked into these other matters and was simply a tremor within the political and economic upheaval of the time, but it is that tremor that would reverberate for centuries across the Atlantic Ocean.

The code of gentility brought forward the question of labor, celebrated the elite's ability to escape it and yet live a life of unimaginable and irresponsible consumption, and to stigmatize as unworthy and lowly those who worked with their hands and, with less disdain, those who worked with their minds. The gentility doctrine was an outrageous insult to any who earned an "honest living," who obtained a skill and brought some useful product into existence. It is hard to imagine any other idea that the aristocracy could have propagated—save slavery for all who were not noble born—that could have more outraged men and women about the worthiness of their lives, especially if they obtained sufficient income to live in modest comfort. The reason for this reaction by commoners who obtained an income (probably greater than twenty-five pounds annually) is an understanding of how "mastery" operates among human beings. For the elite, "mastery" meant mastery of other men: to command subordinates, extract the appropriate behavior, and receive the expected deference. "Mastery" among commoners meant mastery over nature, over some art or craft, some knowledge that permitted a person to achieve an outcome that others could not (growing crops, constructing homes, fashioning metals, putting together furniture). It was and is from this sense of mastery that men and women developed pride; they accomplished feats that could be touched and admired. This mastery of contribution—of production, of manufacture, of skill—was debased by the code of gentility in favor of—of what? Of mastery of dress, of mastery of riding in a carriage, of mastery of manners, of mastery of reading—but not of a mastery of anything at all representing skills that created a useful product. It was simply a celebration of wealth and a belittling of those who in this case had not inherited wealth.

Of course, there were Englishmen and women who might have agreed that common labor was degrading. Perhaps one-half or more of the English population was engaged in brute labor. For peasants having little income, for agricultural workers hired weekly by tenants, for London wage-earners, labor may have been a physical curse. It is hard to dignify the hoeing of plants to keep weeds out of the wheat and rye; reapers using the sickle had to bend over for hours to harvest the grain; construction workers used a shovel and their hands; in London, hod-carriers emptied buckets of human excrement and took them out to the fields for fertilizer. Such labor may have been necessary, but few would find the actual work to be barely more dignified than that performed by an ox.[25]

This description of the different experiences of labor for the English commonalty evokes a long-standing argument about seventeenth-century English society: whether a middle class existed. Early advocates of a middle class lumped together yeomen, craftsmen, manufacturers, merchants, and the lesser gentry. Christopher Hill used the idea of the middle to foist an interpretation on the age that the aristocracy and high gentry faced a new class, the bourgeois middle, that resented medieval regulations that constrained their pursuit of profit. In its extreme version, he argued that there existed a class conflict not so much between the aristocracy and the middle class but between workers and the bourgeoisie because the bourgeoisie wanted to impose upon England a society of capitalists who employed wage workers.[26] Of late, historians have concluded there was no middle class as is currently defined and that the gentry in no sense should be considered a part of the middle class. Yet at the same time authors recognize that a "middle" did exist, consisting of small property-holders (yeomen and some secure copyholders) and of modest and even high income-earners (merchants and skilled craftsmen in villages and London).[27]

In the generic category of the English middle in the seventeenth century, most authors have emphasized how small property-holders, when compared to the gentry, feared the growing number of beggars and vagabonds who seemed to swarm over England. The years between 1550 and 1640 saw peasant uprisings over enclosures and famines. The middle elements in English society denounced those who were not and seemingly did not want to be employed, believed paupers were unworthy of charity and salvation, and were willing to pass repressive laws to stop the impoverished from upsetting property relations.[28]

The stress on the fear of the propertied toward the wage-earning English masses and the totally dispossessed has obscured the anger of the modestly propertied Englishmen at the presumptions of the gentry and aristocracy. Louis Wright was undoubtedly correct when he wrote that the middle class "found that trade was not a tarnishing, shameful thing, but something to be proud of." In particular, the middle cast aspersions on being idle because of income derived from landownership. The Puritan preacher William Perkins sermonized, "The idle body and the idle braine is the shoppe of the diuell." The middle, wrote one Robert Reyce, "though hee striveth ordinarily well, yett he laboreth much." One Puritan said, "God sent you unto this World as unto a Workhouse, not a Playhouse." And sometimes the attack on the gentry lifestyle was explicit: "miserable and damnable is the state of those that being enriched with great liuings and reuenewes, doe spende their daies in eating and drinking, in sports and pastimes." Wealth because of industriousness and activity had a rationale, but those who inherited wealth could point to no meritorious deeds, behavior, or actions.[29]

Two groups within the Elizabethan middle have attracted the most attention, the first being the yeoman farmer. The years 1560 to 1640 have been called the "golden age of the yeomanry."[30] The oddity is that most of the time the yeoman farmer was nearly statistically invisible in English society, and after 1660 he started to become invisible in reality. The main reason he has been noted with so much fervor is evidently that he was the exception to the rule of concentrated landownership by the aristocracy and gentry, the only Lilliputians among the Brobdingnags.

The numbers of the English population in the categories offered by Gregory King in 1688 have been cautioned against numerous times, but they remain nearly the only source available to scholars. By King's number of families, the category of "freeholders" (two of them, for what reason remains unclear) constitutes only 13 percent of all families in England (total families equaled 1,370,586); if the farmers are added, which means tenants of whatever sort, the percentage only rises to 24 percent. The laborers, cottagers, and common soldiery represented 62 percent of the people.[31] Yeomen in Tudor times had a measure of comfort unknown to peasants and laborers, but they rarely knew luxury. The acreage they owned varied from as little as 4 acres to 120, but unless they obtained some 300 acres or more and rented land out to tenants, they could only obtain moderate incomes of less than a hundred pounds.[32]

The relationship between labor, industry, and idleness for the yeoman farmer arose from the circumstance of crop production on a family farm. While they needed the labor of sons and a year-round farm laborer (then known as a farm servant) and perhaps some wage-earners during the harvest season, the yeoman farmer did manual labor year-round. He did not direct labor, he performed it. By avoiding the usual "sins" of idleness, intemperance, gamesmanship, socializing, and hunting, he enabled the earth to yield crops necessary for survival, and he earned income to maintain himself and his family. As a landowner, the yeoman farmer is often lumped into the category of property-owning elite. However, according to the code of gentility, the yeoman farmer belonged to the category of those who labored with their hands. The latter condition pits the yeoman against the gentry, even if some yeomen had aspirations to be gentry. The reputation of the Tudor-Stuart yeoman was the "honest, industrious" yeoman; it was his labor that made him angry at the leisure and idleness of the gentry and aristocracy. Albert Schmidt summarized, "The yeoman prided himself on his industry and ridiculed the gentry, in particular, for their idleness."[33]

In historical accounts, the yeomen are incorporated into the other great component of the English middle, the Puritans. However, the Puritans now represent an interpretive muddle. In the histories of the first half of the twentieth century, the Puritans were quite plainly presented as spokespeople for the middle in terms of industriousness, avoidance of idleness, stress on moral behavior, cultivation of thriftiness, and strict self-control—the bourgeois values.[34] The elements of society Puritans drew upon, however, have been questioned. Puritans had some gentry and aristocratic participation. Much of the study of Puritanism is connected to the origins of the English Civil War and controversies over church structure and Catholicism.[35] In modern historiography, the Puritans are represented as unusually fervent religious believers who were rabidly anti-Catholic. In most generalizations about Puritanism, the radicalism has been diluted and historians have emphasized the Puritans' stress on obedience to civil authority, acceptance of social hierarchy, and justification of wealth inequalities. Most scholars now see Puritanism as a reaction to the social unrest at the lower end of the social scale, especially to the numbers of vagabonds and beggars.[36]

Nonetheless, Puritans leveled distinct criticisms at the gentry and aristocracy, all directed at idleness and conspicuous consumption. They took direct aim at the code of gentility—at least at its main props if not overtly mentioning the code itself—and did so with the doctrine of calling. The code of gentility and the doctrine of calling were polar opposites, and at its basic level this was a disagreement about the act of labor.

The Puritan doctrine of the calling has been inextricably connected with the rise of capitalism. Both R. H. Tawney and Max Weber made the initial analysis and thus provided intellectual fodder for argument ever since. A calling was one's occupation, or one's place in a working society; by being laborious, attentive, dutiful, and absorbed in one's occupation, one contributed to the well-being of society and spent one's time so engaged as to avoid sin. Weber pointed out that the act of labor had little status in early Christianity, St. Paul and Thomas Aquinas merely saying one labored only to obtain food. In much Christian theology, labor was a punishment inflicted on humans for the sins of Adam and Eve; for disobeying God, they were driven out of Paradise, where work was unknown. The calling evidently did not have much favor with Martin Luther. Calvin on the other hand elevated the doctrine among the reformed religious sects, making it a prominent feature of life among certain groups, such as Baptists, Congregationalists, Quakers, and Presbyterians. Weber stressed the creation of an ascetic mentality: all work and no play was the religious obligation of a true believer, and one continuously worked all life long. Original Puritans decried the urge to obtain wealth and to hoard earthly goods. Their purpose was to commit adherents to a godly life that would lead to salvation (predestination permitting). For Weber, the behavior demanded was perfect for a capitalist world: it produced human machines who automatically kept on frantically working, reinvesting, and economically growing, a behavior that had not been previously sanctioned by Catholic theologians. This was how Puritanism produced modern capitalism.[37]

Most of the writing on Puritanism, or at least a great deal of it, has centered on the vicious attitudes of the Puritan clergy to the poor, to wage-earners, and especially vagabonds.[38] Much of that attention should be redirected to the Puritan attitude toward the gentry because that is where the doctrine of calling clashes so dramatically with the code of gentility. The code of gentility celebrated a life of conspicuous consumption without labor. The doctrine of calling demanded labor, self-control, and abstention from luxurious display.

Puritanism had an odd egalitarianism about it that sat uneasily with a concurrent sense of hierarchy. The equality came out of a belief in virtue, which any person might possess (but most would not), but those with virtue obviously should have higher rank and responsibilities than those without virtue. The Puritan attack on the gentry and aristocracy was that neither birth nor wealth entitled an individual to respect because neither reflected virtue. Moreover, as the minister Richard Baxter said, "God is no respecter of persons." Among the radical Quakers later on, this religious social leveling would lead Edward Billing to say that he approved of those who said it "would never be a good World so long as there was a Lord in England . . . for the whole rabble of Duke, Marquesse, Lord, Knight, Gentleman by patents; I find no room . . . in scripture."[39]

Puritans attacked the conspicuous lifestyle of the great magnates and the extravagance of the Stuart courts. Minister John Rogers said, "for those that can work and will not, let them starve"; another equivalently famous line of St. Paul was, "If a man will not work, neither shall he eat." These expressions can easily be seen as a criticism of the gentry, not only the poor, for they, by the code of gentility, did not work.[40] The Puritan complaint against "idleness" and people without a vocation seems almost directly a slap at the gentry. That the gentry had no calling was a violation of God's strictures: "none must be idle but every man must diligently imploy himselfe in such affaires and business as God (his Master) hath committed to his care." Leisure was wrong: God "commanded men to labour six days, not to live idly, slothfully and unprofitably."[41]

Indeed, historians of sixteenth- and seventeenth-century Puritanism have focused on the obsession of these ministers and laypeople with labor. Out of their speeches and writings, says Christopher Hill, came a proclamation of "the dignity of labour" which in some instances became "almost a democratic doctrine." But Hill interprets the "dignity of labour" ideal of the Puritans as an attack on the working class.[42] Given the time period and the surge in the rush to gentry status, the target was as likely the code of gentility and the supreme status given leisure and conspicuous consumption. The doctrine of calling and its emphasis on labor was a direct challenge to the aristocratic standard of a leisured life.

Synthesis: The Free Labor Ideology

Out of the clash between the code of gentility and the Puritan doctrine of calling arose the emphasis on the dignity and usefulness of labor, which by the nineteenth century would be noted as the celebration of free labor and later called by American historians the free labor ideology. Certain basic features need to be underlined. First, the free labor ideology aimed at the pretensions of aristocracy. Second, the doctrine of calling and the free labor ideology had their roots in the countryside. Freedom from aristocracy meant freedom from the domination of landlords, freedom from tenancy, and freedom from a life as a wage-earning farm laborer. This is why the yeoman, otherwise a minor part of English social history, looms so large. Third, the appearance of the doctrine of calling and its elevation of a life of labor reveal the other path English history could have taken. One direction was the consolidation of landownership in the hands of the aristocracy and gentry. Yeomanry was the alternative: it placed landownership in the hands of the multitude.

An element was absent in the doctrine of calling and its insistence upon the dignity of labor that stopped it from being the free labor ideology: a celebration of social mobility, a joyous song about the ability of common people to rise to property ownership and even wealth. The doctrine of calling did not necessarily include the idea of "rising" to higher social levels. Rather, the Puritans simply insisted that one had an occupation and one labored faithfully at it and that was the will of God. In their hands, the doctrine of calling could still support a hierarchical, stratified, immobile society. The free labor ideology when it emerged with full force in the nineteenth-century United States would be—with a few qualifications regarding rural life—democratic, leveling, and socially upwardly mobile.

However, the Puritans had one economic view that would easily mutate into social mobility but at the same time would actually limit the "capitalist" ethos. The Puritans, in accord with their disapproval of the code of gentility, sought *competency* in economic life: "For to seeke for aboundance is not lawful, nether dooth it stand with good conscience"; "we have no warrant to pray for aboundance: for things necessarie & sufficient, we haue warrant to pray." The word that describes the attitude is "*competence.*"[43] In material terms, one worked in the secular world to achieve competence, and the accumulation of goods stopped there. This is a mighty restraint on the capitalist ethos of eternal and never-satisfied self-aggrandizement.

The Levellers and the Distribution of Land

In 1640 began a series of confrontations in English politics that led to open warfare between the king and Parliament. A forbidding controversy over the causes of this explosion squats down among interpretive schemes and proves to be immoveable: whether the civil war and its aftermath was the creature of large social forces and class antagonisms, or the unanticipated consequence of having a politically incompetent king.[44] An unresolved problem remains. For all the confusion about the placement of people of different classes among the parliamentarians and royalists, and the lack of overt land distribution questions in the early years of the conflict, Christopher Hill is correct to point out that for centuries the English people had been ruled by kings, nobles, and bishops, and that no group had seriously challenged this structure. Yet, in just a few years following the breach between king and gentry, the nobility was stripped of its role in the government, the king executed, the church reshaped, and an outburst against hierarchy voiced. England had plenty of political upheavals before, the most notable being the War of the Roses, and yet no such manifestation of popular distrust of the social order had appeared. Why then did it appear in the English Civil War? Why was there a Leveller movement at all?[45]

One could offer this conceptual understanding of why the English Civil War took its peculiar form. First, one would assume that any society with a large population numbering at the least in the hundreds of thousands would experience political problems and upheavals about who had the governing power. Such battles have always occurred as elite groups vie for power and divide amongst themselves. Not all of them, however, involve the other elements in the society. What mates a social upheaval with a political upheaval is the experience of an economic/social transformation that has upset some traditional norms or has introduced novel social relationships or has simply injured a large number of people. To put it in the nineteenth-century historiographical terms of Leopold von Ranke, the socioeconomic transformation produces a "zeitgeist," a spirit of the times. Exactly the form and the content of the zeitgeist may be problematical, but it contains a disposition to want to change existing institutions and establish some ideal of "justice," the form of the justice depending on the cultural background of the society. In the case of England, the zeitgeist was a vast discontent with the new landed relationships that allowed accumulation of land for a few persons and enabled them to live luxuriously without labor. At the same time, such individuals held political power and used it to enhance themselves, while using their social power to belittle others. The alternative was also present: the emergence of the yeoman and the possibility for a society of equals that consisted of small independent producers who worked for their sustenance. In short, the zeitgeist was an animus to the wealth, privileges, and status of the aristocracy and gentry, and an anger at the abasement that commoners had to display to them. For these ideas to materialize, to go from spiritual discontent to behavioral action, the zeitgeist needed a moment of crisis. Charles I gave them that moment.

The convoluted political history of the reign of Charles I and his problems with Parliament need not be repeated here. Beginning his reign in 1625, Charles upset the equipoise between the gentry and the monarchy that the Tudors had established, seemed to be driving toward an absolutist form of rule, tried to govern without Parliament, raised taxes by questionable means, and allied himself with religious leaders who seemed to lean toward Catholicism. Because of his military failures against Scotland, the gentry in the House of Commons, becoming the Long Parliament that lasted from 1640 to 1660, moved against Charles and attempted to restrict his powers. In the midst of this activity against the king, the London populace became active, sending to Parliament the "Grand Remonstrance" that condemned the king's usurpation of power. By August 1642, two military forces appeared; one for the king, the other for Parliament. The civil war approached. The actions of Parliament and its members' animosity not merely to Charles but to monarchical power altogether frightened enough of the aristocracy and gentry to realign themselves in favor of the king out of apprehension that social hierarchy was being undone. The moment of Cavaliers and Roundheads had arrived.[46]

For four years the two sides fought battles, but the superiority lay with the parliamentarians under the leadership of Oliver Cromwell and Lord Fairfax once they created the "New Model Army"—which was less a new model than a combining of forces under astute leadership. Charles I was taken prisoner and held in London; he tried to escape and was captured. A different Parliament would finally resolve Charles's fate. In December 1648 Colonel Thomas Pride arrested or dismissed a majority of the members (143) trying to enter the House of Commons (Pride's Purge), resulting in only 78 remaining to transact legislation. This Parliament, called the Rump, was the institution that in January 1649 tried Charles I for treason and had him executed on January 30. In March, Parliament then abolished the monarchy and the House of Lords. Oliver Cromwell, who had been engaged in brutally suppressing revolts in Scotland and Ireland, returned to London in 1653, dismissed the Rump, and set himself up as Protector of the Commonwealth. He ruled until his death in 1658. After two years of indecisive leadership and a freefall into anarchy, General George Monck reestablished a full Parliament by allowing the members ousted by Colonel Pride to sit again. They overcame the Puritan diehards of the Rump and invited Charles II to reclaim the throne. The restoration of the monarch as well as the House of Lords began.[47]

Within this mass of confused movements and events erupted unexpectedly in 1647 a "democratic" movement. Parliamentarian spokespeople had indicated a desire for a government with a much more restricted monarchy and a more powerful Parliament. Some intimations of increased suffrage were there. The flowering of the sentiment for a politics guided by the votes of the mass of Englishmen came from a group lodged within the New Model Army called the Levellers. The name indicated that this faction desired to place all people on the same social and political level, essentially eliminating the political power of the aristocracy and gentry. Between October 28 and November 1 occurred the famous meetings between Oliver Cromwell and his son-in-law, Henry Ireton, standing for the generals of the army and actually for the belief in gentry leadership, and the spokespeople for the army rank and file, John Lilburne, William Walwyn, Richard Overton, and Colonel Thomas Rainsborough. Levellers had numerous demands, but the question of suffrage grabbed the most attention. Levellers at first demanded the extension of suffrage to all Englishmen, basing their claim for near-universal suffrage on "natural right"; they also attacked monopolies, the House of Lords, and the monarchy. They seemingly were advancing the idea of a republic. Ireton rejected the idea of suffrage to all because it undercut the basic gentry insistence that political power derived from landownership, and to dissolve that connection was tantamount to repealing property rights. The Levellers backed down from support for universal suffrage but nonetheless wanted it widened far beyond what existed in England in 1647. As the Levellers tried to put more pressure on Parliament and the generals, Cromwell determined in 1649 to break the movement, and he suppressed it.[48]

Scholars have long realized that the Levellers constituted a profound change in English constitutional thought. "Unquestionably during these years 1646 and 1647 a democratic movement unique in the history of England was making swift headway," wrote Maurice Ashley in 1958, while G. E. Aylmer added in 1986 that the Levellers represented a "giant stride forward towards democracy." Not the least remarkable doctrine voiced in these years was the assertion that all men had natural rights, and a few have intimated that the Levellers virtually wrote the U.S. Constitution.[49] The radicalism of the Levellers, however, has been questioned as they never attacked property rights and whom they actually represented remains questionable.[50]

Within the Levellers' assertion of human equality was an open rejection of the claims of aristocracy, and behind that a concealed questioning of the amount of land the nobility possessed. The attack took several forms. The first was a defiance of the conventions of hierarchy. John Lilburne, a born pamphleteer, attacked the House of Lords in 1646. He was then arraigned by them for libel and brought in for interrogation: "Lilburne came into the House [of Lords] with his hat on, refused to kneel, and stuffed his fingers into his ears to avoid hearing the charge." Later in 1649, a group began using land for planting crops on St. George's Hill in Surrey, taking land that was not theirs. They were called the "Diggers." When two Digger emissaries went to Lord Fairfax to explain their activities, the men refused to take off their hats because all humans were "fellow creatures."[51] This was a rejection of the deference upon which gentry society depended.

Another obvious manifestation of the anti-aristocratic spirit was the elimination of the House of Lords. The Levellers disliked the institution for the political privilege it gave to the aristocracy, but probably its termination came from the domination of the House of Commons by the Rump after Pride's Purge and the members' determination to execute Charles I. The members of the House of Lords refused to sign the execution order. The House of Commons abolished it on February 6, 1849, for being a useless branch of government. The abolition of the House of Lords has nowhere claimed the attention that the execution of Charles I has, yet it is perhaps more symbolic of the broad-scale social movement swirling in England than anything else; it was a direct attack on a social system, not merely on an incompetent and perhaps dangerous monarch.[52]

Puritanism in the years from 1560 to 1640 may have been primarily about religious temperament, but the movement planted seeds of equality that sprouted in radical directions, thereby producing a striking number of critics of aristocracy by the late 1640s and early 1650s: the Diggers, Ranters, Fifth Monarchy Men, and Quakers.[53] Besides the religious arguments, these critics employed a particular argument based on labor. Their attack on aristocracy was directly aimed at the code of gentility. George Fox struck out at those who owned great estates: "You wallow yourselves in the earths treasure like swine in the mire." John Lilburne in 1649 wrote that the rich gentry lived in rebellion to God "first, because they live without a calling, and so are idle, being vagabonds, and wasters of the creatures, by drunkenness, pride, gluttony." People, he continued, needed to "learn to work with their hands." Radicals at Chelmsford stated that "it is now fit that the nobility and gentry should . . . work for their own maintenance; and if they will not work, they ought not to eat." One of the critics of the House of Lords charged "which of ye before this Parliament, minded anything so much as your pleasures—Plays, Masques, feastings, huntings, gamings, dancings, with the appurtenances." The Digger leader and writer Winstanley denounced the miserly wages that workers received while others who were idle had wealth. Lilburne did defend property—but not the inherited property of the aristocrat but the property earned by labor. He wanted a man to "injoy the fruit of his own labour, industry, and sweat of his brow." Propertied gentlemen often drew the conclusion that the London radicals and their religious fellow travelers desired to destroy property rights as well as the aristocracy and gentry.[54]

Despite their attack on aristocracy and privileged positions, the Levellers handled the question of land distribution with caution. They offered no real program of redistribution, but they did protest enclosures and advocated making freeholders of copyholders and peasants. As well, they wanted to alter the laws of inheritance so that all land did not go to the eldest son.[55] Radicals on the question of landed property existed in the English Revolution, but, as interesting as they were, they were only a minute element in the political ferment.[56]

Out of this mass of criticism about England's social, religious, and political system came a treasure of political theory. Within this treasure, two works have relevance: James Harrington's *Oceana* and John Locke's *The Two Treatises*. Harrington was something of a privileged Englishman who really did not want to upset entirely the English social order, but he saw property distribution as the problem in England's turbulent post-1640 history. An imbalance had occurred due to the sale of monastic lands and monarchical holdings to the aristocracy and gentry. Because political power flowed from economic power, and in the seventeenth century economic power was distinctly landownership, the best way for a commonwealth to exist was to have an equitable distribution of landed property. To obtain equitability in landownership, he proposed an "agrarian law" that no estate worth more than two thousand pounds in landed property nor two thousand acres in extent be obtained via inheritance. By limiting the size of inherited wealth, he hoped land might circulate more freely, thereby halting land monopoly and allowing industrious commoners a fair opportunity of obtaining a freehold.[57]

It would appear that John Locke made the final statement, or perhaps compromise, about the worthiness to be accorded labor and its relationship to aristocracy and the code of gentility. Locke certainly is the center of contention, being seen on the one hand as the ideological father of possessive individualism and on the other as a pure advocate of human freedom and equality. From the vantage point of this work, however, it is interesting—casting aside all the other disputes about religion, governmental forms, governmental powers, and political participation—that the chief doctrine Locke is known for is the labor theory of value. Value was created by the act of labor; that value was then imbedded in the product and became the property of the individual. On the one hand, this definition immediately empties the code of gentility of any worthiness because gentility praises leisure while Locke affirms what most of the Levellers and other dissidents argued: labor creates true utility. On the other hand, Locke then went on to justify the right of transmission of property to progeny, thereby finding a way to sanction the great propertied lords in England. That was his compromise: he gutted the code of gentility yet still found a way to approve of inherited wealth, a position that the dissidents in the English Revolution would not have accepted.[58]

The economic dissenters in the English Revolution lost entirely. The Commonwealth fell apart. Parliament was reconstituted with the members who had been purged, thereby ending the Rump's majority; it reestablished the House of Lords and the monarchy, and it called upon Charles II to take the throne. As many have written, the social establishment after 1660 very much resembled the one that existed in 1600; the great landed magnates reasserted their position and political strength. They then made their status nearly unassailable in the Glorious Revolution of 1688–89 when they ousted James II and brought in William III and Mary under conditions that inhibited the monarchy from challenging the supremacy of the gentry and aristocracy. The great magnates and their families would rule the House of Commons and the House of Lords until the beginning of the twentieth century.[59]

New England and the Radical Impulse in the English Civil War

The radicals of the English Civil War—the Levellers, Puritans, Diggers, and others—never managed to put their alternative society into effect. Part of the reason for that failure was that English society was complicated by groups pulling in different directions, some more interested in political forms, others in religious structures and doctrines, others in economic arrangements. The direction the radicals wanted to take can be clarified by reconnecting the New England Puritan experience between 1629 and 1660 with the English Civil War, 1640–60, by seeing it as a part of English history rather than as the start of a North American colony. Too often the New England Puritans have been treated as a story unto themselves.[60] From another angle, the New England Puritans provide a glimpse into the type of society that the English Puritan movement and its radical spinoffs were trying to nudge their countrymen into adopting.

The Pilgrims made the first settlement of the New England landscape, but their numbers were tiny. The great influx came after 1629, when mainstream English Puritans flooded into Massachusetts Bay under the leadership of John Winthrop, a member of the gentry. In the 1630s, they drew from England 13,000 to 20,000 immigrants, and by 1660 the number in New England had climbed to about 23,000.[61] After some confrontations with the tribes in the area and the acquisition of land, the Massachusetts Puritans set up their society.

The first task of the Puritans had to be the disposition of land and eking out survival from the grumpy New England soil. It is first and foremost the matter of land that revealed the difference in perspective of these dissenters from the social establishment in England. Land disposition was done by colony leaders, usually church leaders and those appointed by the Massachusetts Bay Company. When a group wanted to form a new church outside of Boston, they were given a land grant, and the leaders were then given the power to dispose of the land to the church members in accordance with need. These leaders took into consideration size of family, variation in soil types, position of land, and status—they gave ministers more land, as well as (eventually) selectmen and other government officials. Because of the desire to keep settlers near to the church, all were given town lots. Land allotments could vary from 80 to 300 acres, depending on status; the Massachusetts gentry obtained perhaps 200 acres, in some cases up to 500 or 700 acres. John Winthrop got initially about 600 acres and for his services eventually about 5,000. The farms of the earliest settlers had actually only 4 or 5 cultivated acres because the task of clearing the land was so arduous. It would take years before arable farms of 30 or 50 acres could appear.[62]

All students of the New England migration emphasize the general equality of land holdings in New England prior to 1660, while at the same time stressing that hierarchy and status had their say. No one worshiped democracy or equality of possessions in Puritan Massachusetts; the word of God needed no majority of votes to sanction its truthfulness, and the Puritans believed in an ordered society. But there were also no landless people. Kenneth Lockridge called Dedham, Massachusetts, "in many respects a peasant utopia."[63]

The Puritans were gifted wordsmiths, but in the distribution of land their behavior spoke far louder than any sermon. They rejected the hierarchy of squires, gentry, and aristocracy.[64] They easily could have given out tracts of 5,000 and 10,000 acres to selected individuals and have created a powerful landed gentry like the one in England. They purposely did not. In their behavior involving land grants, the New England Puritans demonstrated that the dissident mind in England between 1600 and 1660 found the basic flaw in English society to be the distribution of land and its concentration in the hands of the aristocracy and gentry. If they had been convinced that such concentration were necessary for social stability, order, and religious rectitude, they would have copied it. While historians of England have often explained the Puritan movement as a reaction to the exploding disorder among the lowest classes, what David D. Hall has written about the New England attitude is striking in its implications for Puritan causal reasoning on social affairs: "No one in New England wanted to re-create the vagrants and idle poor seemingly omnipresent in England and everywhere a strain on local taxes. Handing out land was a practical means of keeping poverty at bay and caring for the disadvantaged."[65] The logical implication is that Puritans did not necessarily believe vagabonds and unemployed were in their condition because of bad character traits. Given that the New England solution to those suffering unemployment and impoverishment was to provide land for them, then the problem in Old England was the great magnates' ownership of nearly two-thirds of England's arable land. It was the aristocracy and gentry who produced unemployment and vagabonds by denying so many a share in the land.

The hostility to England's aristocracy and gentry in the New England Puritan experiment also surfaced in the not-so-subtle rejection of the code of gentility. The New Englanders faced the problem every colony in North America would face, the fundamental condition that ruled their existence. After dispossessing Native Americans of their land by force or treaty, the reality was that land was abundant and labor scarce.[66] The English gentry needed an abundance of people to make the tenant system work; a mass of cheap labor had to be found for tenants to employ so the tenant farmers could be rewarded at the same time large rents were extracted. Without cheap labor, the tenant system fell to pieces.

New England never developed a servant culture the way Virginia did and did not resort to slavery. They instead turned to family labor and thereby created the family farm. However, the family often was not large enough, and so farmers had to be assisted by youth who were apprenticed to families as farm laborers. Interestingly, this virtually condemned the husbandmen of New England to subsistence living. The New England farm simply had no opportunities for commercial dealings until the middle of the eighteenth century, and even then those opportunities were limited.[67]

In particular, the establishment of the family farm as the model life of a New Englander meant rejection of the code of gentility for two reasons: manual labor was absolutely necessary, and a life of leisure was deprecated. Author after author has emphasized the doctrine of calling and the "Puritan work ethic" among the New Englanders. They have not been quite as quick to acknowledge that this elevation of working as a commandment given by God was an absolute rejection of the aristocratic ethos.[68] The life of leisure of the English gentry rested on the acquisition of landed wealth that produced a rental income enabling one to live lavishly without labor. In New England the goal so often repeated was to obtain "competency," not wealth. Reverend John White put the idea in religious terms: "Nothing sorts better with Piety than Competency." Although John Winthrop had many acres of land, and though he was honored and prior to 1630 was indeed a member of the gentry, it is recorded that he had to resort to manual labor on his Massachusetts farm. And when he did that, he was no longer a gentleman.[69] Of course, this discussion deals with agricultural New England; Boston and the seaport towns were a different economy and a different society.

One episode in the life of the early Massachusetts Bay colony revealed precisely how strongly the Puritans wanted to be rid of the hierarchy of aristocracy and the code of gentility that had engrafted itself on the mother country. Puritan nobles Lord Say and Sele, Lord Brook, and others, were interested in the New England experiment, but as they looked over the political institutions of Massachusetts and Connecticut in 1636 they were disturbed that the religious brethren in America had not created a ruling house for the aristocracy. John Cotton explained to them that the Puritan colonies would welcome the gentry of England to their land, but the Bible did not sanction hereditary rulers; only the virtuous were to be allowed governmental power. This was a cautious answer, but the implication was that the hereditary principle was not held in esteem in New England. The nobles were upset at the prospect of life without guaranteed political power, and so they attempted settlement in the West Indies.[70]

New England Puritans *chose* to jettison aristocracy and the leisured life, and it must have been a conscious choice, regardless of the written record. The reason for the assertion is the history of Virginia. The comparison is stark. William Berkeley, governor of Virginia, hated Puritanism, stood for the royal forces during the civil war, and desired a gentry for his colony. He, too, faced the land-labor dilemma of all the colonies, but it made no difference to him whether individuals could farm huge tracts of land. Under his governorship from 1642 to 1677, he handed out tracts of land containing thousands of acres to his favorites and members of the council. Huge land grants continued as a tradition into the eighteenth century.[71] One could hardly have a clearer contrast between the ideals of royalism and those of Puritan dissenters than the way Virginia and Massachusetts distributed land in North America.[72]

Certainly the differences between the New England Puritans and the larger movement in Old England loom large. Nonetheless, the New Englanders can be seen as indicating the direction that the dissidents in Old England wanted to take during the civil war. They sought some restructuring of the agrarian economy, some diminution of the power of the aristocracy, and some appreciation and honor for those who labored. New England was able to jettison structurally the main features of the gentry and to repudiate their gentility ethos largely for one reason: no entrenched landed elite had established itself.[73] But the aristocracy and gentry in Old England were no figments of the imagination. Their members had military power, a tradition of political leadership, and a social hegemony by which they extracted deference from the mass of the population. Thus the momentary surge to egalitarianism was defeated in the era of the English Civil War, and England would settle back and live under aristocratic rule for the next two and one-half centuries.[74]

There is an old interpretive tradition that splits the American colonies and early national experience into Cavaliers (southerners, supporters of the gentry ideal) and Roundheads (northerners, the Puritan challengers to Charles I). Not many scholars have accepted the division, and in terms of society, ethnic groups, and behaviors it is of dubious worth, but viewing American history from the aspect of the disposition of land, the old perspective has some merit. It was a common idea in the nineteenth century, ably expressed by one of Great Britain's preeminent prime ministers, William E. Gladstone. In his 1878 essay "Kin Beyond the Sea," Gladstone proposed that England was not changed by the English Civil War and in it "inequality lies embedded in the very base of the social structure," while inequality was minimized in the United States. What accounted for the difference was Puritanism; American character and institutions did not derive from the Cavalier. American Puritans severed "all those elements of political Anglicism, which give to aristocracy in the country a position only second in strength to that of freedom." The English Civil War transferred democracy to North America: "All that was democratic in the policy of England, and all that was Protestant in her religion, they [American Puritans] carried with them in pronounced and exclusive forms, to a soil and a scene singularly suited for their growth."[75] The anti-aristocratic attitude, with all its views of labor, failed to take root in England but succeeded in North America.

2

The Standard

BRITISH AGRICULTURE, 1660–1870

Upon the collapse of the republican movement in England and the restoration of the monarchy and the House of Lords in 1660, coupled with the gentry's dominance in politics assured by the Glorious Revolution of 1688, British agriculture moved on a particular trajectory that would lead to startling records of output, productivity, and income—and to the birth of the large farm nestled within the estate system. It also produced an amazing concentration of landownership by the 1870s, one that produced the wealthiest aristocracy in the world, an agriculture characterized by tenant farming and an absence of small property-owners. This agrarian setting had vast significance for the American colonies and for the United States. It paved the way for the divergent paths of southern and northern agriculture. British farming methods became as well a standard by which to judge the sophistication of American farming, much to the latter's detriment. The British-American agricultural comparison fed directly into the debate over equality versus inequality, and this argument in turn became deeply rooted in sectional quarrels involving slave versus free labor. Finally, any survey of agriculture in Western Europe can only conclude that the large farm enterprise existed in only two places: Great Britain and the southern United States.

British Agricultural Developments, 1660–1880

The failure of republicanism to take institutional root in England had major results for British agriculture. Because the great landholders now ruled England, the obstacles to land consolidation by enclosures of common land were removed. To keep the great estates in familial possession, the magnates rewrote the rules for entail by creating the legal device called strict settlement. In its simplest form, the strict settlement was an agreement between three generations (grandfather, father, and son) that the estate would be kept whole under the direction of the male heir, who became in effect a "life steward." How the strict settlement affected agriculture has been a cantankerous problem, but it seems safe to generalize that it certainly aided the drive to consolidate landholdings and did nothing to hinder it.[1]

Over the decades, the major developments in English agriculture were a new enclosure movement, the reduction in numbers of yeomen and copyholders, and the mutation of cottagers into wage-earners. The agricultural system evolved into one in which the great landholders established large farms operated by tenant farmers who employed wage-earners to do the physical labor of farming. While there is general consensus that these trends occurred, there is much less agreement about the timing of the changes and their extent. The landed magnates were in place by 1640, but some would have the large farm appearing by 1720 or so, others not until 1790. Some have argued that the yeomen started to diminish by 1710 while others maintain that substantial numbers of yeomen continued not only throughout the eighteenth century but throughout most of the nineteenth. Debate over the enclosures has tilted to the view that they had less impact on most rural dwellers than had previously been believed, but that they undoubtedly injured cottagers.[2]

Several outside pressures buffeted English agriculture, the outlines of which are presented in table 2.1. The first factor obviously confronting British agriculture was population expansion. While the population grew slowly between 1660 and 1750, so that a mere expansion of land under cultivation could have fed the people, the rate of growth after 1750 was remarkable. Feeding this exploding population after 1750 was the triumph of British agriculture. Agricultural output in cereals doubled between 1600 and 1800, but the number working in agriculture only modestly increased. Compared to other nations, English agriculture was far more productive—one-third higher than France, and twice as great as Russia.[3]

During the eighteenth century, the sectors that would challenge the supremacy of the landed interest were beginning to emerge. Expansion of trade built up a wealthy interest, merchants and financiers, that over time could modify the agricultural bias of the landed magnates in governmental policies. Then there was urbanization, the growth of London and, in the nineteenth century, a number of provincial cities.[4] Finally, industrial manufacturing commenced after 1750, the area that has by far attracted the most historical attention. It is clear from table 2.1 that manufacturing and trade

TABLE 2.1 English Population and Agricultural Statistics, 1100–1901

Year	Population, England and Wales (Millions)	Density of Population (per sq mile)	London Population (Thousands)	Percent Rural	Percent in Occupation		
					Agriculture	*Manufacturing*	*Trade*
1100	1.2–1.6	20.6–27.4					
1350	5.0	85.8	50				
1450	2.1	36.0					
1500	2.4	41.2	90				
1550	3.0	51.5					
1600	4.3	73.8	200				
1650	5.2	89.2	400	80			
1700	5.8–6.0	99.5–102.9	500–570				
1750	6.3	108.1					
1800	9.2	157.8	959	71	36	30	11
1831	14.0	240.1			25	41	12
1851	17.9	307.0	2,360	50	22	43	16
1861	20.1	344.7					
1901	32.6	559.1			9	46	22

SOURCES: Population: Edwards, *Making of the Modern English State;* 8; Smith, *This Realm of England,* 233; Wright, *Middle-Class Culture in Elizabethan England,* 10; Wrightson, *English Society,* 128; Chambers, *Population, Economy, and Society in Pre-Industrial England,* 1208; Underdown, *Riot, Revel, and Rebellion,* 18; GBCensus1851, 8; Mitchell, *European Historical Statistics,* 24; Mitchell with Deane, *Abstract of British Historical Statistics,* 8–10. Rural percentage: Allen, *Great Britain and the United States,* 41. Occupational distribution: Deane and Cole, *British Economic Growth,* 143; Overton, *Agricultural Revolution in England,* 8.

would overwhelm the importance of agriculture by the middle of the nineteenth century. However, the political victory of industry and trade over aristocracy and agriculture would await the early twentieth century.[5]

British agriculture continued its trajectory of landed concentration in the latter eighteenth century and throughout most of the nineteenth. The great disturbing element was the French Revolution and the Napoleonic Wars. The conflict produced large gyrations in the price of wheat, and, when the war ended in 1815, an immense depression in the countryside. Returning soldiers flooded the rural areas, depressing wages and producing social conflict for several decades, most notably in machine-wrecking and rural riots. At the same time taxes to pay off the debt from the Napoleonic Wars and the extra expense of the poor rates drove many tenant farmers to distraction. A considerable out-migration of farmers from England occurred between 1815 and 1830. English agriculture stabilized in the 1830s, and by the 1850s, it entered a period of prosperity that would last until 1874. Farm population reached its apex in terms of numbers in 1851 when some 2.1 million worked in agriculture.[6] For the next two decades, the aristocracy, peers, and many tenants reveled in high incomes. Entering the 1850s, the decade that saw sectionalism in the United States reach its political climax, the English gentry and aristocracy sat unrivaled in their grandeur. It was also the moment of glory of English farming—of its productivity, efficiency, and beauty. Most of the achievements that English agriculture had reached were ascribed to the organization of agriculture into large farms, managed by tenants and worked by waged labor. But within the garden was a growing man-made blight.

British Agricultural Techniques

Improvements in cultivation occurred throughout eighteenth-century England, but how extensively they were adopted remains questionable. Labor productivity—output of agricultural produce per hour worked—seems not to have advanced much during this time, and in the nineteenth century it evidently surged because agricultural labor became more male than female.[7] It would not be until the middle of the nineteenth century that machines had any impact upon British agriculture: the tools employed on the farm in 1800 were hardly different from those in 1600. The amount of land used for agricultural purposes grew from 21.0 million acres around 1690 to

29.1 million acres in 1800, but the division of this land into areas devoted to crops versus pasture is uncertain. It is known that English landowners were using drainage systems and reclaiming some wasteland.[8] Yields per acre for grains probably did increase as knowledge of Flemish methods of cultivation spread throughout the British Isles, rising from ten bushels of wheat per acre to thirty or more. The fallow period—letting fields lay idle for years—was abandoned as farmers began using turnips, legumes, and other nitrogen-adding plants to the fields and allowed animals to graze upon them. Better seed selection also aided yield per acre. In addition, manuring, especially the use of sheep dung, became a widely adopted practice to preserve soil fertility. Manuring was in fact a formidable word in British agriculture in the eighteenth and nineteenth centuries, and a word used more than any other to determine the sophistication of farming practices. It was intimately connected to other economic aspects of British farming, especially to the need for cheap labor. Manuring was vital to obtain good crops, but it took much labor to spread the manure properly over the fields, and thus one obtained the English farmer's insistence on a supply of cheap agricultural labor.[9]

Historians in the late nineteenth and the early twentieth centuries linked certain names to improvements in agriculture and talked of an agricultural revolution. Jethro Tull's (1674–1741) work in Oxfordshire and the use of clover was lauded as was his insistence on eliminating weeds. Lord Townshend (1674–1738) labored at reclaiming land, using turnips to restore nitrogen to the soil, and helped establish the "Norfolk rotation" on the Norfolk estates; he did not fallow land and used turnips and other legumes to feed sheep. Robert Bakewell (1725–1794) began the improvement of sheep stock, followed by Charles Colley (1751–1836), who worked on improving cattle. Last of the great names of English agriculture in the eighteenth century was Alfred Young (1741–1820), the secretary of the First Board of Agriculture, a promoter of improved agricultural practices, and an advocate of large-scale farming (farms of 150–300 acres). They were the apostles of manuring, clover crops, draining and reclaiming, and applying marl, but how much impact they actually had is now doubted. Nonetheless, they reflected changes occurring in British agriculture.[10]

While the total of agricultural output and the forces producing it remain the focus of economic historians, others have been more absorbed by the subject of landownership. The pattern is clear: English agriculture was characterized by the large-scale farm, the disappearance of land-owning peasants, the growing prevalence of tenancy, the emergence of the agricultural wage-earner, and the seemingly inexorable elimination of yeoman farmers. Unfortunately, there are few numbers to indicate this transformation, especially of peasants, wage-earners, and yeomen. The only statistics are those offered by Gregory King in 1688, and almost everyone cautions that they are unreliable—but no one has yet offered an acceptable alternative.[11] The numbers in agriculture that King produced are these: aristocracy (temporal and spiritual lords), 60 families; baronets, knights, esquires, and gentlemen (the gentry), 16,400 families; freeholders (two undifferentiated categories), 180,000 families; farmers, 150,000 families; cottagers and paupers, 400,000 families. The category of laboring and out-servants consists of 364,000 families, some of whom must have been in agriculture. Robert Allen insists that the 150,000 families labeled farmers, which normally would mean tenants, made copyholders almost equivalent to yeomen. One may assume that 100,000 families in the farmer category were copyholders and 50,000 were tenants; and that of the 400,000 families among cottagers and paupers 300,000 were peasants owning or using less than 20 acres of land and 100,000 were wage workers; and of the 364,000 laboring and out-servant families, 200,000 were agricultural wage workers and the rest were urban and village workers. By 1851, the number of landowners dropped to about 300,000, of whom 217,000 were peasants, 100,000 were yeomen, about 15,000 were gentry, and 1,460,000 were farmworkers. The bulge in the laboring category is unmistakable, and by 1850 one obtained the well-known result that some 15,000 landowners ruled English agriculture—and the situation was even worse in Ireland and Scotland.[12]

The erosion of the "middle" in English agriculture during the eighteenth century, accelerating into the nineteenth century, was due to several forces. The first and foremost was the extreme social status given to landownership. It not only conferred economic and political power, but a prestige English society bestowed upon those living independently in the countryside and engaging in rural sports—even if the magnates spent two-thirds of the year in their London homes. That desire to be a landowner meant that primogeniture and strict settlement were vitally important legal devices to thwart the sale of land according to market principles.[13]

Previous historians had believed the largest culprit for amassing land into a few hands was the second round of enclosures between 1750 and 1850, the greatest activity being in two periods, 1760–80 and 1795–1815. Enclosures raised a howl from historians in the first half of the twentieth century, but modern scholarship now finds fewer social dislocations resulting from it. The largest group affected were probably cottagers who lost their small holdings once they had no access to the open fields and could not meet subsistence needs; they had to rent cottages (for about two pounds per year) and were turned into a propertyless farm proletariat.[14]

Many now believe the fate of the small propertyholders was determined by price movements, weather, and inaccessibility to financial services. For those who owned less than twenty acres, life was precarious. They had enough land to feed the family and perhaps generate a minimal surplus, but saving for difficult times was out of the question. A series of bad harvests could throw cottagers and peasants into desperation, leaving only the possibility of selling the few acres they possessed. Because of their minute size, they could not arrange bank loans. Over time, the larger landowners simply bought them out when they fell into indebtedness.[15] Copyholders, in contradistinction, lost rights to their land because at various times the copyhold had to be transmitted to a new generation; then the landlord could impose fines that proved too heavy a burden and the copyholder lost the ability to keep and transmit the land to others. Smaller yeomen, possessing less than 50 acres of land, faced some of the same problems as cottagers, but an alternative may have changed their status. Yeomen, accustomed to independent thinking and having managerial skills, may have decided on becoming tenant farmers if the land they were to manage was significantly greater than what they owned. Simply stated, ownership of a few acres was going to equal a small income no matter how intense the cultivation; if a steward or bailiff offered a yeoman the chance of managing 100 to 200 acres, the chances were much better that his income would significantly increase.[16]

Financial capabilities of the peers and gentry expanded markedly after 1690 and were largely responsible for the consolidation of landholdings. A mortgage market emerged in the late seventeenth century that gave the landed magnates income beyond their rents. Thus, when land became available through some misfortune of the small landowners, the magnates had no difficulty in getting the finances to purchase the land. It is well known that the gentry and aristocracy were eager to expand their holdings in the eighteenth century. The result was the amassing of land into a few families until by 1873 or earlier some 7,000 families owned 80 percent

TABLE 2.2 Landholding Pattern of England and Wales

	Percent of Land Owned (Selected Years)			
Group	*1436*	*1690*	*1790*	*1873*
Magnates	15–20	15–20	20–25	24
Gentry	25	45–50	50	55
Yeomen	20	25–33	15	10
Church	20–30			
Crown	5			
Church and Crown	5–	10	10	10

SOURCE: Overton, *Agricultural Revolution in England,* 168.

of Great Britain.[17] Table 2.2 gives the landholding pattern of Great Britain over time; it was assembled primarily through the researches of G. E. Mingay and F. M. L. Thompson in the 1960s and has not been altered since.

The agriculture in England that was emerging by 1750, and evident as early as 1600, was land amassed in the hands of a few families who then subdivided the often scattered lands they owned and found tenants to work the farms. The farms ranged in size, but at some point in the eighteenth century the great landowners wanted farms exceeding 200 acres because more rent could be obtained. They used a supervisor to get the tenants; the tenants then recruited wage-earners by the week from cottagers and villagers. The main characteristic of English farming became the large farm, the tenant manager, and the mass of wage-earning farmworkers. The small farmer continued to exist except he worked a small farm (under 100 acres) by renting it. Yeomen had not entirely disappeared, but increasingly England was known as a land of tenant farmers.[18]

The Aristocracy and Gentry: the Great Landholders

Eighteenth-century England and Wales made up a rural, agricultural nation only beginning its momentous journey into industrialism and urbanization, and it was particularly known for the magnificence of its aristocracy and high gentry, who exhibited their superiority by a dazzling display of wealth—outside of a few merchants, they had no competitors in shows of materialist grandeur. In table 2.3, the estimates of yearly income for various occupations are provided as a backdrop to comprehend the heights the landed magnates of England attained.

Around 1800, there were perhaps 300 peers, 890 knights and baronets, 6,000 esquires, and 20,000 gentlemen, for a total of perhaps 25,000 people in the ill-defined category of gentry.[19] Accounts about the aristocracy's annual income were almost as awe-inspiring as their expenditures. This was at a time when it was estimated that an income of £100 to £150 meant a comfortable life. It was generally conceded that to be a "gentleman" an individual needed an income of £300 and to possess 300 acres—the lowest rung of the leisured class. (Probably two-thirds of the English population earned less than £50 per year.) The Duke of Newcastle in 1715 obtained from rents £32,000 per year. In the nineteenth century, the American traveler and agricultural observer Henry Colman reported incomes of two earls at £100,000 and £150,000. The exchange rate between the pound and the

TABLE 2.3 Examples of Annual Income of Selected English Occupations, 1600–1870

Occupation	*1600–1700*	*1700–1800*	*1800–1870*
Agricultural Laborer	£10–15		
Artisan	£25		
Tenant	£44		
Yeoman	£50	£50–100	
Greater Yeoman	£84	£100–200 (up to £700)	
Gentleman		£300 minimum	
Clergy		£500	
Steward		£300–400	£200–1,500
Lower Gentry	£500		£1,000
Upper Gentry			
Squires		£1,000–£3,000	
Knights & Baronets		£3,000–£4,000	
Peers	£5,000	£5,000–£10,000	£5,000–£10,000

SOURCES: Agricultural laborer: Mingay, *English Landed Interest in the Eighteenth Century,* 9, 19–24. Artisan: Kulikoff, *From British Peasants to Colonial American Farmers,* 20. Tenant: Porter, *English Society in the Eighteenth Century,* 68. Tenant, laborer, and yeoman: Clarkson, *Pre-industrial Economy in England,* 221–24; Beckett, *Aristocracy in England,* 51. Gentleman: Porter, *English Society in the Eighteenth Century,* xv, 14. Clergy: Porter, "Development of Rural Society," 841. Steward: Porter, "The Development of Rural Society," 844. Lower gentry, upper gentry, and peers: Thompson, *English Landed Society in the Nineteenth Century,* 27, 110; Beckett, *Aristocracy in England,* 51.

dollar hovered for two centuries around $5 for £1; thus the above incomes would be $160,000, $600,000, and $750,000 in the years 1715 to 1850. Roy Porter suggests that, to translate prices between the mid-eighteenth century to 1990, one can multiply the 1700s figure by 60 or 80. The above numbers thus translate into annual incomes of tens of millions of 1990 dollars (times 60 in the above cases would yield $9,600,000, $36,000,000, and $45,000,000). In the eighteenth century, this income was derived principally from rent. And it was well known that to amass this kind of income from rent, the magnates needed to own thousands of acres of land. In an 1856 article in the New York *Saturday Evening Post,* the writer jotted down the size of several estates: the Duke of Sutherland owned 96,000 acres; the Marquess of Breadalbane could march a hundred miles from his house to the sea and never leave his property, the Duke of Norfolk had a personal park in Sussex that was fifteen miles in circuit; and the Duke of Richmond had 40,000 acres at Goodwood and 300,000 at Gordon Castle. The eighteenth- and nineteenth-century presses may have been guilty of some exaggeration, but historians have for years confirmed the magnitude of these holdings.[20]

The conspicuous consumption of the magnates took the form of a building program of breathtaking proportions: estate mansions and vast holdings in the city of London. One such mansion was built in Kent with 365 rooms; the Marquess of Rockingham had two hundred yards of foreground before entering his house; the Duke of Buccleuch had eight country houses and two palaces in London. The Duke of Marlborough built Blenheim castle for £250,000 and Eastbury for £125,000; the Earl of Leicester built Holkham Hall for £90,000, and the Nevilles spent £100,000 for Audly End; one nobleman spent £1,219 on the trimmings for his velvet bed. To keep these huge establishments in operation, an army of servants had to be employed. The owners had huge libraries and possessed a treasure trove of art.[21]

Two other initiatives in the countryside had a pronounced effect on how British travelers would view rural America. The artistic quality of the estate mansion and its surrounding area became a focus of attention in the seventeenth century both in Europe and England, and the result was the creation of sweeping manicured gardens; trees were groomed and trimmed to exhibit the magnificence of the environment. Trees could grow to immense girth and height and be admired for their overwhelming presence. Part of the reason for such growth was the lack of forest. Great Britain had over the centuries converted woodlands into cropland and pasture, resulting in trees not having to battle other trees for sunlight and water. The forests had been reduced to about 5 percent of the land and were reserved for the gentry's hunting activities. However, the English needed a barrier to stop domestic animals from invading the crops—they required fencing. They had insufficient supplies of lumber for wooden fences due to the lack of forest land. The solution was the English hedge, most often the Hawthorne hedge. The hedge grew to a height of six to nine feet and a thickness of three to six feet; it was so thick at the soil that it prevented animal entry into the crop areas. And the Hawthorne hedge had white blossoms that the Swedish visitor Peter Kalm described in 1748 as an "incredible ornament . . . because wherever one turns his eyes it seems as if the whole country were a beautiful and delightful garden."[22]

The impact of the hedges, gardens, huge trees—often elms and oaks—upon the foreign visitor was immediate and breathtaking. Agricultural England looked like a perfect garden. Wrote Frederick Law Olmsted upon his entry into the English countryside: it was "green, dripping, glistening, gorgeous! We stood dumb-stricken by its loveliness." Peter Kalm found England full of villages with well-built brick structures and the countryside laid out into ploughed fields, meadows, orchards, and gardens. Indeed, he characterized his journey as being in a continuous garden—"The country was everywhere divided into small *inclosures,* with hawthorn and other hedges round them, so that one could only suppose that he was travelling all the way through a garden."[23]

Writing of the splendor he saw in the country mansions, Henry Colman remarked that it was "very difficult for New England men even to conceive of such wealth," while Olmsted wrote that the life of the country gentry was "beyond any conception I had previously had of such things." One item that visitors noted was the striking handsomeness of the high gentry and aristocracy: they were the original "beautiful people"—they were "incomparable." Their habits and manners were a subject of some comment, but the overwhelming impression was beauty. As Washington Irving wrote, the gentry had "elegance and strength, a robustness of frame and freshness of complexion." Nobody forthrightly mentioned it, but this beauty came from the leisure class doing no enervating manual labor.[24]

The peers and high gentry required modern management staffing to run the farms on their lands. Part of the problem was that usually the magnates owned disparate patches in several counties, making coordination difficult. The landowners turned to estate stewards who were the active managers of the tenant farmers—they were the individuals who recruited tenants, oversaw their activities, took care of contracts and other legal matters, and maintained the accounts. For the great estates, the management could become complicated, with a chief steward, several accountants, agents for each separate estate (a chief agent); the agents might employ numerous bailiffs (usually a tenant farmer who oversaw the activities of four or five other tenants). Landowners might wish to be free of the chores of management, but they could not; they had to pay some attention or risk the infliction of financial improprieties.[25]

For all that has been written about the large farm in England and the estate system, some basic information is still absent. How many tenants were enlisted is not common knowledge. David Spring shows that the Duke of Northumberland's bailiwicks—146,000 acres—in 1849 had 372 farms averaging 392 acres per farm. Henry Colman described Lord Yarborough's operation of 60,000 acres as having 150 tenant farmers (but then oddly mentioned 600 in all).[26]

Also at issue is the size of the farm. The acres composing a farm provide an important yardstick to measure eighteenth- and nineteenth-century American agriculture. Contemporaries considered England the home of the "big" farm, so some understanding of its dimensions are necessary when comparing it with northern and southern farms. One practical matter has to be understood, however: the acreage given for English farms is in American terminology "improved acreage"—that is, land being cultivated either for crops or grass. Unimproved acreage in the United States was wooded land not under cultivation. The English made no such distinction; it can be assumed that, in American lingo, their farms were all improved acres. The size distribution of English farms and definition of size is given in table 2.4.

The size of the large farm in the estate system is uncertain. Colman said usually the farms were 300 to 400 acres, though he saw some containing 800, 1,000, and even 1,200 acres. Arthur Young put the large English farm at the beginning of the nineteenth century at 200 to 500 acres, while Robert Allen believed that farms above 200 or 250 acres were modal for the magnates. Given the huge number of small farms, however, it may be guessed that the great landlords must have permitted some smaller tenants to persist on their lands.[27] Considering only farms greater than fifty acres, 50 percent of the land could be said to be in large farms, 27 percent in the farm

TABLE 2.4 Definition and Size Distribution of English Farms, Census of 1851

PART A: DEFINITIONS

Designation	*Size*	*Source*
Cottager	<5 acres	Allen, *Enclosure and Yeoman,* 57–58
Peasant	5–30 acres	Allen, *Enclosure and Yeoman,* 57–58
		Clarkson, *Pre-industrial Economy,* 66
Subsistence Farmer	50 acres	
Family Farmer	50–99 acres	
Capitalist Family		
Farmer	100–299 acres	Allen, *Enclosure and the Yeoman,* 92–93
		Daunton, *Progress and Poverty,* 48
Large Farm	300+ acres	Mingay, "The Size of Farms in the Eighteenth Century," 470

PART B: ENGLISH FARM SIZE DISTRIBUTION, 1851

Acreage of Farm	*Number of Farms*
Cottagers, peasants, lesser yeomen, 63.7%	
<100	142,358
Greater yeomen, commercial farmers, 28.7%	
100–199	45,752
200–299	18,401
Large Farmers, 7.5%	
300–399	8,061
400–499	3,585
500–599	1,971
600–999	2,372
1,000+	771
Total farms =	223,271

SOURCE: GBCensus1851, 60.

size category of 100 to 199 acres, and 22 percent in the category of less than 100 acres.[28]

Why English farms grew in size—and everyone agrees that they did—is an important question. There are three general answers. Robert Allen believes the growth was due to landlords who wanted to charge more rent, not because of a quest for greater productivity. Increasing the size of the farm reduced the number of farmers, and so the landowners saved on costs and could charge more rent. A second justification for large farms came from the magnates: they insisted that agriculture could only be improved by the application of capital, and capital outlays produced the best results on large farms. This rationale certainly was true in the twentieth century when mechanization and chemical fertilizers became the means of augmenting yields, but was less so in the eighteenth and nineteenth centuries when animals supplied the needed power. Capital applications became important when ditching and draining were necessary.[29]

However, one particular argument connected with the large English farm has application to the question of the efficiency of the southern plantation. It is the matter of the existence of economies of scale. Several English agricultural historians insist that the large English farm developed not because it increased yield per acre, but because it lowered costs per acre. The cost per acre dropped on the large farm compared to the small farm because the same amount of capital was capable of working over more acres than composed the small farm—the capital being in this case plows, hoes, animals, harrows, and other devices. As G. E. Mingay wrote in 1962, "with a large acreage, he [the tenant] found it possible to utilize more economically his teams, wagons, carts, ploughs and other implements." In other words, the capital costs needed to farm 50 acres were nearly the same as 200 acres. To this was added specialization of labor in doing certain tasks, such as plowing. The end result was not that the large farm produced more product per acre; it produced the same amount of product per acre—maybe even less—but did so at a much lower cost. The lowered cost meant the landlord could charge more rent.[30]

Rents differed throughout England, but overall they increased with the advent of the large farm. Robert Allen has found that rents in the midlands were about ten shillings per acre in the early eighteenth century and then doubled by the late eighteenth to one pound. In the 1870s they would rise again to nearly one and a half pounds.[31] Generally, it could be said that rent in the nineteenth century was averaging between one and two pounds per acre. G. E. Mingay has surmised that the magnates were making 3 to 4 percent return on their investments in land, not a particularly good rate.[32] However, until the twentieth century land had such social and political prestige attached to it that the landlords were not overly concerned about the marginal efficiency of their investment. For the great magnates, the ones owning 10,000 acres and more, the rate of return may not have had much meaning. After all, they were making over ten thousand pounds per year (or fifty thousand dollars in nineteenth-century American terms, three million dollars in 1990 dollars) and could live quite handsomely on that amount. Percentage gains or losses may mean something in comparative economics and to psychotic financiers, but absolute dollar amounts for an individual may be all that matters.

It is also vital to understand one glaring fact about the large English farm: although there was a crop rotation and several grains were produced, basically the large English farm was engaged in wheat cultivation. Economies of scale in England were obtained on large wheat farms. The same condition was as fully applicable to North America except for one element that was lacking: cheap labor. And there was a way to maneuver around that barrier.

The Tenant Farmers

To run their large farms, the magnates turned to tenants. The subject of tenancy runs into problems with an American audience because of the bias in favor of family farm ownership. When Sung Bok Kim investigated the eighteenth-century New York patroons, he had to take time to point out "our ingrained ideological animus against landlordism" and its "inevitable corollary . . . [the] worship of yeomanry." Among American historians, the cry of degeneration and inhuman inequality rises as soon as a tenant is discovered on American soil. England probably had a 33 percent tenancy rate in 1600 or 1640; by 1790, it was estimated that 75 percent of Great Britain's farmers were tenants, or, in qualitative terms, England was a land of tenant farmers, not owner-occupiers. But England had a vastly different opinion of tenants than did Americans.[33] Roy Porter summarized that in eighteenth-century England the tenant farmers were the "pride of English agriculture" and that they were "one of the success stories of Georgian England, and they marked their good fortune by claiming their share of the good things in life."[34]

English tenants had to possess a considerable amount of capital to engage in farming successfully. Usually the landlord constructed buildings and any permanent improvements on the land—often forbidding tenants to do certain things, such as clearing trees or draining ponds. Tenants provided animals, implements, seeds, and equipment. Manures were an important expense for the tenants; Henry Colman noted that thousands of pounds were spent annually on it. The rule of thumb was a tenant needed capital that was three or four times the amount of rent; thus a rent of £2 per acre meant a tenant needed capital of £6 to £8 per acre. Tenants often had thousands of pounds of capital—very large sums of money. In the eighteenth century, tenants also had an idea that a farm produced three rents: one rent to the landlord, one rent for costs and taxes, and one rent for individual profit. The profits could mount up. Henry Colman related his finding on Lord Yarborough's land that one tenant raised eighteen thousand bushels of wheat, and many of his tenants paid rents of £1,000 and £1,200 yearly—considerable sums of money. Colman also found that these tenants lived like nobles and rode in carriages. A report in the *Farmer and Gardener* said the Earl of Yarborough paid one of his tenants £54,000 a year, or $290,000—at a time when the income of the president of the United States was $50,000 per year.[35]

Tensions flared up between the landlords (or their agents) and tenants over three precise matters. They argued about the amount of rent and the length of the rental agreement—which by the nineteenth century was becoming yearly instead of nineteen years as it had been in the eighteenth century. The third area of dispute was improvements to the land. Over time it became undeniable that tenants improved the value of the land by their operations, but they had no way to claim the value of the improvements. In the last quarter of the nineteenth century, Parliament began attacking the problem, trying to enable the tenants to claim the value of the improvements they made.[36]

The British Agricultural Laborers

When Henry Colman visited England to investigate its husbandry, he noted that the word "farmer" meant something different from its meaning in the United States. The English farmer did not "labor in the field; they rarely do any personal labor whatever." Indeed, the social aspect of English farming "strike[s] a visitor from the United States with much force." The nation had clear class boundaries between landowners, tenants, and laborers. The agricultural laborers were the ones in England who did the actual farming. The whole of English agriculture rested on their activities, and in the picture of beauty painted of the English garden they were the blight spreading over the land.[37]

Wage labor in English agriculture had existed since the fourteenth century, but it was then a minor feature. Most agricultural labor was done by copyholders and yeomen with the added help of farm servants. The latter were young men and women who lived in the house of the farmer and who did agricultural chores; they were called servants. Ann Kussmaul has estimated that, prior to 1700, 60 percent of people aged fifteen to twenty-four were servants. The enclosure movement, the rise of the large farm, the ending of the commons, and the establishment of wastelands and forests as preserves for the gentry and nobility dispossessed many of their small holdings, and they began to rely entirely upon wages for survival. Some of the excess population in agricultural labor arose from the failure of youth to move out of the countryside and into urban areas; for some reason, perhaps relating to the poor laws, England's agricultural population seemed immobile. By 1770 or 1790, it was apparent that a propertyless wage-earning agricultural population inhabited the countryside, a "rural proletariat" in the phrasing of Peter Mathias. They worked for weekly wages, rented cottages from landowners, and, except for beggars and the urban unemployed, were the most wretched element of English society.[38]

According to a table of occupations for Great Britain in the census of 1851, there were 1,460,896 persons who were male agricultural laborers, farm servants, and shepherds, and 227,000 female agricultural servants; the next largest category was domestic servants (1,038,791), then cotton and calico workers (501,465), then building craftsmen (443,000), followed by undefined laborers (376,551); a long list of other occupations ensued.[39] As of 1850, the category of farm laborer contained by far more people than any other. Because farm servants most frequently were treated differently from weekly agricultural farmworkers, the farm servants should perhaps be separated from the weekly farm wage-earners. Those wholly dependent on wages probably numbered between 600,000 and 1,000,000. Given the existence of some 225,000 farmers (yeomen, tenants, and landowners possessing more than fifty acres of land), the ratio of laborers to farmers (yeomen, tenants) was about 4 to 1, but perhaps for wage workers the ratio was closer to 2.5 to 1. Whichever ratio is correct, the obvious inference is that nineteenth-century rural England was a land of wage-earners, not owner-occupiers.[40]

Wages paid to agricultural labor varied considerably in England and Wales. The lowest were in the southern part of the nation and in Wales; they improved as one moved north. Another condition was the proximity of farm laborers to industrial towns where an alternative employment could be obtained. For men, wages ranged from 6 to 12s. per week, also dependent on whether the season was harvest time or not, wages rising during harvest because of the heightened and urgent demand for workers. Women's wages were far less, ranging from 7 to 10d. per day in the spring, and perhaps 1s. during harvest time. These wages did not include room and board. Peter Mathias gave an illustration of how much lower agricultural wages were than manufacturing/artisanal ones for the years 1815 to 1850. He found that farm wages ran from 8 to 10s. per week; the average for unskilled industrial employees was 18s., for London skilled workers 25 to 40s. per week, and for cotton textile laborers 20 to 25s. per week. In the countryside, farm laborers put their children to work because their employment could raise family income considerably. By 1850, a single man at Norfolk and Suffolk could make £25 per year; if he had four children over age ten and put them to work, he could obtain £50 11s. per year (or in U.S. terms, roughly $255 per year). However, after subtracting the cost of rent and food, the farmworker made barely over £4 to £5 per year (or $20 to $25). At the same time, a male American farm laborer was making beyond the expense of room and board roughly $150.[41]

A hierarchy existed among farmworkers. The lowest was the unskilled worker hired for ditching, hoeing, manuring, and harvesting. Certain workers became specialized and were often hired yearly as servants: shepherds, horse tenders, ploughmen, wagoners, and cattlemen, with shepherds occupying the highest rung on the ladder. But the mobility stopped there by the middle of the eighteenth century. Prior to 1700, many of the farm laborer positions, and especially farm servants, permitted saving and slow entry into landownership—perhaps to the status of a cottager or a peasant, but the English farmworker had not been absolutely frozen in his or her occupation. That mobility disappeared in the countryside. In 1881 George Brodrick wrote, "Still, the fact remains that he [English farmworker] scarcely ever becomes the owner of land, and seldom rents enough to call forth his full energies, or to serve as a stepping-stone of a cottage to the occupation of a farm."[42]

The comparison of British large-scale farming with the American slave plantation rises almost immediately. The usual definition of a member of the planter class is ownership of twenty or more slaves. Besides ownership and distribution patterns of wage labor in English farming and American plantations, one other feature in Great Britain invites the comparison: the development of the "gang" system in British agriculture. The gang system appeared on the large English farm after 1815, though some date it to the 1820s or 1830s. Arthur Young saw gangs of six to ten men, used in hoeing, harvesting, and manuring. By mid-century, there were gangs of up to seventy, men and women, often migrant Irish workers. They were brought together by an overseer, who took them from farm to farm to get jobs; sometimes they ranged seven miles from their homes and were taken on wagons when work was at a longer distance. The overseer set the wage rates low; the work was done on a piece-rate basis, thereby encouraging a maximum of effort and enabling the overseer to weed out the inefficient. Henry Colman saw the system at work and was appalled at the condition of the workers. They were "starved, ragged, and dirty beyond all description," especially the Irish. He also said prostitution ran wild among them. Colman wrote that he had no objection to women working in the fields—they did so in Connecticut during turnip harvesting—but the gang system in England resulted, one told him, in 70 percent of the women becoming prostitutes. Eventually Parliament moved against the gang system by legislation in 1867–68. Historian W. A. Armstrong wrote, "The gang system was a self-evident scandal to everyone excepting those with an interest in its perpetuation." The American plantation gang system also used women and children and worked slaves in groups of fifteen to thirty. According to Stanley Engerman and Robert W. Fogel, productivity and economies of scale in the cotton fields came out of the gang system, but they never referred to it as a "self-evident scandal."[43]

Historians have fastened upon the factory system and the wretched conditions of English industrial workers so much so that they have excluded the British agricultural workers from their narratives. That has created an unbalanced perspective on the nineteenth century. Putting agricultural activities and particularly the condition of agricultural laborers back into the flow of events produces a different understanding of nineteenth-century viewpoints on the comparison of political systems and of free labor and slave labor.

No one has painted a pleasant picture of the farmworkers of England after 1790 or even after 1600. They were desperately poor, destined to remain that way, and they and their families may have represented, said Henry Colman, one-fifth of the kingdom.[44] Americans were aware of the distressed state of the English farmworker. One writer for a Boston periodical talked of the nonsense of England as a bucolic garden with happy workers tending the fields: they were "bound, as with chains of adamant, to a destiny immutable and eternal of poverty, and hardship, and sorrow, and ignorance, and brutishness; in the very midst of enormous wealth, and overflowing superfluidity, and inordinate, unbounded luxury, and refinement of self-indulgence, such as the world has never seen." This combination of the extraordinary luxury of the nobility with the awful poverty of the farmworkers was typical of American commentary. In it one reads the condemnation of the code of gentility, which in its economic essence meant some must live laboriously without reward while others lived luxuriously without labor. In 1885 American observers were still struggling with the social debacle of English farming. Adam Badeau, an American diplomat, summarized how the destitution of the farmworker was tied to the fantasy lifestyle of the aristocrat: "But it is upon this poverty, degradation, and misery [of the farm laborer] that the grandeur and luxury of the aristocracy are founded. One is the direct cause of the other."[45]

The personality traits of what the English farm laborer had become were part of the observations commentators made. Historian Pauline Gregg described how "their low condition . . . showed itself in their small stature and slow gait," and they seemed "dull-witted as well as slow of speech." Adam Badeau wrote in 1885 that they were "dwarvish, their gait slow and sluggish, like their minds." When visiting in 1850, Frederick Law Olmsted called the farm wage-workers he met "the most degraded, poor, stupid, brutal, and licentious that we saw in the kingdom"; the workers looked down, having a "quiet, sleepy, stupid expression." Henry Colman reported the same thing; their "gait [was] very slow" and they grew old early.[46] Connected to all these traits was also something very unsettling to Americans: servility. As Badeau said, "They have, of course, no independence, and are in reality serfs of the soil." The Americans witnessed a gaping separation between the classes: the farmworkers bowed and scraped before tenant farmers and gentry; they always doffed their hats; they never entered a farmer's house through the front door, and the custom of eating at the same table had by 1860 been absent for at least six decades. In short, the English farmworker was treated as part of the livestock.[47]

Farmers and landowners seemed not overly disturbed by the debased condition of the people who actually worked the soil. Several historians early in the twentieth century believed the tenants and landowners wanted laborers deprived of any landed property so that they would be compelled to work for wages or face starvation. Others simply said this species of human being deserved no more than animal existence. Historian E. L. Jones quotes one as saying that "farm wages should be unrelated to rent or profit levels and that a bare living wage, by the standards of the day, was all that the labourer had a right to expect." Olmsted decided that the English gentry "regard their laboring class as a permanent, providential institution, not to be improved in any way, educated, fitted to take an equal share with all Englishmen in the government of the commonwealth of England"; they were "tyrants." He added a few pages later, "God keep us evermore free from a powerful conservative landed gentry."[48]

One of the features that greatly disturbed American visitors and English reformers was the absence of education for farmworkers and their children. Olmsted was amazed at one worker who could not read or write and did not know his own age. Some education services were available for a fee, and several noted the existence of Sunday schools. The illiteracy festering in the English countryside was starting to become troublesome economically as machinery and knowledge became important to farming after 1870. Any number of English visitors to the United States noted with astonishment the high educational level of American farmers and farmworkers because of the public school system. An English clergyman traveling to St. Louis in 1867 recorded his amazement at a discussion among farmers during a railroad trip: "Men of the roughest exterior, of the class we might call 'clod-hoppers,' and from whom one would expect no coherent reply to any question on matters beyond the daily toils and necessities of life, were discussing domestic and foreign politics with great astuteness."[49] Until 1860 the gentry attitude toward education was probably that it was a taxpayer burden and money wasted on people who belonged to the lowest rungs of society and were not capable of improvement.[50]

Citizens of the United States focused on conditions in Great Britain because of historic ties to that land, but they were aware of agricultural developments in other European lands. England was unusual in its large-scale agriculture; most other nations (save Italy and Spain) moved to protect small farmers (the peasantry). Observers from the American republic, however, had to admit that, as wretched as the English farm laborer's condition was, the condition of the European counterpart was probably worse. The British farmworker at least had a cottage; the European farmworker, according to one report, slept in the stables with the animals.[51]

The Gentry's Ideology of Inequality

The English agricultural example until the middle of the twentieth century was an aristocratic one, and as such it was founded explicitly on the ideal of inequality. The spokesperson for this viewpoint was Edmund Burke, who became the standard-bearer for aristocracy, inequality, and tradition.[52]

Everything in eighteenth- and nineteenth-century England was so socially, economically, and politically lopsided—and had been so since the 1500s—that a defense of inequality naturally arose. The justification derived from religion and the natural world. Britons argued that both God and nature had made men unequal in talents and abilities. In the case of religion, it was a reflection of the Great Chain of Being and the belief that God created certain people for certain tasks. In terms of nature, it was simply an observable fact that different people had different talents and levels of intelligence. Thus the ideal of equality defied the rules for the human world fashioned by both God and nature.[53]

Given the fact of unequal talents, the group who had talents had to be permitted to exercise those abilities for the betterment of humankind. These kinds of talent were intellectual, and so to foster them this group required leisure—a position of luxury that enabled them to live without labor. A leisure class became vital in politics because leisure enabled that group to study problems of human relationships with a supposed lack of passion and bias because they were not consumed by straining all their energies to earn basic subsistence. Thus politics was to be restricted to those who had the talent and the leisure to do a proper job; the masses, doomed to grinding labor, simply lacked the ability to think beyond their immediate physical needs.

The leisure class in the English vision needed to be composed of the great landowners. Supposedly rural life produced a conservatism best for the community. Moreover, as property was the basis of civilization and property ownership required the power of the state to insure its preservation, only the propertied should operate government. The constitutional benefit of an aristocracy was also ascribed to its ability to check the arbitrary rule of the monarch and thus preserve liberty, the word "liberty" becoming the political standard to which the British aristocracy rallied after 1689. Explicit in this rationale for the existence of a pampered upper class was a negative assessment of common people: passions and immediate needs ruled them, and they lacked the wisdom and intelligence to produce stability and civilized life.[54]

British landowners prized most highly the policy of low taxes, especially on land. Their luxurious lifestyle was to be maintained at all costs, even at the cost of impoverishing the people inhabiting the countryside. By their control of the county courts, the churches, and county officials, the squirearchy made sure that they could evade local taxes.

In their insistence on the inviolability of their property rights in landownership and their vigilance in maintaining low taxes on themselves, the high gentry and aristocracy, aided by the other landowners, made no bow to any doctrine that held that their holdings were sacred because of any act of labor. The English aristocrat or the English gentleman was known in the Anglo-American world as a person who "live[s] without labor," and who never had to suffer the biblical dictum, "in the sweat of your brow shall you eat your bread." Their economic magnificence drew its justification not from contribution to social wealth but from family longevity and the code of gentility.[55]

The British ideal of inequality produced a society that prized hierarchy, patriarchy, and class differentiation. It extracted deference from the commoners—often a brutal deference. Supposedly the inegalitarian society would work well because the lower orders had little possibility of rising to higher positions—a few did, and society should accept those with talent—and the problems of the lower classes with life would be smoothed over by paternalism, charity, and the church. More to the point, the miseries of the poor were ignored—it was so easy to ascribe their condition simply to the will of God and the decree of nature. Luxury felt no need for compassion or concern for the lower class, only protection of its own lavishness. By implication, low wages were also a part of the divine order of things.[56] By implementing the inequality principle in politics, one obtained reasoned discourse instead of demagoguery and outlandish debate. Aristocracy would protect liberty. And while progress would not be denied, it would come slowly because tradition and family would rule most matters.

Human nature did not change. This was the lofty, idealistic concept of life in an aristocracy. Sometimes parts of it were obtained. But in many ways, the aristocracy's and the gentry's rule in Great Britain could almost be reduced to a control of government to insure that their positions were not threatened, their sway over locality was never challenged, and their material magnificence never interrupted.

Agrarian America would be a challenge to aristocratic England. The two were in one philosophical sense opposites: England stood for inequality, while Americans stood for equality—although Americans frustratingly never bothered to define the term. Most historians and commentators have not found equality, with the possible exceptions of universal white adult male suffrage and the demise of deferential behavior. The economic side of the American belief in equality, however, has been hotly rejected. But it was there. It was there in rural America and in the free labor ideology.

PART II

Northern Family Farming and the Southern Plantation Complex

English migrants to North America brought with them their practices of farming, tastes for certain crops, use of animals—and ideals about society and economy. What happened in North America was determined by a combination of five factors: availability of land, availability of labor, climate and soil, the native population, and attitudes toward wealth. Together these five factors would shape the first two hundred years of North American existence.

The essence of the British experience was that, in order to be wealthy, to have status, to live in mastery over others, one needed to possess land by the thousands of acres. But there was a condition on having this much land: vast amounts of acreage meant little unless one also had access to cheap labor. To obtain the profit that enabled one to live in gentility, one had to possess the *combination* of vast acreage and cheap labor. It was not the landlord-tenant relationship that made the magnates of England fabulously wealthy; it was the existence of cheap agricultural labor. Without a source of such labor, English agriculture would have been a mass of small tenant farmers whose farms would have averaged thirty to one hundred acres; they would have been family farms, but the families would have borne all of the cost of the rent. (To point this out exactly, a system of tenancy in England in which families ran small tenant farms would have been nearly equivalent to southern sharecropping after the Civil War.) Rent would have sapped the initiative of the family farmer, and English farming would have been little better than the husbandry of open-field farming under the manorial system. The magnates would not have earned the vast monetary sums they obtained from dividing the land into large farms operated by tenants. An oversupply of labor opened up the ability to manure the land cheaply and to obtain a division of labor that the family farm could not possibly duplicate.

Labor conditions determined the choices that the English in North America would make. Given the abundance of land, a re-creation of the English land system was possible, and certainly numerous individuals carried that motivation with them to North America. The amount of labor necessary to make a magnate system operate successfully did not exist. So the choice became either to accept the dictates of the existing conditions of land and labor and let agriculture develop into small farms run by families, or to create artificially a supply of cheap labor and establish great estates. Thus one obtained the slave-plantation South and the family-farming North, to be known in the nineteenth century as the slave labor South and the free labor North.

3

English Husbandry on Another Continent

NORTH AMERICAN FARMING, 1607–1815

When the English began colonization in North America in the early seventeenth century, they brought over their customs, their values, their religion—and their land laws. In particular, except for the Puritans, they brought over the English estate ideal. They did not bring over feudalism because feudalism had been dead in England since at least 1500. What the upper class carried mentally was the ideal of family ownership of vast acres that were cultivated by tenant farmers who employed cheap labor. In this way the leadership could obtain wealth without manual labor and be eligible for entry into the gentry class and, with some connections and luck, into the aristocracy. Settlement in North America followed a pattern that rested squarely on European ideas and practices. To obtain settlers, some authority (monarch, corporation, proprietor) gave out huge land grants, containing tens and hundreds of thousands of acres, to highly placed individuals who then bore the responsibility for placing settlers on the land. This relieved the government of the expense of handling colonization itself, but it gave tremendous authority to individuals to provide orderly settlement. Some tried fiefdoms, most sought a system of tenancy, and some, like William Penn, allowed easy acquisition of land by small farmers.

All North American landowners ran into the basic dilemma of the vast amount of land—once the Native Americans had been deprived of it—and the insignificant number of laborers.[1] The English wanted to establish the estate mode of farming. Tenancy under certain conditions could work well for landlord and tenant, depending on the level of technology and farm productivity, but at the crude level of agriculture in the seventeenth and eighteenth centuries, tenant farming done solely by a family simply could not generate enough output to earn sufficient income for both tenant and landlord. Landlords could not grow rich, in short, by owning vast amounts of land divided into tracts of fifty-acre farms worked by family-farming tenants. Thus the origins of North American slavery: it arose out of the desire of English leaders to reproduce the estate system of English agriculture in the American colonies.

The attempt to replicate the estate system in North America led to land "engrossment," the term frequently used by contemporaries to describe the consolidation of landholdings into estates of several thousand acres. This process took place in all the colonies, but most markedly in the Tidewater area. It had disastrous consequences because the consolidation of land into large holdings created an artificial land scarcity and sent the humbler members of society, who abhorred the idea of being poverty-stricken tenants, in search of new western land. With them went the sons of great landowners in search of their own little kingdoms. The engrossment of land in the Tidewater produced one of the forces of colonial western expansion, which in turn produced continuous war with the Native Americans after 1750.

Land engrossment in the Tidewater was not the only force promoting the westward movement. Population growth was the other, although it arose from a surprising source: the yeoman farm. To obtain subsistence from a family farm, the family had to include at least five members—three males and two females. To satisfy the labor needs, farm mothers and fathers needed to procreate numerous children. Family sizes of five and six members thus resulted in a population growth that would equal or exceed 3 percent per year, doubling the population about every twenty-five or thirty years. When the children grew up, they needed land to farm, but they could not find it within the constrained boundaries of New England. The New England egalitarian example might have fallen into the condition of overpopulated Old England had it not been for the West. This became a second source of expansion.

Colonial Land Distribution

The equality in landholdings established by the Puritans in Massachusetts, Connecticut, and Rhode Island continued to be the primary characteristic of agriculture of these colonies throughout the eighteenth century. Of course, there was a catch to this equality: only religious dissenting English Protestants were eligible for landownership and citizenship in the New England colonies. The New England equality in landholdings has to be understood as a form of "herrenvolk democracy," a term usually reserved for matters of race in which democracy becomes the condition for members of the master race but servitude or slavery for the social "others," that is, people of a different ethnicity. In the New England case, the intense tribalism of the Puritan descendants limited equal treatment to people of their own ethnic stock and religious point of view; others were outsiders to be banned from their communities.

Great Britain introduced some landholding inequality in New England after the Restoration and the Glorious Revolution. New Hampshire had been given as a proprietary to John Mason while the territory of Maine was given to Fernando Gorges in 1639. However, Maine was too cold, and few migrated to it until the last half of the eighteenth century. New Hampshire almost became incorporated into Massachusetts until the Lords of Trade split it off into a separate colony in 1679. New Hampshire followed the Massachusetts settlement pattern, but Maine had a more troubled experience because it was dominated by proprietors looking for income and perhaps estates. The proprietors of Maine wanted tenants while the settlers wanted freeholds. Thus began a struggle that ended only when Maine was separated from Massachusetts in the Compromise of 1820. Battles over appropriate land dispersal also involved New Hampshire and New York over control of land between the Connecticut River and Lake Champlain. By 1783 this became the state of Vermont, recognized by the Union in 1791 as the fourteenth state. Although there were more struggles over land in the eighteenth century than had been the case in the seventeenth, the family farm continued to win.[2]

The middle colonies of Pennsylvania, New York, and New Jersey proved the power of the estate ideal and underscored how unusual the land pattern of New England was. These states operated under the common belief that the best path to settlement was to give or sell land in large amounts to high-status individuals who then undertook the task of acquiring settlers. Some of the purchasers attempted to set up the English estate system of landowner and tenant, but they had dubious success. Slavery did not root as deeply in the area as it did in the Tidewater, and without a cheap source of labor the landlord-tenant farming system did not operate so profitably. The result was a continuous pressure to sell lands instead of holding on to them. The exception, at least for a while, was the patroon system of the Hudson Valley.

William Penn in 1681 obtained from Charles II the tract of land that would become Pennsylvania, about 45,000 square miles. In laying out plans for the colony, Penn turned to the mechanism of dividing the land into large lots to give to friends and investors who would then find settlers. He sold 300,000 acres, in lots from 250 acres to 10,000 acres, to 250 people. To receive income from the transactions besides the sale price, Penn included a quitrent (basically, a small state tax on the land). Penn also realized that the prosperity of his colony depended on obtaining settlers, and he advertised in Europe that he sold land cheaply. The offering of cheap land when added to religious motivations brought a surge of immigrants from Germany and Scotland into Pennsylvania during the eighteenth century. Although the settlement plan could have fostered large estates, in practice it did not; cheap land led to family or community farming.[3]

Although Pennsylvania was potentially a large colony, the land available for agriculture from 1680 to 1750 was actually quite restricted. The Allegheny Mountains cut across the state diagonally, and settlers soon ran into the Blue Mountains not far west of the Susquehanna River. If one can assume that the available improved and unimproved acreage of 1850 approximately represents the amount of available farmland in 1750, then the southeast corner of the state had potentially slightly over 3.6 million acres for settlers, a serious limitation. Family farming predominated in Pennsylvania in the eighteenth century, but the ingress of immigrants led to tenancy and wage-earning agricultural laborers, researchers finding that the county of Chester had nearly 50 percent of its population without land. Although the numbers are not yet conclusive as to either the extent of landlessness or the collapse of the ability of poor people to obtain property over time, the obvious result of population pressure led to expansion into other parts of the proprietary.[4]

To the north of Pennsylvania slowly grew the colony of New York, which by most accounts had the most notorious land distribution of all the colonies. Under the direction of the Dutch East India Company in 1629, the Dutch created a settlement plan that gave huge tracts of land to the person (a patroon) who would transport fifty families into the colony. Over time these became the patroonships along the Hudson River—huge estates with tenant farms that dominated the colony of New York when the English took over. The estates were Rensselaerswyck (1,000,000 acres), Livingston Manor (160,000 acres), Cortlandt Manor (186,000 acres), and Philipsburg Manor (92,000 acres). At the end of the seventeenth century, things were made worse as governors in the eighteenth century gave out large land grants; two were for 100,000 acres each, eighty-three were over 10,000 acres each.

The crown was not favorable to such grants, but colonial governors when it came to land did not much heed their London superiors. The patroonships made land concentration highly visible in the colony of New York, and therefore a matter of some comment and attention.[5] To farm the land, the patroons used tenants.

New York's development until 1750 followed the Hudson River from Long Island up to Albany. The reason was geography: the area to the west was filled with mountains populated in the valleys by tribes, in particular the Iroquois. After 1750, more Europeans ventured into the Mohawk Valley (running from Albany to Rochester) and into the Wyoming Valley of the Upper Susquehanna River. Then tensions rose with the tribes. Yet, considering the geography, New York, which looks like a large state, was in colonial times confined to the Hudson River Valley. Again, using the counties that surround the river and the improved and unimproved acres enumerated by the U.S. Census of 1850 as a means of calculation, the available land was about 4.7 million acres.[6] Both Pennsylvania and New York had significant obstacles for land-hungry settlers, and thus both had less available land than one would think.

The Tidewater region of Virginia and Maryland was destined to become the land of the great plantations, and soon South Carolina and Georgia would follow their example. These colonies attempted to replicate the British estate system of agriculture, but it took a long time to accomplish. For much of the seventeenth century, farms were more modest in size, resembling a yeoman society. Neither Virginia nor Maryland had the large-scale farm as the model type of farm in the seventeenth century. If newcomers (indentured servants) survived the Tidewater environment, they had a good chance to acquire land and become yeomen.[7]

From the beginning, however, the Virginia Company tried to attract settlers by offering land to immigrants. This was the beginning of the "headright" system by which a planter obtained fifty acres from the government for each individual he brought into the colony, and of "indentured servitude," in which individuals in England, for whatever reasons, decided to trust their luck in North America by having someone (a planter) pay their passage across the Atlantic in return for becoming a farm servant for four to seven years. The climate took many immigrants to an early grave. It is estimated that servants comprised three-fourths of the mid-seventeenth-century population of Virginia.[8]

The headright system was one means by which large acreage fell into the hands of a few individuals. Another was the practice of the Virginia Company of giving immense tracts of land to company officials. The most noteworthy effort to establish a gentry style of landholding pattern in Virginia, however, came from the designs of Governor William Berkeley (governor 1642 to 1677, except for 1676). Berkeley belonged to the English gentry, believed in hierarchy and rule by the betters, and during the English Civil War supported the Royalist cause. The importance of Berkeley in Virginia agriculture was his activity in recruiting the sons of English gentry to migrate to Virginia by enticing them with large land grants. It was in the last half of the seventeenth century that the First Families of Virginia were established. Moreover, the role of politics in the creation of a class of great landholders has to be stressed. A tight clique ran Virginia and sat in the Governor's Council. These individuals saw no injustice in rewarding themselves and their friends with handsome land grants, a practice that continued into the eighteenth century. Altogether, some 2.35 million acres were patented between 1650 and 1675.[9]

Maryland experienced the same pattern, and there was no outlandish William Berkeley there. Part of the reason was the cultural heritage. To live a life of gentility without manual labor, one needed a huge estate and cheap labor to cultivate it. Once the Tidewater had a measure of safety and promise, the people with money in England were attracted, and they had the means to purchase vast tracts cheaply. Some historians have looked upon the purchases as an early example of land speculation. A prime motivation, nonetheless, had to have been to amass land and seek entrance into the gentry. The other route to land engrossment was similar to the means in England: wait for small farmers of fifty acres or less to run into debt due to difficult times and then buy them out.[10]

As the large plantation began to take shape after 1700, a land scarcity appeared. Tenancy was beginning to replace yeomanry, and sons of yeoman farmers had trouble obtaining land—seemingly there was no land in the Tidewater to purchase. The reason that yeomen could not get land was that it was being engrossed. By 1705 in Accomack County, forty-six persons held 43 percent of the land, and in Northampton, twenty-one persons (3 percent of the tithables) held 39 percent of the land.[11]

The eighteenth century saw no letup in the willingness of the Tidewater gentry and political leaders to grant huge tracts of land to themselves and their friends. Large plantations moved into the Piedmont and partly into the Shenandoah Valley. Most famous was the grant of 5.3 million acres—called the Northern Neck—to the Fairfax family. Likewise, during the middle part of the eighteenth century, the area known as the Southside (extending from the James River south down to the North Carolina boundary) attracted settlers.[12] The landholdings of some of the plantation lords were stunning (see table 3.1).

TABLE 3.1 Examples of Individual Land Holdings in the Tidewater Area, Eighteenth Century

Name(s)	*Acres Owned*
John and Isaac Van Meter	40,000
Alexander Ross and Morgan Bryan	100,000
Jacob Stone	10,000
John Parke Custis	15,000
Charles Carter	15,000
Richard Blackburn	13,813
Lord Fairfax	5,282,000
David Ross	100,000
19 people, each	>20,000
45 people, each	>10,000
George Mann	8,852
Edmund Randolph	7,463
Robert Beverly	50,000
Harrison Family	651,000
Henry McCulloch	72,000
John Lewis	120,000
Beverly Manor	118,091
James Patton	120,000
Alexander Spotswood	83,000
William Fitzhugh	50,000
Robert (King) Carter	332,000
John Tayloe	3,000
James Patton, syndicate	100,000
Edmund III Berkley	10,000
John Baylor	6,000
Richard Corbin	6,000
John Mercer (Ohio Co.)	60,000

(continued)

TABLE 3.1 Examples of Individual Land Holdings in the Tidewater Area, Eighteenth Century (*continued*)

Name(s)	*Acres Owned*
Robert Goldsborough	2,900
Richard Tilghman	15,000
James Carroll	5,421
Charles Carroll	13,400
Nicholas Meriwether	13,762
Thomas Nelson	11,000
Matthew Marable	7,500
John Wilson	7,000
George Adams	6,000
Richard Bennett II	6,760
Philemon Lloyd	7,430
Daniel Parke Custis	20,743
Robert Bristow	9,750
Edward Lloyd	43,000

SOURCES: Voorhis, "Land Grant Policy of Colonial Virginia," 155–56, 168–70, 183–89, 191; Brown and Brown, *Virginia,* 36–46; Dowdy, *Golden Age,* 9, 16; Main, "The One Hundred," 358–60, 363–64; Gray, *History of Agriculture in the Southern United States* 1: 391–92; Fischer and Kelly, *Bound Away,* 85–90, 95, 100; Lorena S. Walsh, *Motives of Honor, Pleasure, and Profit,* 246, 250, 251–57, 271, 298, 307, 324, 489–90, 491, 499, 515, 519, 552, 557; Billings, Selby, and Tate, *Colonial Virginia,* 210.

To comprehend the Tidewater elite's unquenchable thirst for land, the example of Great Britain must be understood. This was the time when the great estates of England were growing, the number of English yeomen contracting, and the supremacy of the high gentry and aristocracy unassailable. The Virginians simply wanted to participate in the creation of an astoundingly wealthy elite that governed socially and politically without opposition; they wanted to join in the creation of an inequality that could never be breached. The American plantation lords openly mimicked England. Throughout the eighteenth century, the code of gentility stretched its tentacles from England to North America, and southerners rapturously became ensnared in its embrace, striving to replicate the lifestyle of the English gentry: they entertained lavishly, provided for education for male progeny, copied the style of English dress, rode in carriages, built brick mansions, and established elaborate gardens. From the other whites they demanded deference. They dominated politics, especially county government, where they were appointed as commissioners, justices of the peace, and to other offices. They also tried to establish family longevity by implementing the practices of entail and primogeniture.[13]

Through headrights, land grants, or purchase, the plantation was becoming established in all the southern colonies. North Carolina, being for many years after its separation from South Carolina in 1729 a small farming colony, escaped some of the consolidation going on elsewhere, but as trade routes improved the land magnates moved in and the monarch handed out land with an alarming largesse. South Carolina took the commercial route early in colonial life and by 1720 found its export staple in rice; in the lowlands the large plantation took form. Georgia was just beginning its colonial experience and had laws to favor small settlers, but nonetheless the ability to employ slave labor in tropical staples evoked the plantation. Maryland did not directly participate in the great Virginia land rush, but consolidation of holdings into larger estates took place there as well.

Whether the planter class deserved to be called an "aristocracy" is a matter of some debate. Several historians, such as Jack P. Greene, Aubrey C. Land, David Carlton, and Peter Coclanis, describe the planters more as business managers, overseers of large enterprises, rather than the stereotypical "rentier."[14] Yet they seemed an aristocracy in the making. In England, the low gentry needed a minimum income of £300 from at least 300 acres. Most yeomen farm-owners in North America had farms 300 to 500 acres, so it would seem that a great number of Virginians met the gentry requirement of acres owned. However, they failed the income test. Amongst a welter of numbers to describe rents (which could be money or crop shares or a stipulated number of pounds of tobacco), we will employ those of Lewis Cecil Gray. He said rents ranged from 10s. to £10 per 100 acres in 1700; by 1750 a medium plantation (assume 200 acres) rented for £5 to £8 per acre. That yields a rent by 1750 of about £0.025 per acre, or half a shilling per acre. Rent in England varied from 10s. to £1 per acre before 1800. Thus to equal the minimum income requirement for a member of the British gentry a Virginian would have to rent out 6,000 acres. For the high end of the gentry, £1,000 per year from rents was needed. For a Virginian, that would equal the renting of nearly 21,000 acres. The great aristocracy, over £10,000 per annum was probably beyond the ability of any but a select few Americans.[15]

Yet all the pieces were in place to transform this landholding elite into an aristocracy. From table 3.1, it is clear that some landholdings were large enough to qualify for aristocratic titles; the income levels could rise over the years as population increased.[16] Most of the aristocratic families of England, moreover, had not entered the aristocracy in one fell swoop. Families earned their aristocratic status by rising to the top of the income heap and purchasing land, educating their sons, holding the estate together, and waiting two or three generations to produce the leisured rentiers that led them to be considered pure gentry and eligible for elevation to nobility. These conditions were all in place in Virginia.

The disturbing feature in the scenario was slavery. Slavery had trouble fitting into the British aristocratic pattern, because American slavery was the slavery of Africans. British aristocracy sat squarely on the inequality of people of the same race, and the mastery the aristocrats exercised was a mastery over people of their own ethnicity. American racial slavery could undermine this principle of inequality among people of the same ethnicity by requiring a bond between higher and lower classes of whites to keep Africans in servitude—this is from another angle the same interpretation as that offered by Edmund S. Morgan.[17] The use of racial slavery for the plantation labor source made the American plantation gentry potentially different from the English gentry.

Slavery vs. Tenancy

In the settlement of North America, few if any of the promoters and planners of the American colonies, except possibly for New England, envisioned a yeoman society arising without some type of elite direction. The leaders in North America wanted the estate system of England, and that meant landownership concentrated in a few hands, the farms managed by tenants, and work done by wage laborers. What made the estate system of agriculture profitable was cheap labor. That was the problem in North America. Cheap labor did not exist but had to be invented. Southerners solved their labor dilemma ultimately by instituting African slavery, but for a variety of reasons the northern colonies could not.

Most northern colonies had problems with tenancy. They did not have the population density to drive down wages, and they did not have the unique staples that gave the southern colonies an edge in overseas markets. They attempted tenancy by using family farmers on small acreages. These small farms could not possibly generate enough income to satisfy both the landlord and the tenant. The only way a landlord could acquire enough income from tenancy was to have hundreds or even thousands of tenants working for small rents; Van Rensselaer had 2,100 tenants, Livingston Manor had 470 tenants, and the Cortlandts and others had 430. Then, a sizeable income might have been obtained, but even these were insufficient for the consumption desires of the patroons.[18]

Sun Bok Kim investigated the New York patroonships and found that rents were minor. He gives as example a Cortlandt manor rent of £3 19s.—call it £4—for 238 acres. He used Sir William Johnson as another example; he charged in the 1750s a rent of £2.5 for 100 acres. That works out to be £0.017 and £0.025 per acre, respectively. That is considerably beneath the standard English charge of £0.5 to £1.0 per acre. Nonetheless, most colonial historians stress conflict between tenants and landlords, not the diminutive amount of the rent charged.[19]

Historians have found tenancy in early Pennsylvania, yet the northern colonies failed to acquire a permanent pool of low-wage labor. People simply migrated out of southeastern Pennsylvania to the South by the Great Valley Road or pushed into land owned by the Shawnee, Delaware, and Susquehanna, thereby sparking off wars with Native Americans. The great landed estate that operated by the employment of tenants did not become characteristic of Pennsylvania agriculture.[20]

All the problems of estate agriculture in the northern colonies were forcing the elite to conclude that the estate system was not working in North America the way it did in England. Fortunes were to be made in cities, in mercantile activities, and not on the land. Rather, fortunes could be made by selling land to actual settlers. In many cases this became a matter of pure speculation. However, many purchasers of vast tracts, such as William Cooper and the Holland Land Company, figured out that waiting for settlers to purchase uncleared land was a losing proposition. To get settlement, the first settlers had to be encouraged, aided, and given easy terms; those settlers would then undertake the arduous task of clearing the land and erecting the first buildings. As more land became improved, the rest of the land lots could be sold at a higher price. Sometimes the procedure worked, sometimes it did not. But the rise of land companies financed by city dwellers and overseas investors, from the Ohio Company to the Yazoo land companies, demonstrated one fact: the British system of landlords and tenants simply was unworkable in North America.[21]

Because the Tidewater staple of tobacco had some monopolistic qualities in the European market, the potential for large profits for tobacco planters appeared good. Nonetheless, the planters needed a cheap labor source. Initially they found that source in indentured servants. Starting in the 1660s, however, the real wage rate in England rose and, as it climbed, indentured servants were less willing to risk the American climate and forgo the advantages of settled English community life. As the source of English indentured servants dried up, the alternative for the Tidewater planters was trying some form of tenancy among the free (ex-servant) population. However, tenant farming in England was advantageous to landlords precisely because the tenant had access to cheap labor to do the physical labor. In North America, tenant farming could only be a variant of family farming. Tenant farming in North America would have none of the profitable advantages for the landowner that it had in England, resulting in Tidewater landlords having to charge less rent; and less rent meant the income of the landlords could never approach gentry standards.[22]

The choice made here needs to be made clear and simple. Rising labor costs outmoded the large plantation in North America because it made production so expensive that it could not turn the landlord a sufficient profit. The large landlord could have accepted the fate given to him, and the landlord could have scaled back activity and sold land to others, opening the door for the continuation and expansion of a yeoman society. But the great landlords would not accept the idea that they would not achieve the status of gentlemen who received large annual incomes without manual labor. So they cheated by using physical violence to get the low-cost labor they wanted. The Tidewater planters turned to slaves imported from Africa. The solution was slavery of someone, and the someones were Africans because the trade had been in operation for decades, the racial justification for African slavery already enunciated, and the use of large numbers of slaves in agricultural operations already established.[23] In North America, slavery came to exist and continued to exist only because it was a system of cheap labor.[24]

Lorena S. Walsh has found evidence that large Tidewater planters were turning to African slavery as early as the 1630s, but it was after the 1660s, and especially after Bacon's Rebellion in 1676, that importations of Africans rose dramatically. In 1660, the population of Africans in Maryland and

Virginia was barely 1,000; by 1680 it was nearly 4,300; by 1720, 30,600; by 1740, 84,000; by 1760, 190,000; and by 1780, 304,000.[25] Slavery's superiority in economic organization, many have argued, lay in the cultivation of semitropical staples, such as tobacco and sugar and later cotton, because of the length of the growing season for each crop. Given the longer time needed to cultivate the semitropical staples, planters needed a labor force almost continuously. The grain staples (wheat, oats, corn), however, only needed intermittent work equal to perhaps four months out of the year. Thus, slave labor as an institution for the entire year was cheaper than hiring "free" labor for the same time frame.[26]

Beyond the planting and slaveholding class, Virginia and Maryland society exhibited a confusing trait: a combination of independent family farmers, tenants, and great planters. Much of the study of Maryland and Virginia has taken the route of regional or county history, and one is never certain how applicable the generalizations of those studies are to areas beyond their boundaries. Taken together, however, the existing studies indicate that in the two colonies by 1770 about one-half of the white male adult population were landowners and the other half were landless; the landless were perhaps composed of 30 percent agricultural laborers and 20 percent tenants. Thus one does get a "yeomanry" in Virginia and Maryland, but its significance is difficult to assess.[27]

Even at this early stage one can see the reason why laborers and independent producers—small farmers—feared the institution of slavery. Slavery devoured the American landscape because it required such vast amounts of land to reward the owner with a luxurious lifestyle. By doing so, it pushed nonslaveholders off the land.[28] Out of the planters' decision to institute African slavery came potential conflict between the great landlords and the nonslaveholders.

Expansion Explodes

After 1750, North American Europeans surged to the West. This push created the Great War for Empire between Great Britain and France from 1754 to 1763, finally resulting in British sovereignty over the internal area of the continent. It also provoked a war with the tribes who were increasingly angry at the inroads American colonials were making in the Appalachians and beyond. In 1768, the Iroquois and British officials—George Croghan and

Sir William Johnson—negotiated the Treaty of Fort Stanwix, in which the Iroquois basically ceded western New York and Pennsylvania to the British Empire. The tradeoff was a pledge on the part of the British to stop colonial squatters and speculators from invading Native American lands. The British wanted peace; the colonials wanted land. The colonials satisfied themselves at the expense of both the Native Americans and the British.[29]

The timing of colonial expansion is telling. New Englanders, having little area, did not immediately spill out to the West. Prior to 1700, they moved to Long Island, Maryland, and parts of Virginia. Otherwise, the rising New England population swelled over in the 1750s into the areas of Vermont, Maine, and New Hampshire. By the 1760s, Connecticut was struggling to get a foothold in the Wyoming Valley of Pennsylvania—the two states finally settled the dispute peacefully—while various New Englanders were invading the New York manors and stirring up rent riots. Because New England bordered the entire eastern boundary of New York, the westward movement for them meant flowing into the backwoods of that colony. In the 1790s, a massive flow of New Englanders, especially from Massachusetts, started settling the Mohawk Valley. But it was not until the first decade of the nineteenth century, and really after 1815, that New Englanders began their famous westward march into Ohio, Michigan, northern Indiana, northern Illinois, Wisconsin, and Minnesota. Until 1800, the most densely populated portion of the colonies satisfied their ravenous land hunger by staying east of, essentially, the western boundary of Pennsylvania.[30]

Pennsylvania, and to a lesser degree New York, earned fame by being aggressively expansive against the Indians, and the continuous battle in the Carolinas with the Cherokees and the Creeks earned the English colonists there no fame for being peacemakers. However, the individuals who caused the most disruption were the Tidewater inhabitants, the Virginians, Marylanders, and to some extent North Carolinians. By 1750, Virginians spilled over into Kentucky and Tennessee, and incited war with the Shawnees that last until the 1790s. Virginians, led by John Sevier, marched into Tennessee and countless battles with the Cherokees.[31]

The fighting stopped in Kentucky but continued in the Southeast between the Georgians and North and South Carolinians against the Creeks, Cherokees, and Seminoles; in the northern areas, the battleground moved to Ohio and Indiana. Ultimately, the tribes were defeated. The major victories for the United States came in the battle of Fallen Timbers (1794), and then coincident with the troubles with Great Britain after 1807 in the battles of Tippecanoe (1811) and Horseshoe Bend (1814). These defeats ended the power of the tribes east of the Mississippi River to thwart the advance of Euro-Americans onto their land. The stage was set for Andrew Jackson's policy of Indian Removal, passed legislatively in 1830 and put into practice 1831–40. The quelling of the Native American's capacity for war unleashed the flood of pioneers into the Old Northwest and the Old Southwest. Why did that flood appear?

Scholars have offered a number of explanations for the vigorous westward expansion after 1750. Chief among them is the growth of the European population in the colonies. Another way of expressing the same interpretation is by affirming that the colonists had run out of land, either by a swelling farm population that simply could not obtain small farms or by depletion of the amount of arable land by soil exhaustion. The population-pressure argument makes some sense for the New England colonies (and then states) because of the odd population dynamic of the family farm.[32] When the farmers' children became adults, the farmer in the seventeenth century could bequeath the land to his children, but this was not the case by the middle of the eighteenth century. The children would have to seek their livelihoods elsewhere, and this became the main source of outward expansion.[33]

Other researchers point to the obsession of elites for land engrossment, either to build estates or to make fortunes by speculation. A considerable number of writers ascribe expansion out of Virginia to ruined land because of abuse of the soil. An older interpretation stresses how pioneers wanted democracy and sought escape from eastern elites.[34] Running strongly in these explanations of colonial expansion is the theme that planters were pushing yeomen off the land. Closely allied to this interpretation is the belief of many historians that a large number of nonslaveholding Virginians sought to escape the effects of the plantation by heading to regions where slavery was nonexistent. At least one million Virginians fled their native state between the American Revolution and the Civil War, and a hefty portion went to Ohio, Indiana, and Illinois. Edmund Morgan sized up the situation: Virginia had an "artificial" land shortage, and the reason for the shortage was the plantation. Ultimately, it was the attempt to engraft the estate system of England onto North America that caused Anglo-American expansionism between 1750 and 1800 and nearly seven decades of war with Native Americans.[35]

The form the estate system took in the colonies, however, was plantation slavery. Because of land engrossment into plantations, nonslaveholders could not obtain land. They cast their eyes westward and saw land for themselves once the Native Americans were displaced. The alternative was to stay along the seaboard and become poverty-stricken dayworkers—which, because of population growth over time, was going to be their fate. Whether they could even become tenants was dubious because of the existence of African slavery. The plantation owners had already determined that slave labor produced more profits than white tenancy. Ultimately North America, if stopped at the Appalachian Mountains, would come to resemble something like but oddly different from England: a few landed magnates, some successful tenants, slave agricultural workers, and a horde of farthlingless agricultural workers. Whether Tidewater yeomen thought this way or not, the result was this: better to war against the Indians—who could always be cast as a social "other" and stigmatized racially—than suffer the fate of the bulk of England's rural population.[36]

The numerical dimensions of Virginia's loss of population and the surge in numbers for Ohio and Kentucky are given in table 3.2. The numbers show a population explosion in Kentucky from 1780 to 1810, and then the rate of growth slows down. For Ohio, the migration begins in 1800 and becomes a torrent after 1810. Both Maryland and Virginia had healthy rates of population growth until 1790, and then a stagnation appeared.

A comparison of population density between England, Virginia, the middle colonies, and the New England states does not reveal a population problem in Virginia. England and Wales contained 58,355 square miles (minus the islands); Virginia (including the present state of West Virginia) had 64,284 square miles, while the New England states (minus Maine) covered 32,083 square miles (with Maine it had 61,978). With the exception of Maine, Virginia in 1770, 1790, and 1800 had fewer people per square mile than any of the New England states (table 3.3). But New Englanders did not start migrating en masse until the 1790s (into New York along the Mohawk River), and did not start populating the Western Reserve until the 1820s. New England did face a "Malthusian moment," starting in Rhode Island and Mas-

TABLE 3.2 Population of Kentucky, Ohio, Virginia, and Maryland, 1700–1830

Year	Kentucky	Ohio	Virginia	Maryland
	(in thousands; decade percent change in parentheses)			
1700			59	30
1750			231	141
1760			340 (53.7)	162 (16.0)
1770			447 (30.3)	203 (22.5)
1780			538 (22.4)	246 (18.9)
1790	74 (64.9)		692 (39.4)	320 (26.7)
1800	221 (195.1)	45	808 (17.2)	342 (3.3)
1810	407 (80.0)	231 (408.9)	878 (7.5)	381 (8.8)
1820	564 (34.0)	581 (152.0)	938 (9.5)	407 (10.6)
1830	688 (19.3)	938 (60.8)	1,044 (14.9)	447 (11.9)

NOTE: Total population is African Americans and whites.

SOURCE: HStatUS76, 28, 33–367, 1168.

TABLE 3.3 Comparison of Virginia and New England Colonies/States with England, 1770–1830

	Population per Square Mile (Selected Decades)			
State/Colony	*1770*	*1790*	*1800*	*1830*
NEW ENGLAND				
Connecticut	38.1	49.4	52.1	61.8
New Hampshire	6.9	23.5	20.4	29.8
Massachusetts	29.3	47.1	52.6	75.0
Rhode Island	54.5	64.7	64.7	90.9
Vermont	1.1	9.3	16.9	30.8
Maine	1.0	3.2	5.1	13.3
VIRGINIA: TOTAL POPULATION				
(Slave + free)	7.0	11.6	13.8	19.0
White only	4.0	6.9	8.1	10.9
ENGLAND AND WALES	(1750 = 108.1)		157.8	183.0

SOURCES: England and Wales from table 2.1. Population numbers from HStatUS76, 24–37, 1168; land area (Virginia area includes West Virginia), from HStatUS76, 39.

sachusetts, but the explosion out of New England into the West had to wait until the 1800–1820 period.[37]

The distribution of landownership marks a vital difference in the expansionist drives between New England and Virginia. The family farm permitted the New England population to remain longer in the home states or along the eastern coast because the farm size was small and an additional farm population could be accommodated simply by making more small farms. Of course, a limit was reached, and then those who wished to continue as farmers would have to migrate elsewhere.[38]

This was not the case in Virginia. Because of the engrossment of land into "gentry" estates in Virginia, a growing farm population could not be accommodated. Those who wished to become "independent" landowners had to leave the colony. The situation is illustrated in table 3.4. Using the censuses of 1850 and 1860 and hoping they reflect the potential amount of improved and unimproved acres in farms in 1790, one can compare the farm acres per capita for the New England states, for the counties of Virginia that existed in 1771, and for England and Wales. The results show the squeeze on Massachusetts, Rhode Island, and English agriculture—but certainly not on Virginia's agriculture. The average family size in the United States in 1800 was about 5.5 to 6.0 persons, so in terms of the number of acres per family unit in 1770, 1790, and 1800, the acreage in table 3.4 can be multiplied by 5.5 or 6.0. If, in American and English farming, the farmer possessed under 5 acres, he or she had to engage in wage labor to survive (as a cottager); that condition probably extended up to 20 acres. From 20 acres to 50 acres, the farmer could produce a subsistence lifestyle but only in good weather conditions. Above 50 acres the farmer could engage in some commercial farming, and more successfully when above 100 acres.[39] On the basis of acres per assumed family unit (5.5 persons per family), only Rhode Island in 1780 was falling into the subsistence category (52.8 acres per family), while Virginia had, in 1780, 118 acres per family (counting free and slave) and 200 acres per family (whites only); in 1750, on the eve of Virginia's push into the West, it had 505 acres per white family. Even when using the white population of 1800 and dividing that into the acreages that existed for the counties of 1750, Virginia averaged 122 acres per white family—a size of farm that equaled the average number of acres per farm in Illinois and Indiana in 1860.[40] From this vantage point, Virginia had no lack of acreage to support good-sized commercial farms for all its family units. The same analysis can be applied

TABLE 3.4 Farmland Availability, Population, and Population per Acre in England and Selected States, 1750 and 1780

State	*Population 1750 (Thousands)*	*Population 1780 (Thousands)*	*Potential Acres (Millions)*	*Ratio Acres per Person 1750*	*Ratio Acres per Person 1780*
Maine	—	49.1	5.7	—	116
Mass.	188	268.6	3.338	17.8	12.4
N.H.	27.5	87.8	3.684	131.6	41.9
Vt.	2.0	47.6	4.160	—	89.0
Conn.	111.3	206.7	3.503	31.6	16.9
R.I.	33.2	52.9	.511	15.5	9.6
N.Y.	76.7	210.5	4.671	60.7	22.1
Penn.	119.7	327.3	3.609	30.1	11.0
N.J.	71.4	139.6	2.753	38.8	19.7
Va. (all)	231.0	538.0	11.495	49.8	21.4
Va. (whites only)	129.6	317.4	11.495	91.9	36.3
Md. (all)	141.1	245.5	4.564	32.4	18.6
Md. (whites only)	97.6	165.0	4.564	46.6	27.7
England & Wales	6,253	9,168 (1800)	29.1	4.7	3.2

SOURCES: Acreage based on reported farm acres, improved and unimproved, given in the *Eighth Census of the United States* 2: *Agriculture;* for New York, Pennsylvania, and New Jersey, the *Seventh Census;* both the New York and Pennsylvania counties used are given earlier in the narrative and based on the maps in Hawke, *Colonial Experience,* 483, 392–93; the Virginia acreage (11.4 million) based on counties existing in 1771 as given by Jewel Spangler, *Virginians Reborn,* 107. Population numbers from HStatUS76, 24–37, 1168; acres in England from Mingay, *English Landed Society in the Eighteenth Century,* 12; and Allen, "Agriculture and the Industrial Revolution," 103.

to New York; by the numbers, New York and Virginia had no shortage of land and no just reason for a land mania. What they had was a land system characterized by engrossment.

Because Virginia and other societies growing semitropical staples aped the example of the large English farm, they sought with "demonic force" (Stanley Lebergott's phrase) to glom onto large tracts of acreage. But in North America, the only way large-scale farming could be made enriching in the style of the English gentry was to use slaves because recourse to tenants yielded so little. The engrossment of land either squeezed out the small landholder or his children, sending them flying to lands where they could reestablish a freehold. The plantation, in short, was the enemy of the small farmer's dream of landownership.[41]

This circumstance of the plantation crowding out the nonslaveholding southern farmer, especially in the years 1770–1820, explains the antislavery animus that arose in the Northwest Territories. Yeoman farmers were running explicitly away from the plantation lord. They feared the Virginia gentry. They understood that there was a connection between land engrossment and slavery, but their focus was hardly on the condition of the slave but more on how the slaveholding gentry ruined their dreams of landownership. Thus in the battles over slavery in Ohio, Indiana, and Illinois—and partially in Kentucky—one sees an attack on the slavemasters rather than on slavery itself.[42] This fear of the engrossment tendency of the plantation lords would soon be called the slave power's hostility to free labor and would after 1810 play a key role in the future debate over western territories.

Population pressure would eventually push New Englanders into Ohio and into conflict with the Native Americans, but had land been more equitably distributed in Virginia and Maryland, the land lust of the Euro-Americans might have been delayed until 1800 or 1810, instead of emerging in 1750. If the land had not been engrossed, an outcome different from the "dark and bloody ground" of Kentucky might have been possible. Because of the landholding pattern the plantation generated, the Indian Wars of the Revolutionary period can be attributed to the institution of slavery.

4

Dimensions of Antebellum Northern Farming

In the northern states of the antebellum era, the agricultural dreams of the Levellers became reality. No great landlords dominated landownership, small farms overwhelmed the landscape, and farm laborers, a minority of the rural population, were not imprisoned in a caste. How this state of rural economy came into existence requires some explanation, for the actual heritage of British agriculture was the preponderance of large estates owned by the aristocracy or the gentry. The British yeoman was more a creature of a nostalgic imagination than a social reality, but the U.S. northern yeoman was not.

The Statistical Preponderance of Agriculture in the North

The United States in 1860 was a rural, agricultural nation. This was obviously true of the slaveholding states but no less so of the nonslaveholding states. In table 4.1, the northern states, excluding the far West and the territories, were overwhelmingly rural and not greatly different in 1860 from the norm of the country (the nation was 80 percent rural while the northern states were about 74 percent). Comparing the rural/urban status of the northern states in 1850 to their status in 1860, it becomes obvious that much of the urban surge in the North came from the tide of immigration. The only urban states in 1860 were the relatively small ones of Rhode Island and Massachusetts. Moreover, the northern states, save lower New England, continued to grow in rural population. It was the immigration wave after 1880 that pushed the northern states into urban life. The numbers characterizing the urban-rural residence of northerners in 1860 do not depict an urban society; it was overwhelmingly a rural society.

Given the rural nature of the antebellum northern states, it should come as no shock that the occupations of farmer and farm laborer dominated all

TABLE 4.1 Rural Characteristics of Northern States, 1850–1920

	1850	1860	1870	1880	1890	1900	1910	1920
New England								
Total Population (Millions)	2.7	3.1	3.5	4.0	4.7	5.6	6.6	7.4
% Rural	71.2	63.4	55.6	47.6	38.4	31.4	26.7	23.8
Great Lakes								
Total Population (Millions)	4.7	7.8	10.3	13.6	16.7	20.00	22.6	26.3
% Rural	91.1	86.5	78.6	74.5	64.3	58.0	50.6	43.0
Mid-Atlantic								
Total Population (Millions)	5.9	7.5	8.8	10.5	12.7	15.5	19.3	22.3
% Rural	74.5	64.6	55.8	49.8	42.0	35.0	28.8	24.6
New England, Great Lakes, and Mid-Atlantic								
Total Population (Millions)	13.4	18.3	22.6	28.1	34.1	41.1	48.4	55.9
% Rural	79.7	73.7	66.1	61.4	52.4	45.7	38.6	33.1
Entire United States								
Total Population (Millions)	23.8	31.5	39.9	50.3	63.1	76.1	92.4	106.5
% Rural	82.5	80.0	71.9	71.6	64.7	60.2	54.1	48.5

NOTE: New England states: Maine, New Hampshire, Vermont, Connecticut, Massachusetts, Rhode Island. Great Lakes states: Ohio, Michigan, Indiana, Illinois, Wisconsin, Iowa, Minnesota. Mid-Atlantic states: New York, Pennsylvania, New Jersey.

SOURCES: Standard definition of urban area being an incorporated area of population 2,500 or larger; computed from HStat76, series A 6–8, p. 8, series A 57–72, pp. 11–12, and series A 195–209, pp. 24–37.

others in the census of 1860. Problems about definitions plague the tabulating of people who were farmers and rural residents. While the census lists the specific categories of farmer and farm laborer, the occupation list for each state also includes items such as farriers, curriers, drivers, drovers, fruiterers, gardeners, and others. How many of these belong to the farmer and farm laborer category is not apparent from the census totals. Many craftsmen were needed to enable the rural economy to operate, but people of that description could as easily be urban dwellers. The division of merchants into urban and rural categories is unknown. Many of those listed as laborers could have been agricultural workers who simply gave a bland response to the census taker. Given the limited urbanization of the nation in 1860, it is probable that some 70 percent of the northern population had strong links to the farm economy (see table 4.2).

Two characteristics of the northern farm stand out. The first is that the farms were owner-occupied. This was a land of yeoman farmers. Travelers

TABLE 4.2 Agricultural Occupation Characteristics of Northern State Populations, 1860

State and Region	*Percent Rural 1860*	*Percent of Total Occupations Farmer/ Farm Laborer, 1860*	*Percent of Male Occupations Farmer/Farm Laborer, 1860*
Maine	83.4	39.1	43.1
Vermont	98.1	52.8	59.4
New Hampshire	77.9	38.6	42.1
Massachusetts	40.5	13.8	15.8
Connecticut	73.5	26.1	29.1
Rhode Island	36.6	16.5	18.8
New England	63.4	25.7	30.0
New York	60.7	27.7	33.1
New Jersey	67.3	22.9	26.9
Pennsylvania	69.2	27.9	31.6
Mid-Atlantic	64.5	27.1	31.2
Illinois	85.6	50.7	55.0
Indiana	91.4	59.3	62.8
Iowa	91.1	61.6	64.3
Michigan	86.6	52.6	56.6
Ohio	82.9	46.5	50.2
Wisconsin	85.6	53.7	57.4
Minnesota	90.7	52.3	55.3
Great Lakes	87.2	52.4	56.1
All Northern States Listed Above	74.3	36.2	41.0

SOURCES: Occupation totals by state given in *Census of 1860* 1: *Population*, total numbers and farm and farm laborer categories. The census has numerous categories that are obviously female; the reason for the third column is that nearly 11 percent of total occupations listed in the census were held by women, but their participation varied considerably state to state, women being employed more in the eastern than the western states. In the cotton South, the percentages are usually less than 5 percent. A comparison that does not adjust for the presence of women in the occupational listings of the census thus makes the North less agricultural than the South.

from England and Europe throughout the nineteenth century seized upon the ownership of land by northern farmers as the chief social and economic characteristic of the republic. Alexis de Tocqueville wrote in the 1830s that in the United States there were, "properly speaking, no farming tenants; every man owns the ground he tills." His observation was affirmed time and again; Michael Chevalier added that "the cultivator is the owner of the soil; in the West, this rule is without exception." In the 1850s a visiting member of Parliament, W. E. Baxter, recorded his view of New Hampshire farmers: "Nowhere in the world will the stranger find a more thriving, comfortable, and independent, a better educated, or a more moral and religious peasant proprietary, than inhabit this New England district."[1]

This startling revelation of a vast country filled with owner-occupiers led to discussion of two traits of the northern people that seemed inseparable from yeoman farming: the importance of physical labor for successful cultivation and the social equality that invariably accompanied a community in which almost all owned 80 to 120 acres. After discussing American farming, Hugh Tremenheere in the 1840s plunged into the reality of Tocqueville's "equality of condition" of northerners; he described Massachusetts, with a few exceptions, as being "composed of one great middle class."[2]

Part of the motivation for making these assertions was the common practice of comparing the rural society of the northern American states with that of England's. The northern states had 1,523,000 farmers in 1859–60, and 535,000 farm laborers, yielding a ratio of farm laborers to farmers of about 0.33 to 1.00. For England and Wales in 1851, there were, for those possessing more than five acres, 307,000 farmers—owners and tenants combined—and 1,461,000 agricultural laborers (laborers, ploughmen, herdsmen). Depending on who is included in the laboring group, the ratio of laborers to farmers in England and Wales was between 2.5 and 4.75 to 1.0. The common rural person in England was an agricultural laborer; in the northern free states, he was an owner-occupier—a yeoman.[3]

Modern investigation of the nineteenth-century American farm has revealed in the antebellum years a tenancy rate of at most 20 percent; otherwise, the person who farmed the land owned the land. The tenancy rate went up after the Civil War, but the rate for the succeeding century held steady at about 30 percent. Most studies of the agricultural North find tenancy was used as a step on the ladder to farm ownership and was not permanent for most tenants. To put the matter in perspective, the landowning characteristic of the northern farmer has to be compared to Great Britain in the nineteenth century, and there tenancy in 1860 was approaching 75 to 90 percent.[4]

The second noteworthy characteristic was the average size of the northern farm and, associated with that observation, its longevity. The numbers in table 4.3 are averages and thereby mask much information about the miniature farms (3 to 20 acres) that could not sustain a family's subsistence needs, and the large farms over 300 acres, which were obviously commercial. Nonetheless, extremely few northern farms reached the size that could be put into the "high gentry" class in England, farms over 1,000 acres. Excluding California and Oregon, the northern states in 1860 had only 467 farms larger than 1,000 acres, and most states (save Illinois and New York) had fewer than 15. For comparison, South Carolina alone had over 482 farms greater than 1,000 acres—greater than the total for all the northern states (exclusive of California and Oregon). One of the facets of the size of the American farm was how long it endured.[5] As given in table 4.3, the average size of northern farms in the states actually declined after 1880—in marked contrast to the budding consolidation that was going on in manufacturing. The decline in size of farm was most likely due to farmers passing on portions of their holdings to their sons and daughters until a minimum size was reached (about 80 acres which, theoretically, could still sustain the family and earn some money on the market).

Farming in these northern states between 1850 and 1900 had reached a balancing point between population growth in the rural areas and the formation of farms. Rural population grew by 68.7 percent between 1850 and 1900 while the number of farms increased by 127.6 percent. (computed from tables 4.1 and 4.4). The ups and downs of the business cycle made life difficult, of course, but the situation in the last half-century was different from the half-century after 1750 when population pressure in New England pushed people west to find farms. What accounted for the difference between 1750–1800 and 1850–1900 was farm technology—the need for labor on the farm changed.[6]

The remarkable feature of northern farms is the degree to which they persisted well into the twentieth century. Urban populations swelled, as did the number of workers in industrial, financial, and mercantile occupations, and as a result the percentage of the population in rural areas and in farming declined—but that is a percentage. In absolute numbers, the number

TABLE 4.3 Average Acres per Farm, 1850–1960

	1850	*1860*	*1870*	*1880*	*1890*	*1900*	*1910*	*1920*	*1930*	*1940*	*1950*	*1959*
New England												
Maine	110	109	108	104	104	107	104	109	114	99	122	164
N.H.	116	123	122	116	119	123	120	127	132	109	128	172
Vt.	139	136	134	138	135	143	143	146	157	156	185	243
Mass.	99	94	103	88	87	83	78	78	78	61	75	102
R.I.	103	96	94	83	85	83	84	81	84	74	74	99
Conn.	106	100	93	80	86	85	82	84	87	72	82	107
Mid-Atlantic												
N.Y.	112	107	103	99	97	100	102	107	113	112	128	164
N.J.	115	108	98	85	86	82	77	77	69	73	70	89
Pa.	117	109	103	93	87	86	85	87	89	86	96	119
Great Lakes												
Ill.	158	146	128	124	127	124	129	135	143	145	159	196
Ind.	136	124	112	105	103	97	99	103	108	107	118	145
Mich.	129	113	101	90	86	86	92	97	101	96	111	132
Ohio	125	114	111	99	93	89	89	92	98	94	105	132
Wis.	148	114	114	114	115	117	119	117	120	123	138	161
Minn.	184	149	139	145	160	170	177	169	167	165	184	211
Iowa	185	165	134	134	151	151	156	157	158	160	169	194

SOURCE: HStat76, Series K 17–81, p. 461.

TABLE 4.4 Number of Farms in Northern States: New England, Mid-Atlantic, and Great Lakes

Region	*1850*	*1860*	*1870*	*1880*	*1890*	*1900*	*1910*	*1920*	*1930*	*1940*	*1950*	*1959*
						(In Thousands)						
New England	168	184	181	207	190	192	187	157	125	135	103	57
Mid-Atlantic	322	381	421	489	469	486	468	425	358	348	297	198
Great Lakes	368	587	762	985	1,009	1,136	1,123	1,085	967	1,006	885	666
Minn. + Iowa	15	79	163	277	369	384	373	391	400	410	382	321
Totals	873	1,231	1,527	1,958	1,987	2,198	2,151	2,058	1,850	1,899	1,167	1,242

NOTE: New England: Maine, Vermont, New Hampshire, Massachusetts, Connecticut, Rhode Island. Mid-Atlantic: New York, New Jersey, Pennsylvania. Great Lakes: Ohio, Michigan, Indiana, Illinois, Wisconsin.

SOURCE: HStatUS76, Series K 17–81, p. 459.

of farms in these antebellum northern states kept growing until 1900. They then declined by about 100,000 farms in 1910 but held steady at about 1.9 million farms until 1940 (table 4.4). The collapse of rural America in these states, though differing by region, was a post-1940 phenomenon. The same pattern is illustrated in the average acreage of farms in these states; they held constant from 1910 to 1940, and then in 1950 the average acreage per farm began to shoot up. For rural America, change in farm size and organization began after 1960.

Of course, in the years following the American Revolution, large farms existed. Many of these were located in New York and attracted considerable attention because they were on the route to one of the scenic wonders of North America, Niagara Falls. The route took visitors right into the land once occupied by the patroons. Two individuals garnered the most attention. The first was Stephan Van Rensselaer, who owned (supposedly) in the 1820s some 750,000 acres on the Hudson River, had five thousand tenants, and was rumored to possess £7 million. People recognized the oddity of this estate; it was a "remnant of feudalism in a young democracy of North America." The Van Rensselaer estate became famous for the tenant riots beginning in 1839 when the inheritor of the estate, William, raised rents to pay off accumulating debts. Ultimately the government of New York stepped in to mediate the fracas in the late 1840s, and the estate was slowly broken up into freeholds by 1865.[7] The other large holding drawing the attention of visitors was the 40,000-acre estate of James Wadsworth. Wadsworth built and ran his own farm of 1,600 acres and rented out the rest. People were interested in Wadsworth because he was reputed to be a reforming, progressive farmer, the best of a rare breed in America.[8]

Large farms eventually emerged in the West. They were often connected to cattle, horse, and hog raising, activities which required more land for pasture and less labor. The Virginia military tract in Ohio spawned some large farms. Indiana attracted the interest of Henry Ellsworth, who eventually owned 110,000 acres and tried to cultivate it with tenants. Illinois had a number of landowners who each held tens of thousands of acres. M. L. Sullivant of Champaign County earned a place in the agricultural press because of the size of his operation: 22,000 acres with 500 head of cattle, 1,100 acres in corn, 300 in wheat, forty plows, seventy-five spans of horses, and seventy-five yokes of oxen. The potential of an estate system on the prairies came to pass in the person of William Scully, an Irish landlord who inherited the family fortune. He came to the United States because tenant anger over harsh policies led to his elder brother's assassination. By military warrants, Scully amassed over 60,000 acres in Illinois in the 1850s, and expanded his holdings in the 1860s into Kansas, eventually becoming the nation's largest landlord with 225,000 acres and thousands of tenants. Besides actual farmers, speculators flooded the Great Lakes states and held a significant portion of the domain.[9]

Despite the presence of mammoth operations, most writers about antebellum agricultural America stressed the smallness of farms and the backwardness of northern farming practices. Robert Russell, the astute Scottish agricultural observer, visited in the mid-1850s the reputed best farm in Connecticut, of some 700 acres, but 500 were in hills and pasture and only 200 were being cultivated; "it was no great affair after all." He ended up thinking that New England did not even have a farm that met the standards of a fourth-rate farm in England. Most of the time foreign visitors looked at farms between 50 and 200 acres.[10]

Northern farms varied in size from state to state and within regions of each state. The most informative source on the subject is in Jeremy Atack and Fred Bateman's superb statistical analysis of northern farming in 1860. In their sample, the average farm in the eastern states was 118 acres, of which 39 were unimproved (forested area). In the western states (Ohio and beyond, excluding California and Oregon), the average was 155 acres, 65 of it unimproved. Farms in the entire North produced on average 80 bushels of wheat, 378 bushels of corn, and 99 bushels of oats, with Illinois, Indiana, and Iowa producing hefty amounts of corn (from 530 to 800 bushels).[11]

The small size of northern American farms has to be stressed because the comparison to Great Britain is so revealing. Table 4.5 is stitched together from various sources indicating the amount of acreage set aside for various crops in England and some northern American farms. The difference in size of crop acreages and output is striking; it is evident in farm journals when northern farmers talk about selling twenty-five bushels of oats, fifteen bushels of wheat, six or seven bushels of potatoes.[12] The large English farm produced an output of thousands of bushels of crop.

Editors and contributors to the agricultural press hotly debated the question of the appropriate size of a farm. Advocates of large-scale farming fixed on two particular aspects of small farms. First, the capital costs (buildings, ditches, and implements) were excessive, and the capital could have

TABLE 4.5 Comparison of Acreage for Various Farm Purposes, United States and England

United States	*Acres in*			
Year, Location	*Wheat*	*Pasture/Meadow*	*Barley*	*Maize*
1860, New York (average farm)	3.83		8.12	3.93
1860, Illinois (average farm)	13.36		3.00	24.79
1860, Ohio (average farm)	2.83		5.23	8.75

England	*Acres in*			
Year, Location	*Wheat*	*Pasture/Meadow*	*Barley*	*Maize*
1860, Olmsted	100			
1860, Butler Abbey	230	150	150	150[a]
1837, Coke of Norfolk	430		130	
1854, Wark Farm	500[b]	250 + 200		250[b]

[a] Entry for turnips.

[b] Entry for wheat, barley, and rye.

SOURCES: New York, Illinois, and Ohio figures from Atack and Bateman, *To Their Own Soil,* tables 7.2 and 10.2, pp. 113 and 166. England figures from Olmsted, *Walks and Talks of an American Farmer in England,* 233; "Editorial Notes Abroad," *Cultivator* 8 (February 1860): 41; "Mr. Coke of Norfolk, England," *Farmer and Gardener* 4 (October 31, 1837): 210; "Address of C. P. Holcomb, Esq., before the Md. State Ag. Society," *American Farmer* 9 (January 1854): 201.

been more efficiently used by larger farms. Second, there was no adequate division of labor. Instead, the owner-occupier had to labor himself and become master of all trades. Such authors argued that everything was done more effectively when the owner managed the labor of others.[13]

However, most editors and writers favored the small farm, and they explicitly advocated a farm of fifty acres. They had a number of reasons. Economically, it was easier on smaller farms to implement intensive cultivation methods and to keep the soil fertile by manuring. The writers took the position of agricultural reformers who denounced the American tendency to farm until the soil gave out and then simply move west.[14] British observers of American agriculture for the first half of the nineteenth century knew that British techniques were superior to the slovenly way they saw Americans cultivating the land, and it was summed up in the magic word "manure." American reformers agreed that agriculture in the United States was backward, and they blamed it on the Americans' tendency to purchase more land than they could intelligently cultivate.[15]

Proponents of small farms had other rationales. They talked up the virtues of the small farm in an effort to stem the desertion of eastern farmers for the West.[16] Many writers focused on the social results of small farming: personal and political independence, a continual respect for the duty to labor, and the community life that came from having numerous farms in close proximity. For a republic, yeoman farmers were required, not landed magnates. One wrote, "to respect himself, a man needs his own stake in the earth—a little dominion under his own control," while immense landed tracts produced a desire for wealth that resulted in "self-conceit, and breeds a reckless disregard of human rights and human happiness." Republics, wrote another, needed the "subdivision of estates and the ownership of the soil . . . vested in the cultivators, and the consequent independence of our yeomanry."[17]

Northern agricultural writers tended to glorify the small farm and deprecate the English tendency to concentration. They also praised the small farm in Europe, especially in France, and recognized that in landholding practices the English were an anomaly. They also ascribed the same set of characteristics to European peasants that they gave to American freeholders. Henry Colman extended his agricultural tour to France and Belgium, and he decided that the farm laborers there had a decent chance of acquiring a small farm. Another writer found the British and French rural classes to be wholly different: the French peasants were "industrious, sprightly and full of talk and merriment, presenting . . . a striking contrast to the stupid, homeless and penniless farm laborer[s] of England." Moreover, owner-occupiers took care of the soil and cultivated it carefully—the mark of initiative, independence, and aspiration.[18]

Antebellum northern farming was small-scale and remained that way for decades. It was the necessary foundation for a host of political and social ideas that flooded the land and gave physical meaning to such words as "equality," the "necessity of labor," and "independence." Large estates, farms of vast size, existed as did ruthless speculators, but not in a major role; the dominance of the small farm above the Ohio River was overwhelming.[19]

Northern farmers between 1820 and 1860 underwent major changes in their economic environment and agricultural techniques. Historians have debated whether antebellum farmers were subsistence, family-first cultivators or market-driven commercial entrepreneurs. Prior to 1820, the desire of farmers to be either mattered little; if they were not situated on a river, near the ocean, or next to an urban area they really had no choice. Without transportation facilities, they had to be primarily subsistence farmers. The current and standard estimate is that 70 to 80 percent of the crops raised on the farm were consumed on the farm, most credit was extended over years, trade was local, and barter was a frequent, perhaps the most common, mechanism of exchange.[20] However, the states after the War of 1812 supported government projects to end landed isolation by creating transportation systems: roads, river improvements, and particularly the construction of canals, the most famous being the Erie Canal. Soon after 1840, states retreated from government-owned enterprises, and private interests created the railroad network between 1840 and 1860. By 1860, some 30,000 miles of track were operating in the United States, 20,000 miles in the North.

The apparatus of transportation altered the economic environment of northern agriculture significantly—this was the agricultural participation in what has been termed either the "transportation revolution" or the "market revolution."[21] First, it permitted most farmers access to a market—that is, to distant places where people were densely packed and could easily be aggregated in a demand for the farmer's output. Most northern farmers seized the opportunity and engaged to some degree in commercial farming. But the effects of the new transportation system on agriculture far exceeded the question of traditional versus commercial husbandry because it brought different agricultural regions into competition with one another. Eastern agriculture could not withstand the low-cost production of cereals beyond the Appalachians, and the output of wheat surely marched to the West over the decades until in the 1880s and 1890s it settled in the central prairie states. Wheat farming along the East Coast withered—in New England it was never strong anyway—and so farming in New Jersey, Pennsylvania, and New York as well as New England switched into cattle and especially dairy and truck farming (vegetables and fruits). Certain areas in Vermont, New Hampshire, and Ohio became centers of sheep raising and wool production. Farmers who could not make the switch in the East often solved their economic dilemma by moving west. Wheat moved into Illinois,

Indiana, and Wisconsin between 1840 and 1860, but it encountered problems with pests and rust; steadily and surely the Great Lakes states centered on corn and hog production, the practice being imported from the Shenandoah Valley and brought to the region between 1790 and 1830 by upland southerners from Virginia and Kentucky.[22]

Railroads produced an engine of macroeconomic growth that has seldom been witnessed since. Two internal trade areas opened up due to the railroad that had not existed before because of transportation problems. The Great Lakes area was settled in its southern areas first, and farmers used the Ohio-Mississippi River system to send surplus production down to New Orleans. The railroad, however, connected western farms at first with vessels on the Great Lakes to produce the grain trade from Chicago to ports on Lake Erie and Lake Huron, especially Oswego, New York. A momentous traffic in goods poured from the West into New York lake ports, from there into the Erie Canal, and then to New York City via the Hudson River. Eastern manufactured goods made the reverse journey. In the middle and late 1850s, consolidation of some railroad lines and extension of others created the trunk lines—the New York Central, the New York and Erie, the Pennsylvania, and Baltimore and Ohio—that could bypass the lake trade and ship foodstuffs directly from the West into the East. The trade between the two regions of the East Coast and the Great Lakes was the engine of economic growth that would power the U.S. economy for nearly eighty years. What made the trade possible was the growth of cities in the East and the lack of exuberant soil. Easterners, especially in New England but increasingly in New York and Pennsylvania, had insufficient food supplies for their populations. Midwesterners supplied the food for easterners and in return purchased eastern goods.[23]

For four decades many northern agriculturalists experienced prosperity, and the years 1815 to 1860 have been acknowledged as a "golden age" of American farming. The economic reason for the prosperity was the east-west trade and the price differential that resulted. Creation of market areas enhanced the demand for foodstuffs, producing a price rise. Economic historian Don Parkerson figures that winter wheat prices between 1840 and 1860 rose 50 percent, but prices of manufactured goods did not rise at all, the reason being the spread of the eastern factory system. The introduction of new machinery generated a productivity that kept the price of manufactured goods stable in the face of growing demand from a multiplying population. Every incentive imaginable was thus present to get the farmer to commit to commercial agriculture.[24]

Arriving coincidentally with a new market structure in agriculture was a new technology. This subject has attracted perhaps more attention than any other feature of antebellum northern farming. Except for the introduction of the cradle in the 1790s, American farm technology had been stagnant for years. With the arrival of the transportation revolution, an incentive appeared for increased crop production. Thus came in fairly rapid succession such implements as the harvester, the steel plow, the self-acting rake, the twine-binder, the thresher, the combine, the seed drill, and the horse-hoe, among others. During the antebellum decades, animal stocks were improved, the varieties of seed increased, and the use of mortgages extended. However, the pace of these changes remains debatable. Clarence Danhoff, an astute historian of agricultural change, finds that adoption of the new technology was slow and that into the 1860s the cradle was still the most common implement used in harvesting grain crops. Agricultural technique remained surprisingly constant.[25] So did farm size and the dominance of the family farm. The reason for this constancy probably resides in one fact: in 1750 the primary source of power was the horse (or ox or mule); in 1850 it was still the horse; in 1900 it was still the horse; but in 1920 the tractor began displacing animal power on the farm. The difference between the farm and the factory in this instance is unmistakable. The factory, or manufacturing production, moved within the nineteenth century alone from handicraft to water power to steam power to electric power. Along with the change in the source of power came change in size of plant. But on the farm, such was not the case: the horse reigned supreme.

The small family farm in the North had a number of consequences that have not been properly emphasized. Because the family farm was usually under 120 acres, there was a vast plentitude of them. For a township of thirty-six square miles (thirty-six sections each having 640 acres or one square mile), a potential of 192 farms could be formed. Northern counties could and did have thousands of farms. These farms constituted units requiring credit, mercantile, and mechanical services, and thus gave rise to the small town. The small towns in turn required services from larger centers, regional cities, and the regional cities were in their turn sustained by metropolitan centers like Chicago, Boston, and New York. But the base that created this urban network was the family farm; without small farms, the network fell apart.[26] Another way to state this phenomenon is that the North's vast number of small towns and regional cities did not arise from a proclivity of New Englanders or others to live in an urban environment. It was the natural result of having agriculture conducted by small farms. Thus much of the town/urban quality of the North derived from the fact that agriculture was organized on the basis of the small farm. For the postbellum years, this would have a striking result: railroads connected cities, and because the North had such a multitude of urban places, railroad construction would assume gigantic proportions, dwarfing the building of the iron horses elsewhere in the nation, most notably in the South.

The family farm also sponsored a curious social characteristic: community activity. The family farm of 120 acres, using this number as a norm for the North, took up surprisingly little space. The acreage put in the form of a perfect square would measure .433 miles by .433 miles. Northern farms, therefore, were not that distant from each other. Assuming a normal person could walk at 3 miles an hour, then in one hour one could walk past nearly 7 farms. This made visiting and community activities—church, school, governmental duties—fairly easy to incorporate into the routines of life. And the diaries of antebellum northern farmers show considerable visiting of friends and especially kinfolk. This capacity of northern farm families to socialize has to be compared with the isolation brought into the South by the large plantation. Assuming the large plantation was 1,000 acres, then a township system would only house 23 of them (as opposed to 192 family farms), and, if 1,000 acres were put into a perfect square, the plantation would measure 1.25 miles by 1.25 miles. No longer would simple walking or even horseback riding allow easy visitation of kinfolk and neighbors. The geography consumed by the plantation militated against the formation of community bonds.

Speculators, Land Companies, and the Public Lands

In the northern states from 1800 to 1940, the family farm containing slightly under 120 acres statistically dominated. Given the nature of firms in the U.S. economy to consolidate, the example of large-scale British farming, and the attempt of various Americans to create vast estates through land speculation, the question arises as to how this state of things came to pass. Why did northern farms retain their small size; why was there so little aggregation into larger units? The answer to this question is largely economic in the various costs of production and the inability of northerners to apply economies of scale on their small farms; it also had something to do with urban America and the prospects of making fortunes there. Nonetheless, the American land law system added to the oddity of the longevity of the small northern farm.

Virginia's handling of Kentucky in the 1770s and 1780s was a national scandal. By 1800 or 1820, propertyless white men were more numerous in Kentucky than almost any other place in the United States, with landed property ownership among the white inhabitants only reaching some 35 percent. Contemporaries knew the problem. To get a title a person had to travel to Richmond, and the process could take up to three months, something yeoman farmers could not afford. Then came the confusion over land titles due to improper surveys; the confusion led to lawsuits won by richer litigants. The situation was made worse by Virginia issuing Revolutionary military warrants to those who served in the army, a situation made perfect for speculators hoping to buy up the warrants for a cheap price, claiming the land, and then reselling it after they secured the title. These facts were known in the 1790s and repeated in the travel literature of the nineteenth century.[27]

Land companies had started to form in the late 1740s, and they proliferated after the Revolution. All the states exhibited the same phenomenon of incorporating companies to direct migration out west: Massachusetts established the Ohio Company; some dissidents later founded the Scioto Company; Connecticut gave birth to the Connecticut Land Company; Virginia extended its land policy to the (Ohio) Military District; three Pennsylvania individuals established the North American Land Company; all sorts of prominent men in Massachusetts, New York, and Pennsylvania were involved in what became the Holland Land Company; and then there were the four companies that became lumped together in the Yazoo land speculation.[28] The travelers accompanying Louis Kossuth in 1851 noted one thing about Kentucky and early state attempts to control land distribution: it produced large estates. States were a dismal failure at a fair land distribution. Wealthy investors too easily manipulated state legislatures into passing laws inuring to their benefit.[29]

Kentucky's land title debacle occurred during the governorship of Thomas Jefferson, and he must have learned much about how legislation could affect settlement. As the proponent of "yeoman democracy" and the virtues of family farming, Jefferson at the Continental Congress was in a position to write a program for land settlement that excluded the British estate system. He actually failed to do so except by happenstance. The real key was to get possession of western lands out of the hands of the states, which was largely done in 1783 to form the Articles of Confederation. The land ordinances of 1784, 1785, and 1787, based loosely on Jefferson's plans, did not exclude land companies and speculators, but by demanding a grid system (the township divided into thirty-six sections, each section being 640 acres) that was surveyed in advance of settlement, the Continental Congress established a generally orderly system of land disposition with nearly imperturbable titles. The Northwest Ordinance of 1787 also added the famous provisions for public education and elimination of slavery, plus the important mechanism by which a territory could become a state on equal terms with the older states, but one of its overlooked virtues was to ensure the mess of Kentucky was not repeated. And indeed, once the federal government got hold of the disposition of public lands, the fraud and corruption that sloshed out of statehouses disappeared.[30]

Nonetheless, the land ordinances of the Confederation period, which were adopted by the new order under the Constitution, did not make any provision for selling land only to settlers or limit the size of acreage that could be bought. Until the Homestead Act of 1862, the land laws usually set a minimum size of acreage that could be bought, not a maximum.[31] Land speculators abounded in grabbing up land when public sales took place, and land companies still operated. So how did family farming take such root when the opportunities existed for agglomeration?

The first answer was economic and somewhat mixed in with public policy. Speculators found that capital gains from speculating in land (that is, expecting the price of land to rise) did not happen anytime soon. To make profits from land speculation, one needed to have people ready to settle, and the United States was just too thinly peopled. That situation abated by the 1840s and 1850s when immigration, combined with a swelling native population, provided sufficient migrants to the West to yield some tidy returns on land speculation, but in the years 1800 to 1840 an abundance of settlers simply was not there.[32]

Economic circumstances mixed in with public policy in a furtive manner to hinder the potential for creating massive estates. The desire to have an estate with tenant farmers existed but it clashed strongly with the one requirement at the time to make it function: cheap labor. For this reason the Northwest Ordinance was important. By outlawing slavery and giving the yeoman farmers of the South the chance to move into a land where slavery did not exist, Congress removed, or at least put a large hurdle in the way, of the creation of great landholdings utilizing slaves.

The second means by which the great landlord was thwarted in the West was a case of democratic politics in the early republic. By excluding slaveholders from the Northwest, Congress did its best to insure the family farm would predominate for decades. Because landownership was still the basis for obtaining the vote for white adult males, and land acquisition was not difficult, all these small farmers were interested in attracting more population and establishing transportation systems. They were not interested in letting speculators hold land for years, thereby thwarting future settlement, and small farmers did not want landed magnates monopolizing the soil. So they taxed land. Foreign traveler after foreign traveler picked up this tidbit of information. Sir Charles Lyell noticed its effect in Ohio. Taxes were laid per acre of land; if the taxes were not paid yearly, the sheriff seized a portion of the land and auctioned it off "so that it is impossible for a speculator, indifferent to the local interests of a district, to wait year after year, until he is induced by a great bribe to part with his lands."[33]

In England, county government was ruled by the gentry, and the gentry had no interest in heaping on county taxes for public education. In Parliament, they used their powers to keep taxes on themselves low—they passed the tax burden on to their tenants. Southern plantation owners were in the same boat as the English gentry for very much the same reasons. Southern plantations grew large, containing thousands of acres. A land tax could be devastating to the economic position of the planter, and thus, like the gentry of England, the slaveholders argued for low taxes, slaveholder control of local government, and no expensive system of public education. The question of taxation of land could well explain slaveholders' highly vocal denunciation of democracy.[34]

Nonetheless, the principal reason for the establishment and longevity of the family farm in the North was economic. The states and territories above the Mason-Dixon line did not have the excess population to justify economically an estate system like England, and an artificial pool of cheap labor could not be manufactured by resort to slavery, indentured servants, or some other mechanism of legalized servility. And the technology did not favor large-sized firms, except possibly in cattle and sheep ranching. The steam engine might work well in a manufacturing plant, but it was less utilitarian in plowing the land.[35]

An oddity in the status of landholding North and South immediately presents itself. Public land sales in the North yielded the family farm. The same procedure in the South resulted in the great plantation. This bifurcated outcome probably originated from one central fact: northerners who desired wealth went to cities, and land speculators were eager to parcel out quickly their holdings to small farmers. Southerners, being influenced by the cotton boom and the status given to a country estate, a derivative of the English inheritance, saw prestige and wealth in possessing thousands of acres. Thus the wealthy of the southern seaboard bought land during southern public land auctions—either themselves, their agents, or from land companies—in huge tracts. It is one of the oddities of the American agricultural land system that it simultaneously produced the proliferation of both the family farm and the great plantation.[36]

Getting a Farm

The number of northern farms grew continuously for eight decades after 1820, so the question arises how families obtained them. Farms were not cheap for young men and beginning families. Clarence Danhof argued that a forty-acre farm cost perhaps $1,100 when taking into account fencing, land cost, breaking the land, building or purchasing a cabin, stocking the farm with animals, and getting the implements to cultivate the soil. Others have found that initial price too high, suggesting that $500 was sufficient in some areas for obtaining a forty-acre farm.[37] Danhof mentioned several ways around the high cost of a small farm, including obtaining credit through a mortgage, but he thought a farm laborer could obtain the capital through seven to ten years of labor. Northern farm laborers annually hired were making about $150 per year in 1860. It may have been possible for them to have saved $500 to $1,000 over a seven-to-ten-year period, but that act seems to require an implausibly heroic effort on the part of hundreds of thousands of individuals, an assumption that evokes skepticism. Ray Allen Billington, historian of the frontier, believed an eighty-acre farm would cost $1,500 and remarked that for farmworkers it was "impossible to save the money to buy a farm from their wages of $150 a year and keep."[38]

From the gross statistics of the lack of tenancy and the impressive increase of the number of farms throughout the nineteenth century, farm laborers had to have means of acquiring farms rather than outright purchase of a fully stocked farm. Most contemporaries emphasized limiting the first attempt at farming to ten acres, and they often figured in breaking and fencing initially at most twenty acres.[39] There is also a question of the amount of animal stock necessary to begin farming. One suspects an 1850s farmer could at first manage with either a yoke of oxen or a span of horses but not need both. Moreover, most authors stressed that the novice farmer should build much of what he needed himself: fencing, huts, shelters, and animal pens. Above all, they emphasized, as did historian Robert Ankli, that the settler should look to build up his farm over a period of five or so years, especially if he did not have the money for a fully furnished farm at the outset.[40]

Writers on American agriculture added a few other interesting methods of circumventing costs. One was to rely on neighbors for building barns, setting up fences, and doing a number of chores. The exchange was made explicit: to accomplish tasks in newly settled areas, farmers had to swap their labor time for others' labor time. D. R. Thomason wrote that farmers would even trade some of their animals' work in return for hours of labor in the future. So necessary was collective effort in northern agriculture that the Northern Pacific Railroad told prospective settlers in the postwar years to migrate in colonies rather than as individuals.[41]

Commentators on how to become an American landowner also consistently offered one piece of advice: if the would-be farmer lacked the money to purchase a farm and set it up, then he should rent a farm or farm on shares. In the midst of this celebration of landownership in northern agriculture was a constant reference to renting and sharecropping. The typical agreement was to lease ten or twenty acres for one year and split the crop equally between renter and owner. It was commonly stated that this arrangement worked best for young, unmarried men. Reporters of the practice clearly presented this as a step to farm ownership, and they emphasized the need for the lessee to save money for eventual land purchase.[42]

While modern readers may think $150 per year is scanty to try to amass savings to purchase forty or eighty acres, one writer, John Regan, said that a farm laborer getting room and board plus $16 per month would have no trouble: "At such a rate of remuneration as this, the labourer cannot but save money, and so he is fairly on the road to enable him to purchase land."[43] Travelers were making a comparison between the English farmworker and the northern U.S. farmworker, and it was this comparison that left them awestruck about the wages of the latter. The $150 figure for northern farmworkers included room and board. English farmworkers had to pay rent for their cottages and purchase their own food. The wages for an English farmworker was about 6 to 12s. per week. The cost of food and rent took 10s. 6d. from the wages, leading only 1s. 6d. per week, which was about $0.37 in U.S. currency (at $5 per £1). That meant, after food and board were subtracted, the British farmworker was making only about $20 per year, while his American counterpart was obtaining $150. No wonder travelers thought the American pay was sufficient to start farmworkers on the road to farm ownership.[44]

Two other factors helped lower the entry price of farm ownership. The first was the obligation of fathers—sometimes a legal obligation—to help their sons obtain land, to provide a patrimony. Their ability to do so became more limited as the land became more settled, but one would expect some paternal assistance in enabling children to become farmers if that were their desire—and sometimes it was not.[45] The other means was to obtain credit. Advisers to those seeking western farms frequently said farms could be bought on fourths. One fourth was paid down at the time of purchase, and the rest paid in equal installments over the next three years. The major risk involved was beginning the venture near a downward movement in the business cycle—such as buying a farm in 1836 or 1856, only to be devastated and bankrupted by an ensuing panic. Another qualification existed. Lenders were unlikely to loan money to people without reputations in the neighborhood and to people inexperienced in agriculture. This is probably a main reason why western land could not be a true safety valve for eastern mechanics and factory operatives; they simply did not know how to farm. But for farm laborers and the sons of farmers, the lenders could make a reasonable assessment that they had a good chance at success.[46]

If British farm laborers and tenants, who encountered difficult times between 1815 and 1850, were convinced they needed to migrate, they had to possess at least £100 or, even better, £500 so they could purchase a developed farm along the East Coast. The reason for this was telling. The prime attribute needed for successful American farming was manual labor. In 1855, Member of Parliament W. E. Baxter wrote, "It is no child's play to cut down American timber, or bring into cultivation the new land of the West."

The gentleman farmers of England would never survive; the tenant farmers who really did no farming but only supervised laborers would fail; only the working laborers of England and the British yeomen had a chance of surviving the American agricultural ordeal. As Isaac Holmes wrote, in the United States all the family "must . . . labour; the father, mother, sons, and daughters must all work; for it is probable that no domestic servants [farm laborers] [can] be procured." Who physically labored on the farm was one of the great divides between British and American agriculture.[47]

Geography, Climate, and Animals

Sometimes there are odd disconnects between the separate fields of historical inquiry, and much of agricultural history and economic history somehow does not get absorbed into political and cultural history. This was not always the case, but after the 1960s and the onslaught of the "new" histories—that is, the New Social History—the locus of historical writing became the town and the city. The watchwords of those writers were (and are) industrialization, modernization, class formation, and social segmentation. This alteration in perspective has diminished appreciation of the physical qualities of American life. To understand some of the cultural norms, the role of technological change, and political division of the country, awareness of the physical environment faced by North Americans is required. For farmers the physical environment was the ultimate shaper of their lives and gave meaning to the words "labor," "slavery," "dependence," and "independence." Above all, the environmental and geographical conditions gave a profound significance in American agriculture to the words "manual labor."

The first geographical condition that shaped nineteenth-century American life was the reality of the forest. At the time of settlement, the land area of what became the continental United States covered 1,903 million acres; 822 million of it—or 45 percent—was forest. From the Atlantic coast to the prairies of Illinois, the nation was, in the words of geographer Michael Williams, "an almost unbroken expanse of forest." For visitors, the American forest was a forbidding place. A variety of trees leapt up from the ground to heights of one hundred to three hundred feet, but they were not picturesque. The trees were thick, the trunks usually only a few feet in diameter because the trees were fighting with one another for sunlight and moisture. The lower branches were barren, and the ground underneath had little vegetation—the sun did not penetrate to the ground. Where there had been floods, activities of beavers, lightning strikes, and purposeful clearing by Native Americans, some thinning of the forest had occurred, and a few giants emerged.[48]

European visitors found the American forest disturbing, and Britons especially commented on its ugliness. The United States was "an immense forest," "an imperious forest," a "gloomy" forest, "impenetrable forests," and traveling meant going from one forest to another, interrupted by patches of clearing by insignificant farms.[49] One pioneer of Michigan said of Saginaw that the "density of some portions of the spruce, fir, and cedar lands exceeded any tropical forest I have ever seen, and my experience in Central and South American states has been considerable." Up to the Civil War, visitors to the country continued to be surprised at the extent of forest in Pennsylvania and Ohio.[50]

Land was cheap for Europeans in the United States not only because the continent had so few people but because it was so heavily forested. To be made suitable for crops, the trees had to be felled. This fact made commentators very wary about recommending that Europeans migrate to the United States.[51] The process of clearing the trees for farming took several forms. At first, colonists copied the Native American method of land clearing. Small brush and vegetation were raked away (often by women and children), and the trees girdled, i.e., a deep cut made in the trunk that extended around the tree's girth. Over time the tree died and branches fell off. What was left was burned. This process took years. Axes became available sometime in the 1720s, and the colonists and then pioneers used the tools to cut down smaller trees. But the larger ones took too much time and labor. Girdling continued to be a common way to remove trees throughout the nineteenth century. Visitors to the United States immediately recognized this massive difference between the country and England. England had no forests to remove; America was all forests. England was a garden; America was a forest wasteland. England adored well-trimmed and pampered trees, but in America "the axe and the plough are waging war with the wilderness." In 1856, the author Robert Playfair put in an aside, "I may as well mention here, that the farmers seem to look upon a tree as a natural enemy." The estimate was that it took one year to clear ten acres.[52]

But felling the trees gave the farmer—the pioneer—the resource of wood. Wood was used in three ways. First, it was a fuel: the countryside continued to use wood as the primary fuel for heating until 1890. As late as 1922 one-third of all timber cut in the country was for that purpose. Second, it was used for buildings. Americans would not construct the brick cottages and buildings so prevalent in British agriculture.[53]

Third, the wood was used to construct fences to keep animals out of the crops. Climate and soil did not permit the Hawthorne hedge to root in the North American continent, not that many farmers were interested in growing small trees for artistic reasons anyway. Several types of fences were possible, but the type that was probably most common was the Virginia fence, also called the snake fence, the zigzag fence, or the worm fence. Felled lumber was at hand; it was split into rails, chunks that were about eight to ten feet long. A set of rails going in one direction were then alternately laid on top of other rails that were nearly at right angles. Where the two sides met at the right angle, a post was built (or some propping mechanism) to stabilize the fence. The zigzag fence rose to a height of about six to eight feet. It was effective, it could last fifteen to sixty years depending on the type of tree the rail came from, it was cheap because lumber was so plentiful, and it gave the supporters of presidential candidate Abraham Lincoln a motto to exhort in 1860.[54] But it was ugly. In 1824 Isaac Chandler complained of the lack of "ornamental gardens . . . while the zigzag wooden fences have no beauty in themselves." Matilda Houstoun was more direct: the worm fence "offended the eye."[55]

For most antebellum visitors the felling of trees led to the next most prominent feature of American agriculture: the burnt-out, blackened tree stump, whose presence utterly ruined any hope of a bucolic vista. It was probably the most remarked-upon feature of American farming—until visitors got to the prairies. Britons had trouble finding words to describe accurately the ugliness of the landscape. The trunks of trees protruded from the earth anywhere from two to four feet. Farmers usually did not attempt to remove them but let them rot; only the smaller trunks could be removed by use of chains and oxen. The number of trunks must have been unsettling. The editor of the *American Farmer's Magazine* in 1850 estimated 40 stumps per acre and believed that one-fourth of the improved land in the country was riddled with 500,000 stumps. The time it took for a tree trunk to rot depended on the wood; it was said the stumps of pine took half a century. One observer said American farms from a distance looked like a graveyard with the stumps appearing as gravestones. Another compared the farming landscape to "earthy butchers' blocks," and another commented that worm fences and stumps made American farms look "more like a valley of desolation than a smiling vineyard."[56]

Visitors recorded such sights constantly in all parts of the Union until they reached the prairies of Illinois, but especially on the well-trod path from New York City to Niagara Falls. As late as 1859 Charles MacKay mentioned that on either side of the railroad he took to Niagara Falls he saw "the stumps of trees that had been cut down by the pitiless axe of the settlers." Although machines were invented to uproot them, none seemed particularly successful. When Charles Lyell on a railroad trip complained about stumps in the fields, his fellow travelers laughed at the removal efforts of English immigrants. The reason was telling: removing stumps required vast amounts of labor, and in a labor-scarce land a good farmer learned to plow around stumps and not waste energy trying to remove them.[57]

Besides the forest, temperatures drew the attention of travelers. England and France had temperate climates; the heat rarely rose above 80 degrees in the summer, and temperatures in the winter rarely reached freezing. What discomforted so many who journeyed to the United States were "the extremes of heat and cold" that could not "be conceived by an European." Frederick Jobson, who attended a Methodist convention in the mid-1850s in Indianapolis, recorded, "I am daily in such a state of perspiration as to require at times several changes in clothes during the day." The anti-corn-law leaguer Archibald Prentice in 1850 said that it was 82 degrees in the shade "and I can get no one to say that it is hot or even warm weather." The impact of temperature on the farmer and the newly arrived immigrant farmer was monumental. A diary of an Ohio farmer in the 1860s exhibits the problem: July 13, "cutting wheat: reapers give trouble 100°"; and July 20, "finished cutting wheat . . . Mercury at 100°." And Ohio was also cursed with high humidity.[58]

Related to the weather, climate, and forest conditions was sickness. From colonial times to the end of the nineteenth century, the United States was not considered a healthy environment for Europeans. The "seasoning" period in seventeenth-century Virginia is well known, the period in which climate and disease killed English immigrants within a few years. Virginia was not unique. Sickness also pervaded river-bottom areas; the lowlands of Ohio,

Indiana, and Illinois were virtually swamps in the early nineteenth century. An Englishwoman, Mrs. Felton, wrote that she had been in the United States for two years and had "nearly two years of ill health."[59] Sickness in the nineteenth-century United States was a constant occurrence.

An impressive fact about all American farms, northern and southern, was the number of animals on them. Nineteenth-century America teemed with domestic animals. The reason was power: power was animal power in the eighteenth and nineteenth centuries. In most cases, the suppliers were oxen and mules, and, when lighter equipment came into being, horses. Oxen, mules, and horses were, of course, used with the plows and harrows, but they were also necessary for transportation. The crops had to be carted back to the homestead or barn, and then, when possible, to a market. Thus horses and mules dominated American agriculture until the 1920s when the tractor began displacing them. Historians have been so entranced by the railroad that they have ignored the reality that the vast bulk of the American population depended on animals. From their sample, Atack and Bateman estimate that the average northern farm had almost three horses and one ox; mules were used in the South, not the North.[60] Having animals as a source of power shaped farming in a dramatic fashion. Animals had to be fed. Thus, anywhere from one-tenth to one-third of the farm's acreage had to be set aside for hay, grass, or meadow.[61]

When the subject comes to animals, the king of the American farm in the nineteenth century was the hog. The pig was the principal source of meat for most farm families. Pigs could be released into the wooded areas of the farm and subsist on nuts and vegetation, requiring little care from the farmer. When killed, the pork could be salted and in the winter stored for months and still be edible. Some pigs were raised for market, and Cincinnati became famous for pig slaughter. The dressing of hogs for the marketplace, however, can obscure what happened on nearly every farm in the United States during December, for the Christmas month was the month of pig slaughter on every farm. All farm journals have short entries like "Butchered three hogs," "helped father butcher a heavy hog," "killing hogs," went to father's "to help him kill hogs," "killed 4 hogs; 5 be killed later," or "help Bryan kill hogs."[62]

Animals required shelters to be built for them, but they presented one additional task for farmers to perform. Being organic creatures, animals excreted waste. Because the nineteenth-century American farm housed so many animals, the dung had to be removed, and the removal was accomplished only by shovel and physical labor. If the dung were not removed, problems could become serious. There were cases in which farmers let the dung accumulate to such an extent that the barn became useless and had to be burned to the ground.[63]

Equipment and Structures on the Family Farm

Beyond the animals, the forests, and the climate, small farmers required other equipment to enable the land to yield crops. Implements were necessary for even the most basic farming, such as an ax, a saw, a hoe, a sickle, a scythe, and a cradle; later would come the plows, harrows, and rakes. In the 1840s and 1850s, the revolution in farm machinery hit with threshers, steel plows, harvesters, binders, and seed drills.[64] Yet the most important pieces of equipment may have been boots and shoes. The landowners and tenant farmers of England may have ridden around their farms on horses, but American small farmers walked. One can calculate that, for an acre of land in a perfect square, plowing furrows thirty-six inches apart required 2.75 miles of walking. As the average northern farm had at least thirty acres under cultivation, plowing constituted 82.5 miles of walking—behind a horse or ox—and usually farmers had to plow multiple times during a season. With this amount of walking, boots and shoes were an absolute necessity.[65]

Carts and wagons were also vital to the farmer's existence. Every farm had to haul manure or produce from the fields to the barn or to some desired location and finally to urban markets. The reason for this is straightforward: weight. The average weight of a bushel of farm produce was between fifty and sixty pounds. Hauling the harvested crop, with largely all its chaff attached, one-half to one mile to the barn by foot would have been physically devastating. Wagons and carts were essential for the American farm.[66]

Northern farmers did not spend all their time in the fields. They had to be on the road a surprising portion of the year. They had to pay off debts, to get loads of lumber, and to travel to market. Seldom did northern farmers dispose of all of their crops at once; they had to make multiple trips. In 1858 Minnesota, Levi Countryman recorded that on October 28 he took potatoes to the man who made boots for one of his workers; took cabbages to Truax for 308 feet of lumber, and on October 29 he went to Hastings to see about his farm mortgage. In one 1859 diary for the months of January through July, one farmer made nine trips to the mill and thirty trips to town.[67] All this traveling interfered with field work.

Northern farmers could make such journeys because the distances were probably not great. Small farms enabled people to be situated closer together and have ready access to villages. The number of trips they made was also related to the size of the wagon. A tradesman, William Thomson, described farmers around Cincinnati driving their wagons to markct. Thcy contained a few bushels of corn, some flour, a barrel of apples, corn meal, a bag of peas or beans, two or three roasting pigs, and other goods. Now, "Suppose all this bespattered over with dust and mud, and you will have a very good idea of an Ohio farmer to market." From the description, and knowing that the average weight of a bushel of produce ranged between fifty-five and sixty pounds, it seems that a farmer's wagon could not haul more than five hundred pounds of freight.[68] Thus farmers had to make multiple trips to a grist mill, a lumber mill, or a market to get the goods they needed and to sell their crops. Northern farmers were surprisingly mobile; staying on the farm was simply an impossibility—they had to obtain supplies, pay debts, and market their produce.

Focusing only on the major staple crops (corn, wheat, cotton, and rice) produces an incomplete picture of the activities of farmers and the requirements of maintaining the farm. Whether subsistence or commercial, American husbandry had a multiplicity of required chores; if not performed, the enterprise would collapse. Much of this excess activity may be described as the need of farmers to create their own operative infrastructure (power, water, utilities, buildings) that in an urban setting would already be in place. These infrastructure needs had to be built and maintained by the farmer. Thus, in the American farm from 1607 to 1920, there was no such thing as a single person being successful. Success in farming was only secured by group activity—in the North's case, the family.

And even if the farmer could successfully clear the forest, take care of the animals, cultivate the land himself, and complete the necessary business trips—he was doomed to failure and death. He was not feeding or clothing himself. The activity needed to maintain a successful farm needed more than one person. Although the division of labor between man and woman (the "gendered division") into outside chores for the man, cooking and household chores for the woman (with dairy duties usually added) has drawn much criticism, some division of labor was absolutely necessary.[69] There was simply too much to do and not enough time to do it. The subsistence farm of thirty acres or the small commercial farm of a hundred improved acres required three or four people to sustain itself.

Manual Labor and the Family Farm

Climate, temperature, soil types, and forests dictated much of agricultural life in North America. European agricultural modes did not easily fit American conditions. To put the basics of European farming in place, the colonists and then citizens of the land had to wage war on the environment, and battle they did. Early historians described it as heroic; later historians have decried the slaughter of forests and animals. Either depiction leads to one conclusion: the effort and the sweat in making farms in North America was heartless, monumental work. Therein lay the special quality of the United States that would emerge in nineteenth-century rhetoric: labor won the victory in the war on the forest. In winning that war, labor earned a special kind of dignity that in the past had been bestowed on warriors—and the achievement was acknowledged both by Americans themselves and by foreign travelers. And the victory produced the fruits of labor: a competency, an elevation from propertylessness to property ownership.

The reality for all nineteenth-century small farmers was constant, unending, strenuous manual labor. For them, farming required a personal battle with geography, climate, and the soil, and their only weapons were the muscles in their bodies. The people who performed and experienced the effect of all this labor were farmers, farm wives, children, and farm laborers. There was a purpose to continuing that type of labor, a vision of what the "good life" consisted of, and from that vision came political ideals as well.

5

The Actors on the Northern Farm and Their Quest for Competency

In the 1830s, New York Quaker Elizabeth Farnham, who had moved to Illinois to be raised by her relatives and later married John Farnham, was traveling on a steamboat on the Mississippi River around Alton when she met a farmer who was traveling back to his Kentucky homestead with a new wife. He told Farnham, much to her disgust: "I lived up *hyur* two years, and had to have another man's woman do all my washin and mendin and so on, and at last I got tired o' *totin* my plunder back and forth, and thought I might as well get a woman of my own." And as for his choice: "I reckon women are some like horses and oxen, the biggest can do the most work, and that's what I want one for." Elizabeth, who possessed Yankee habits of mind, immediately said to herself, the man did not want a wife, he wanted a slave.[1]

This incident shows many features of American society at the time—the cultural differences between upland southerners and Yankee migrants, the situation of women in American society, the impact of the transportation revolution—but it also underlines a truth about American farming: the woman was absolutely indispensable in holding the family farm together. The man Elizabeth Farnham encountered could not make a farm operate by himself—he needed help. The literature on farm women tends to be sparse, focusing on the duties of antebellum farm women and then plunging into questions of cooperation versus the competitive market. Much more attention is given to urban women, the question of the existence of "separate spheres" for men's and women's work, and the associations women constructed to promote reform.[2] The reason for the farm woman's obscurity is fairly straightforward. As indispensable as the woman was to the family farm, her activities were one vast repetition her entire life, and her lot in life until the 1830s was utter and never-ending drudgery.

The family farm relied on children for labor, and so the first duty of the farmwife was to bear children. And she did. Until the beginning of the nineteenth century, the average American woman bore 9 children; by 1850 the number had mercifully dropped to 5 (and by 1900 to 3.5). This rate was nearly double that of Europe. The reasons for children varied somewhat; the need for labor on the farm was the first requirement, but agrarian societies also produce many children for purposes of old-age protection—to have surviving progeny who could take care of aged parents when feebleness overwhelmed them. The rather sudden drop in fertility between 1800 and 1850 has elicited a number of explanations, but a prominent one was the arrival of machinery on the farm after 1830—the arrival of "labor-saving" machinery. The need for child labor became less immediate as the machines came onto the land.[3]

Children labored on the farm beginning at age eight, boys and girls helping in sowing some of the crops and assisting the mother in carding wool, spinning flax, and making clothes. At about age twelve, the boys began plowing and harrowing, while the young girls became more confined to the mother's work. All the family did chores, leaving various historians to term the family farm an engine of child exploitation. One British traveler in Ohio quoted a pioneer saying that children were the "most profitable kind of cattle we can raise in these woods." Children enabled the farmer to lessen the need to hire farmworkers. Children left the farm when they approached age eighteen and then exited in great numbers by age twenty-one. This release of young farm adults provided older farmers the opportunity to hire experienced help when they reached their late forties and early fifties.[4]

Making clothes and sewing used to be among the celebrated chores of farm women, their activities giving rise to a romantic idea of "homespun" in which families relied on their own resources and not on distant markets. In the frontier areas, women did make clothes from flax. However, the rise of transportation facilities and the development of cloth manufacturing in factories enabled women to go to market to get materials instead of using the raw materials on the farm. When ready-made clothing became available in the 1850s, farm wives no longer needed the spinning wheel and other home manufacturing devices. Currently, the belief is that eastern women were involved in market exchanges before the Revolution; the women in Illinois, Indiana, and Ohio would have to wait until the 1830s and 1840s.[5]

The one area of farm production in which women predominated was dairying and cheese-making. The techniques came from England and migrated to New England, and from there into central New York. Western grain competition hurt eastern farmers, and they turned to butter and cheese—dairying—as a mainstay of the farm. Women were the ones who understood the mysteries of cheese-making, and so they occupied a prominent place. Later in the 1860s, dairying was taken over by men, and by the 1880s western competition rose again until by 1900 even that area of farming was lost to easterners.[6]

When needed, women did do field work, although that tended to differ by ethnicity, but the major effort of farm women took place within the farmhouse. The woman had to cook, wash, iron, clean, and care for the children. She probably had garden duties, although men tended the garden as an integral part of the farm economy to produce necessary subsistence foods. These chores were done relentlessly the whole of her life, and cooking food and cleaning dishes were no minor affairs. Not only did the farm wife have to cook and clean for the laborer hired for the year, but in those times when others came to raise barns, help plow, and attend the harvest, she had to prepare the food for them as well. Visitors who stayed for a week or two were a major burden. One woman said that by age fifty she had cooked fifty tons of food for hired men.[7]

Prior to 1850, the farm woman's lot was hard and depressing. The worst was probably the pioneer stage when women were uprooted from communities and lived in the intense isolation of the forbidding forest.[8] That was when the making of clothes took its heaviest toll. Even after the spinning wheel became obsolete because of manufacturing and trade, farm women's toil seemed excessive. For this reason the lack of servants irked such barbed writers as Frances Trollope.[9]

Trollope called farm women and daughters the "slaves of the soil," and Americans, while they raged at her book for plenty of other reasons, did not really contradict her. The life of a farm wife, said one contributor to the *Ohio Cultivator,* "is one of confinement and unremitting toil. From early dawn until late at night, it is nothing but mend and botch [*sic*], cook and bake, wash and sweep, churn and make cheese, wait upon the husband and his band of laborers, bear children and nurse them." Indeed, the farm journals indicated a real problem with happiness on the farm. One said that young town girls looked upon marriage to a young farmer "as a sort of state prison punishment." A contributor to the *Cultivator* rather implied the role of the woman was pure work because what a good farmer needed was "an industrious wife, not too handsome." Another finally said that the farm woman "has just as good a right to 'life, liberty, and the pursuit of happiness' as he."[10] Visitors marked one of the ways that the continued stress of labor changed American women: they were pretty in youth, but then aged fast when married and took on chores. Farm women, it was said, who were fifty looked like they were seventy.[11]

Historians currently have not looked kindly upon the transition from subsistence to commercial agriculture, but in the case of the farm wife it was probably a god-send. Access to manufactured cloth and ready-made clothing along with the reduced need for laborers due to machinery liberated the farm wife from the crushing burden of subsistence chores. It would not seem extravagant to say that the life of the farm wife brightened considerably after the Civil War when villages became more prominent in the West and made life less isolated, when the chores of family farming and the need for child-bearing were reduced enough to give the woman some spare time to enjoy the rural setting, gardening, and visiting neighbors. But prior to these changes, life for the farm woman must have been miserable; life was something to be endured, not enjoyed.[12]

If farm women have had few investigators, farmworkers have had even fewer. David E. Schob has written the only lengthy monograph on the subject, and his assessment was that farm laborers had a modestly good chance to climb the "agricultural ladder" and to become land-owning farmers. Others have expressed more skepticism. A regional variation appeared, and those farmworkers on the East Coast stood more of a chance of remaining propertyless and poor all their lives unless they migrated west.[13] Aspects of farm laborers, however, remain crucial to understanding antebellum northern ideals—lifetime goals, aspirations, the contemporary understanding of how the world worked. Out of many elements of their lives, two would be central for the antebellum decades: the meaning of the "wage" payment and the idea of "rising" in the world.

Northern farmers due to the small size of their operations did not have exorbitant needs for workers. They attempted to supply their needs with children; assuming one man could cultivate and care for ten to twenty acres, two male children could extend cultivation to sixty or eighty acres, leaving forty or more acres in forest land for fuel. Historians have found that children worked well for farmers between the ages of thirty to fifty, but after age fifty the farmer would have to turn to hired help or possibly turn the farm over to a son. When the children were too young to help, or if they simply had not arrived, the farmer tried to hire one person for ten months or a year (depending on whether terms could be agreed upon, as farmers usually dropped the winter pay considerably because there was less work to do), and as well tried to hire young girls to help the wife (usually at one-half or less the pay for hired males). Beyond that, for the grain crops there were two particular periods when the need for labor was greatest: in the spring for plowing and seeding, and in July through September for harvesting. Then weekly labor could be hired or simply contracted by the task. Certain pools of labor existed: young sons of other farmers, unmarried sons turning twenty-one, hired workers of other farmers, owners of small farms, and village artisans—carpenters, blacksmiths, shoemakers, and leather workers. In the northern farm communities, there was much flexibility in people's schedules to try to arrange workloads to suit everyone's needs.[14] There may have been those who specialized in breaking the sod, pulling stumps, and operating threshers, but few scholars have found in antebellum times a group of farmworkers permanently frozen in their situation as wage-earning laborers.

The system of farm labor, especially those hired out for a year, had distinct characteristics. First, although a farm laborer, the worker was not treated as a servant nor told to eat at a different table nor given demeaning titles; the reigning custom was to treat farmworkers as equals. This aspect upset British migrants, especially tenant farmers, who were not accustomed to such familiarity with the servant class. In the Western Reserve of Ohio, the expression was "If a man is good enough to work for me, he is good enough to eat with me." This custom may have originated in part because often the hired help were relatives or the sons and daughters of neighboring farmers. Also, the arrangement could be broken by the American farmworker at any time. In Europe, the position of farm servant—board and meals—was prized when it could be obtained, and if a laborer "'gets a situation,' he keeps it." But in the United States, the laborers could be working one day, and then gone the next: they "may have left him to get higher wages elsewhere or they may have hired or purchased land on their own account." In antebellum times there were few contracts; agreements were made orally.[15]

While foreign visitors recorded their belief that farmworkers faced favorable prospects, the American agricultural press tended to grump about them. The usual tensions between workers and employers spilled out; the contributors to the farm press, usually farm owners, complained that workers did not exert themselves enough, did not follow directions, and were lackadaisical in their behavior. Largely, though, the journals kept to manuring, care of animals, profits, and the virtues of small farms. The major complaint surfacing was the sudden lack of traditional farmworkers from Anglo-American homesteads and the need to resort to Irish and German workers in the 1850s. In Solon Robinson's thousand-page book on farming, he devoted all of one page to the appropriate treatment of the farmworker.[16]

The system of hired workers in northern farming was probably shaped by one special element: the reliance on the family members of the fathers and mothers to make the farm operate—sons either as teens or young adults or independent farmers themselves, the brothers of the father and mother, and uncles, nieces, nephews, and sisters. Here are some paraphrased samples from one Wisconsin diary in 1855 and 1856: October 9, 1855, worked all day for father on his chimney; Oct. 11, milked father's cow; Oct. 12, helped haul stone for father's hearth (Oct. 13, certified to teach school); May 16, 1856, father gave $2 for work; May 26, 27, ploughing for father; June 3, split rails at father's; June 21, worked in his garden, got $1 from father; July 14, worked for uncle in his harvest; July 30, cut hay for father.[17] Within the diaries, the word used to describe the farmworker was "help," and when aid came from other family members or from neighboring farms, the description was always "help."[18]

Employing sons and daughters of farmers to aid in cultivation has some relevance for the political debates over free and slave labor. In 1838 a contributor to the Baltimore *Farmer's Register* brought up the subject by talking about the uselessness of employing vagrant laborers, saying three out of four were "worthless, absolutely unqualifiedly worthless." This was one area in which slave labor was superior to free labor. He found his conjecture true when he talked to Ohio, Indiana, and Illinois farmers. Their sons and daughters were reliable, dutiful workers because of the family relationship, "but between our hirelings and ourselves, involves us in perpetual difficulty," for they "take offence at trifles, [and] are rude in their manners at table." This source has its dubious aspects concerning hired labor outside the family circle, but the importance of the familial relationship creating an energetic, cooperative, and productive workforce is probably accurate.[19]

Finding workers could be a time-consuming chore. Individuals looking for employment sometimes went farm to farm inquiring about positions;

farmers would go to the villages to see if any men could be spared there. In the nineteenth century, with its lack of communications, missed connections had to be a frequent occurrence. (In economic terms, both sides suffered high transaction costs because information costs were so high: one literally had to prowl the township area to see if labor were available.) Curtis Downer recorded that "Pa went over West this forenoon to look for a hired girl but didn't find any," while a carpenter recorded in his diary for 1856, "Cyrus went to Whitetablet to find work got none."[20]

When the two sides did get together, a negotiation over the nature of the work, the duration of employment, and pay occurred. Because pay rates were so widely known, there probably was not that much leeway in the negotiations, but the interesting feature was the lack of coercion. It was indeed entirely a voluntary transaction in which either side could (and sometimes did) refuse. The entry in the diaries was phrased distinctly as a bargain arrived at by the two sides, not imposed by one on the other. One diary summarized a transaction in 1851 about work commencing that we "agreed Twelve Dollars per month"; another in 1853, "paid John Stapish June 11, 1854 $73.15 and settled with him and agreed to be even." In 1874, Curtis Downer included a line that his father hired "David for $16 per month for 2 months perviding he can't find another man to work longer." At times, the agreement indicated part-time work: "Denny works a.m. for us[,] p.m. for himself."[21]

Once the negotiation ended and work began, farmers kept track of the days worked, any expenses the hired person might ask of the farmer (boots, clothing, and the like), and especially of the days absent (sickness, rain, and other). There was some flexibility in the schedule, as time was taken out for hunting, fishing, and sometimes religious activities—"This Evening we all hands went over to Father Cooks to the Singing School"—but I never saw evidence of workers getting time off for political rallies. Most often workers were docked for sick days.[22]

From the diary entries, it would seem that in the fields, barns, and gardens the hired hands worked without direct supervision. They were either employed for specific tasks, which in the negotiation process must have been largely determined by whether the person had the needed skills, or they were general hires expected to know farmwork. It is nonetheless clear from entries that hired people worked tasks while the farmer was away on some other part of the farm or traveling to the village for some local business. For kinfolk, one would expect a decent work effort from the hired hands so as to remain in familial good graces, but non-kin evidently had a good understanding of the effort required. One of the factors holding back an increasing work demand from the farmer was that he was in the field working himself and knew the limits of human endurance. The cost of supervision is a significant factor in nineteenth-century agriculture as it directly affected profits and productivity; it represents one of the great economic divides between slave labor and free labor.[23]

Getting paid had some interesting variations. Most of the day, weekly, or specific job hires were paid immediately and in cash. Sometimes they were not paid: "Went to S. Cook's after money and got none. Then, to Asa W. Baldwin's after money and got none. Then to Wm Leeman's after money and got none."[24] More intriguing were the number of occasions that farmers or workers when laboring for others agreed to swap labor time. For example, Levi Aldrich in Minnesota wrote in his diary, "A. Click agreed to brake for 3.00 an acre and take work for pay in the fall." Typically corn husking was remunerated with a share of the total husked. William Tebbetts of Illinois started work on twenty-eight acres of land and borrowed a horse to work on another farmer's land "except what I have payed with labor in return and I have several days due me for horse labor." There was direct barter for payment; one Illinois worker was "to be paid in hay" for his efforts.[25] The most interesting way farmers might pay their workers was by animal stock. A Rock Island County, Illinois, farmer took a liking to a young worker, Jonas Peterson, and sought him out as a yearly hire for his farm. At the end of the year, he owed $260, took young Peterson to the corral, and Peterson chose two horses for payment.[26] A worker with access to someone else's farm might be able to build up animal stock over several years and be in a good position to sell his animals for enough money to buy a farm for himself. This was the only such occurrence that I ran across, so there is no indication of how widespread the practice was.

On occasion the farmer fired the worker for incompetence or lack of effort, but the farmer diaries show much more quitting than firing. The writers did not give much detail; the entries were short and to the point: "George Deal quit work." One that was more informative was "Gus Anderson's month was up this morning and he quit work because I would not raise his wages." (Somewhat oddly, Gus Anderson quit, but the next evening he sat at the farmer's dinner table.)[27] On another occasion a farmer released a laborer "because he wouldn't do as [father] wanted him to." When a farmworker quit, the farmer settled with him immediately: "Will Sweeney went away today. I paid him $7.00."[28]

Little information, nonetheless, exists about farmworker's views of their position on the economic ladder. None of the diaries or accounts reveal much about attitudes of these workers or what happened to them. The farm journals also are silent. One can extrapolate one way or another. On the one hand, a high quit rate demonstrates job dissatisfaction. On the other hand, there was never a farmworker movement in the United States until the "Okies" in the 1930s. All of the great agrarian uprisings in American history after 1766—the New York Rent Riots, Shays' Rebellion, the Holland Company riots, the Van Rensselaer rent riots, the Grange, the Alliance, and the People's Party—were movements of owner-occupiers or of fairly well-off tenants—but, unlike England, not of farmworkers. Therefore, an overall characterization of the attitude of farmworkers is ambiguous; discontent there was but not enough to generate a social movement.[29]

Social mobility had to be a fact among the farmworkers in these northern states during the nineteenth century. In 1860 these northern states had 1.2 million farms; by 1900, they had 2.2 million farms. These new farms had to have farmers coming from somewhere, and the most likely source was the sons of farmers who had served apprentice time as farmworkers. How they managed to purchase farms is a troubling question. The one aspect of the nineteenth-century American farmworker that foreign travelers harped upon was their high pay (after subtracting the cost of rent and food, American farmworkers indeed made five to twenty times what their British counterparts did), and thus they had the ability to save. (But where did they put their savings? No diary or ledger gave a clue about that.) In 1823, William Kingdom wrote that American workers could save a "small sum, the saving of two or three years of a prudent working man," and be able to "purchase one or two hundred acres of land."[30] That observation was printed time and again, even as late as 1897. Surely farmworkers must have had family financial aid when they were ready to make the jump from worker to farm owner.

Extrapolating from comments in manuscripts about rates of pay, one could establish these boundaries for the northern antebellum standard of living: for single men, $0.50 per day or $150 per year, subsistence poverty; $1.00 per day or $300 per year, subsistence plus minor comforts; $1.67 per day or $500 per year, middle class; $3.00 per day or $900 per year and above, some affluence. The worth of board, meals, washing, and mending would be $200 in an urban area, perhaps $150 in a rural area. At a pay of $150 beyond room and board, the farm laborer had no affluence, but was considerably above poverty. More significantly, because the farmer provided room and board, the farmworker was in a position to save. For families of five, the boundaries would be roughly $300 per year base subsistence; $500 per year subsistence plus minor comforts; $750 per year lower middle class; and $1,500 and more, considerable affluence.[31]

There is one other possibility to relate. In the ledgers and diaries, one is impressed by how few consumer goods the American farm population bought. They purchased clothes, boots, shoes, coats, and maybe some leather goods; whiskey was popular—and very little else. Farmers bought capital goods for the farm, not consumer goods. There may have been a consumer revolution in America, but it was largely an urban phenomenon; it was a nonentity in the countryside. The accounts farmers kept of the purchases of the hired help reveal sparse and utilitarian choices, not ostentatious ones. The consequence is that perhaps farmworkers were indeed achieving monstrous rates of saving—largely because as single youths, they had little to purchase at all. The end result was that American farmworkers could, and probably did, save out of their wages. This is not a minor finding. It will play a major role in understanding the free labor ideology.[32]

The farm owner is the last actor to consider on the northern family farm in the nineteenth century. Overall, his life from one year to another was an endless repetition. Farm diaries and ledgers show that the cadence of plowing, sowing, harvesting, fence mending, hog-killing, animal tending, and marketing went on year after year, punctuated time and again only by an accident, a sickness, or some community event. Continuity, not change, was the tendency of northern agriculture.

One characteristic above all others described the northern farmer: he labored all the time—he did physical, manual labor with his hired help in the fields. Everyone noticed this aspect of northern farming. Early travelers to the United States wrote, "The farmers live well and work rather hard," or more pithily, "Farming people in Ohio work hard." In 1897 the visitor George W. Steevens noted that on a Wisconsin farm the owner and his sons looked "as rough-clothed and hard-worked as any labourer."[33] Farmers because of their manual labor had no place among those who judged people by the code of gentility. If one wanted the genteel life or to meet its practitioners, one went to the city; on the farm manual labor reigned supreme, and its reign was ruthless.

The major alterations in farm life came with the filling out of the surrounding areas and the taming of the land. Villages grew more numerous, providing farmers with more access to information, merchants, churches, and amenities. Over the first half of the nineteenth century, the intense isolation of early pioneer life dissipated. Most of the farmers had long harbored commercial aspirations and wanted access to markets. The problem of the western states was put into precise form by a Wisconsin settler writing to a brother in Vermont. Why move to Wisconsin? "It is indeed a fact that they [migrants to Wisconsin] can raise wheat and other grain, easier than in your county, but what does it avail them? There is no market for it and but very little sale any way."[34] Writers complained about having to walk or travel twenty or thirty miles to a market. Everyone recognized the problem of transportation, and counties tried to provide some solution by imposing a tax on settlers to construct roads. The tax was one to three days of labor on county roads, to be paid in actual labor, the labor of a son or hired hand, or a money equivalent. This tax prevailed nearly everywhere, and, so far as the ledgers and diaries are concerned, generated no animosity.[35]

As the region matured, farmers completed the arrangements on their farms, the building of houses, barns, stables, and fences. They cleared land until they had enough wooded area left to provide fuel and enough improved land that three adults (approximately) could cultivate. In this process two obvious things would happen. Quite naturally, as the farm took finished shape, a surplus in crop production occurred. At first the farm struggled to provide enough food for family and animals. When the farm neared completion, however, a surplus above family needs appeared. Thus if one takes the traditional number of 60–80 percent of a farm's output going for family needs in the 1760s or 1790s to 40–50 percent by 1860, the change hardly represents a transformation of farm life but an expected happenstance given the evolution of a farm from a primitive state to a finished one. It also meant farm income rose. Atack and Bateman estimate that in 1859–60 owner-occupier farms in the Midwest earned $113 per capita and in the Northeast, $174, the average for the entire North being $135.[36] Assuming the farm family was between 5.5 and 6 members, the farm family income in the Midwest would have earned, on average, $620 to $675 a year; in the Northeast, $960 to $1,045, and for the entire North, $740 to $810 (all figures rounded). This would put most farmers squarely in the middle income level for the nation.

Farmers were earning more, and the value of their property was rising, but work routines did not differ greatly—except they became easier. The impact of labor-saving machinery upon output was not nearly as important as its impact upon a person's physical labor. The toll farm labor took on a human being prior to 1850 must have been staggering. But with machines, corn husking became a thing of the past, the flail no longer made muscles ache, and the sulky—a seating area the farmer rode in, while the horse pulled the plow—spared feet the anguish of walking tens of miles in the fields. Farming became physically easier to perform after 1850 or 1860, but the routines remained pretty much the same.[37]

A striking element about northern farming throughout the nineteenth century was the selling of small amounts of produce all the time. The diaries have entries such as "sold 6 1/4 bushels of oats at 48¢"; description of an Iowa farm having twenty-three acres of corn, oats, and grass and sixteen hogs; sale of 76 bushels of wheat and 10 bushels of rye; hauling 820 bushels of wheat; the results from threshing ten acres of wheat were 70 bushels and of six acres of oats equaling 157 bushels; a crop of 220 bushels of wheat and 125 bushels of corn; the selling of 650 bushels of corn.[38] Although this cannot be proved directly, one gets the impression from diaries that farmers sold some crops to get cash for immediate needs, then stored the harvest in bins, cribs, and basements until they either needed to purchase some new item, pay for a repair, or to retire a debt. Then they took the amount of crop needed to get cash or to use as barter to take care of the problem. Northern farmers seemingly were in no great rush to convert their crops into cash.

The Family Farm and the Exactions of Manual Labor

Northern husbandmen obtained their own plots of land and ran their own lives. Throughout the nineteenth century, orators celebrated the property ownership of hundreds of thousands of northern farm families and the social, economic, and political independence that, allegedly, came in property ownership's wake. But this ownership was no manna from heaven that rained freely on farmers. Weaving its way into and out of all the romantic rhetoric was the understanding that the price to obtain this independence was an excruciating amount of manual labor.

Thomas Lincoln was a typical pioneering semi-subsistence farmer in the late eighteenth and early nineteenth century. Running into the befouled survey problems of Kentucky, he lost his substantial farm (over four hundred acres), and, taking his family with him, he moved to Indiana and then to Illinois. Never earning more than a bare competence in agriculture, he nonetheless found enjoyment in religion, hunting, and telling tales, but his life was one of continuous arduous labor. So it was also for his son Abraham as well as the other boys and girls in the Lincoln household. As Abraham grew up, he learned and did all the chores required for the farm, and to aid the family's income Thomas hired Abraham out to other farmers and had him chop trees and split logs into rails for the infamous worm fence. When Abraham did not work hard enough, Thomas inflicted physical punishment. Abraham had to turn over all that he earned to his father and in return he was supposed to obtain a patrimony—some land, a start in life. Abraham did not receive a patrimony, and young Lincoln abandoned the family as soon as he turned legal age and left farming forever. Some have implied that Abraham did not like the hard work that farming required. That is probably true, but the dislike deserves further consideration. Abraham Lincoln was hardly alone in distancing himself from agricultural labor. One has to conclude when compiling the writings about the first half of the nineteenth century that the quest for farm ownership exacted a heavy toll on the physical and mental health of the seekers.[39]

The amount of physical exertion by the actual laborer in farming between 1607 and 1850 was monstrous, whether done by slaves, tenants, or family farmers. To sow the crop, weed the plants, harvest the crop, tend the animals, construct the buildings, make the clothes, and feed the laborers took an immense amount of work; bending, lifting, hoeing, shoveling, mending, and stitching were all done by hand and with primitive tools. Not everyone had the constitution to keep up this regimen.

Foreign visitors made many comments about American politics, the Lowell mills, cities, and agricultural practices. One aspect they mentioned frequently was the physical appearance of Americans, and their evaluations were not favorable. David Mitchell, a sympathizer with the Confederacy in 1861, commented that the northern farmer of the Yankee type had not a "lazy bone on his body," and was a "rather tall, bony, sinewy, strong man, with very little fat." As to the Yankee farmer's outlook, "there is too often about him a look of being overworked both in mind and body." Mrs. Matilda

Houstoun decided that the Anglo-Saxons had deteriorated in North America because they were smaller in size and weight than their cousins in England. Thomas Hamilton as well thought Americans were not a happy lot as they were "withered by the anxieties of life," while the geologist Charles Lyell remarked about New Englanders that "most of the farmers and labourers had pale complexions and a care-worn look."[40] Women in youth had vigor and color, but soon after marriage all the vitality faded and they appeared old before their time.

Often the physical condition of Americans was related to climate, but others saw in it the onerous exaction of the amount of labor needed to make the family farm operable. One critic said that, when an Ohio pioneer finally cleared the land, "he . . . dies when he should be beginning to enjoy himself." This observation that family farmers were working themselves into an early grave was memorably put into the comment of Anthony Trollope during his visit with a Wisconsin farmer. He admired the independence of the farmer and his family; yet, "His children are sick and he is without comforts. His wife is pale, and you think you see shortness of life written in the faces of all the family."[41]

Robert Russell made a telling observation about the effects of work on the human physique, and at the same time demonstrated why leisure had such alluring charms. When at a fair at Springfield, Ohio, he noticed a set of young Kentuckians making purchases and sales. They stood out from the northern farmers as being larger and better looking. Indeed, they were so "fresh and ruddy" in complexion to make each "every inch an English country gentleman." He decided that the "exemption from manual labor for several generations, it would seem, has not been without its influence on the Anglo-Saxon constitution." In other words, slave labor spared the planters the deleterious effects that accompanied the manual labor of family farming. He continued his contemplation by assessing northern farmers: they "are much smaller men, with a vast amount of activity and energy. All who labour with their hands upon the land in America lose that full habit of body which our agricultural labourers have at home." Family farming required such vast amounts of intense labor that it shrunk its practitioners and stole all youth from them. In one letter, an Ohio farmer let slip that when he migrated to Ohio he weighed 175 pounds; after farming in Ohio he weighed 134 pounds.[42]

European commentary on the physical appearance of American farmers could be dismissed for a number of reasons. Many of them had political axes to grind about democracy and equality versus aristocracy and inequality. Others simply viewed matters from an upper-class snob perspective; they were simply exhibiting their distaste for people who worked with their hands. The oddity, though, is that the nineteenth-century American agricultural press largely agreed with them.

Throughout the North were an oratory and a scribbling that extolled the necessity of labor, the dignity of labor, and the biblical commandment to labor—in the "Sweat of thy face shalt thou eat bread" and "Labor is but the discharge of duty to Him who made the requirement." No shortage of paeans to the hardworking family farmer was to be found in the antebellum decades.[43] Likewise, all sorts of writers and speakers, as well as foreign visitors, recorded their awe at the vast number of independent proprietors that inhabited the rural lands: "We differ from most other nations in this, that our land is divided into *small* parcels or estates, and is *owned* by those who till it."[44] Yet by the side of such writing were complaints about the lack of respect that farming elicited from the rest of society. Farmers had to "live down the old brutal notion that a farmer must be uncouth, uneducated and unthinking—a mere ploddropps [*sic*]." Who got respect? "Professional men—a class of men, who are . . . the drones in the great hive of human society." Rural editors and writers were angry that the "merchant, with bleached hands and well lined purse," looked with contempt upon the "hard toil, brown visage, and moderate gains of the farmer." Young town women recoiled in horror, so it was reported, at the idea of marrying a farmer. Some farmers looked askance at urban dwellers, believing that such people fled the countryside because "they don't like the toil of the farm, and imagine they should be exempt from the mandate, 'In the sweat of thy face shalt thou eat bread.'" The most telling revelation about farm life was given when writers complained that farm children deserted the farm: the youth wanted to exchange "the hard work of the farm for the pleasanter business of selling goods." Being a clerk—earning a wage under the supervision of a wealthy person and thereby being economically dependent—was preferable to the economic independence of a yeoman farmer. One may extract from such statements that, when the amount of physical labor became too high, the reward of republican independence meant very little.[45]

Beneath the tidal wave of hosannas to labor was an undercurrent of discontent about the amount of labor being expended. "Work and slave, indeed, for a living," wrote one, and end up "little if any better than the man who owns no land. What is the use of owning property if it is not enjoyed?" Anthony Trollope, upon watching Americans labor, decided that, yes, labor was good but not to excess—Americans were the most severe taskmasters he had ever encountered. A pioneer of Minnesota did not remember farm labor fondly. He spent one year as a farm hand in his youth which would "never be forgotten, but will ever remain a depressing 'souvenir' in my life's experience."[46]

Most diaries and journals kept by farmers—the few that exist—are not overladen with commentary about politics, work, and society. The authors stuck to the agricultural chores they performed. One exception was the diary of a Minnesota pioneer, Levi Countryman, who came to Minnesota in the mid-1850s, served a brief stint in the army at the end of the Civil War, obtained a master's degree at Hamline College, and lived to ninety-two, passing the last twenty years of his life in California. He kept a remarkable journal between 1858 and 1860, and in it he let loose his frustrations at the workload of small farmers.[47]

Levi Countryman found the life of a small farmer exhausting and overwhelming. "There are so many things to do, and to so little profit on this place that I don't know what to do first." He sodded oats: "Tis a tedious business." He cradled wheat: "The cradling business makes me very tired, and I can hardly get up and down." Even potatoes drew forth complaints: "Dug potatoes all day and am excessively tired to night." Even after all the work was done, "I run 3 1/2 miles, after the [escaped] cow, but did not get her." He noticed other farmers who worked hard and are "nearly worn out"—"*Labor est voluptus.*" He thought he was not suited for farming and hoped to obtain a divinity degree and escape the ardors of rural life, a hope that was realized in the postwar period. But he knew why God created Sunday as a day of rest: "I have worked very hard all week. Oh how blessed an institution is the Sabbath day."[48] "I do not hate work," he wrote dejectedly, but he came to question the amount necessary for survival: "Labor is ennobling truly; but too much labor is degrading, hence that labor *may* be ennobling there must be a limit to its exercise. I believe it to be that amount which is necessary to ensure health: say from four to six hours. That amount is enough to support us, and the balance of time may be spent in accumulating knowledge which is vastly more essential than wealth. Never be idle. There is nothing so painful to me as idleness."[49] The extreme demands of labor on the farm took its toll on the antebellum farm family. Many sons and daughters, of course, learned the routines and remained members of the rural community. Others did not, and some openly told their fathers that they intended to leave.[50]

Farm women for the first half of the nineteenth century bore perhaps the heaviest burden, or at least so it seemed to numerous writers. Prior to the development of ready-made clothing, the pioneer woman made the clothes for the family, including the spinning and weaving of cloth. With the ability to purchase clothes with few alterations to make, the woman was relieved of the chore. A great relief it must have been. Cooking alone consumed the farm woman's life, and putting the added chore of homespun manufacturing on the women and children must have come close to crushing out enjoyment of life. Only later in the century when such tasks were more ornamental to a farm family's existence did homespun and quilting become sources of pleasure.[51]

When William Cooper Howells wrote his reminiscences of life in early Ohio, he hit upon another theme that was important in explaining why some children abandoned the farm for the town. The amount of labor crushed out the realm of mental adventure. Howells wrote, "I liked the work that was to be done; but the quantity of work to be done, and the constant devotion it required, made it slavish to me and irksome"—it consumed all of his time and left none for reading. Howells's complaint was not unlike that of Abraham Lincoln, who also had an active imagination and found that farming detracted from it. The mental vista farming permitted was sometimes quite limited. The noted agricultural historian William N. Parker in a presidential address to the Agricultural Historical Society related a story from the mid-twentieth century about how farming might have provided a competence but not mental satisfaction. Parker once asked Rondo Cameron how he came to work in banking; Parker reported that Cameron said, "'If you'd spent the years from age 9 to 16 walking two feet behind the rear end of a mule in a cotton patch . . . you'd be working in banking, too."[52]

Excessive labor, especially in the heat, also had one other potential effect: it dulled moral sensibilities. Manual labor yielded much exhaustion and so many aches, pains, and tiredness, that it affected the brain and impaired reasoning faculties. There was a reason why commentators so often compared slaves to animals, to the brute kingdom. Although the visitor John Oldmixon found the route to property ownership by working could be traveled, he nonetheless admitted that the world's laboring classes all suffer because the "hard work" they perform "eats into and puts out any little *mind* they may possess." Seldom do diarists and letter writers admit to this, but one did. The editor of the memoirs of Illinois Baptist minister John Mason Peck wrote that the "time, and care, and toil, which he devoted [to] this [farming] seems to have been at times oppressing to him; but though he complains of its deadening his religious susceptibilities, he did not intermit or shrink from these endeavors till his family was made comfortable."[53]

The price of independence obtained through family farming was high indeed, and many were not willing to pay it. They escaped to the city or town and attempted to obtain independence and a higher income by becoming lawyers, doctors, teachers, contractors, merchants, or finding other "middle-class" employments, ones that did not require such debilitating amounts of manual labor in harsh weather conditions. In many cases, they simply became wage-earners of one type or another, sometimes earning an adequate income, sometimes obtaining only enough to stave off poverty. Given the labor requirements on the farm, however, it was no wonder that people deserted it; nor was anyone astonished at why "leisure" was such a prized state of being. As many realized, excessive labor did the human mind and body no favors.

Historians have grafted onto the early years of North American history all kinds of interpretive schemes and ideological frameworks. Perhaps only studies of slave life have preserved the awesome amount of human labor necessary to survive. The cruel reality was that hard manual labor was central to human existence, especially when the population of the society exceeded certain boundaries. It must nonetheless be asked what it was in farming that motivated these people to pursue an avocation that resulted in such physical exhaustion. The answer is an independent competency, and that answer leads to one of the major debates over the nature of antebellum northern agriculture: the capitalistic nature of the family farm.

Northern Agriculture and the Arrival of Capitalism

A divide has opened up in the field of early American agricultural history, and it involves the question of capitalism and the nation's farmers. The general interpretation that has been gathering momentum is that northern farmers became capitalists early. The crux of the interpretation is that, by the time of the Civil War, the North had changed and northern farmers had become capitalist farmers espousing the bourgeois ethos: the character traits of industriousness, sobriety, rational thinking, and individualism. Opposed to these values were the traditional ones that emphasized hierarchy, patronage, obedience to social superiors, community obligations, government direction of life, and kinship relations. Many scholars insist that ultimately the North came to embody the bourgeois values, while southerners, both yeomen and great planters, retained much of the traditional agrarian ethic.[54]

Among agricultural historians, there is a stark division between the economic historians—the "cows and plows" crowd—and the social historians, who tend to come from the New Social History of the 1960s and who emphasize economic systems and class and gender relations. Economic historians have assumed the existence of the market and a market mentality. Their view of the evolution of American agriculture has been focused on subsistence agriculture largely due to transportation dilemmas. They affirm that as soon as the market appeared American farmers joyfully exercised their preferences and produced crops for income. Economic historians' attention has been wholly on output, income, productivity, and typical measures of material welfare. Social historians have been more alive to the possibilities that farmers did not have a market mentality—they had to acquire one.[55]

The question of traditional versus market values was initially proposed by Michael Merrill and James A. Henretta, who looked at the New England family farmer and insisted that their values were not the ones prized by the market. Family was more important than income, and the family farm sought to maintain its stability by providing food and shelter (subsistence economy) for its members and enough surplus production to purchase the few goods not obtainable locally. Trade was conducted locally, often on barter terms, and credit was often carried for years before repayment was demanded. The theme was taken up later by Christopher Clark, who located the New England switch to capitalism in the 1830s and 1840s, when wage work slowly became important both on the farm and as an adjunct to rural income (via outworking).[56]

The foray into the mentality of northern farmers has evoked an explicit Marxist explanation, mostly notable for its confusions. To signal the arrival of capitalism, Marxist theory demands the creation of propertyless wage-earners (the proletariat), a few large property-holders who do not engage in manual labor, and consolidation of businesses into larger units. None of those conditions were present in nineteenth-century northern farming, which is readily admitted by Marxist authors. That does not stop them, however, from deciding that the northern farmer was definitely capitalist and probably the progenitor of American capitalism.[57] Hal S. Barron, a prominent historian of New England rural life, has rejected this characterization of northern farming as capitalist primarily by emphasizing the continuity of traditional practices of farmers. These social historians "have tended to superimpose urban and industrial patterns on the late nineteenth-century countryside."[58]

For a number of scholars, the current interpretations of the northern farmer have been inadequate. The dichotomy made between subsistence farming and commercial farming was too rigid and inappropriate. Several have opted for understanding northern family farming as a compromise between commercial and family-first farming, that most northern farmers followed a mixture of strategies, keeping uppermost the goal of keeping the farm in the family. Much of the literature continues to stress continuity throughout the nineteenth century, the great divide coming sometime after 1880.[59]

Much of the work on northern farming is remarkably well done, attentive, and accurate in terms of family survival strategy, crop routines, roles of youth and wives, market involvement, and use of labor; the economists' determination of output, income, productivity, and demography has been stunning and useful. But the placing of the northern farm family into the framework of general economic systems and culture has been awkward, confused, contradictory, and unsatisfactory. The approach this work takes rests more on the actual physical demands of farmwork than any economic system approach, and it uses for its cultural background not the bourgeois framework but that of mastery.

To evaluate the character of the northern farming community, some definition of capitalism is required. Among social historians the most widely accepted description of capitalism is that the system is evident when the propertyless wage-earner emerges and is frozen in his or her position—it is a permanent condition of impoverishment. To this we will add another partial definition: in the capitalist system, the capitalist always and relentlessly and perpetually reinvests a part of his earnings—that is, a capitalist seeks growth without end. This is part of Max Weber's "spirit of capitalism," the ascetic quality of the capitalist mentality that molds in human form an investment machine whose investment behavior is in perpetual motion not for the rational reason of accumulation for future enjoyment but because that is the moral order of things.[60]

The first task in analyzing the northern farmer is the subject of "bourgeois values." Those are generally considered to be individualism, industriousness, rational decision-making, wealth accumulation, sobriety, self-control, a heightened sense of moral duty in regards to proper individual behavior, and an individualism that stresses that one's actions determine one's welfare. With one or two exceptions, the northern farmer, whether southern uplander or northern Yankee, possessed all of these traits. What is crucial is the determination of *why* all these traits applied to the farmer's life. The answer lies in the nature of the owner-occupier farm and the necessity of labor. More generally, these traits have existed in all farm communities, regardless of ownership type, across the globe. If the farmer or his wife is continually inebriated, the chores will not get done and the farm family will perish. If the farmer does not plant crops with some rational idea of rotation, soil fertility, the needs of animals, the timing of sowing, hoeing, and harvesting, the farm family will perish. Put another way, if one took the opposite of these values of industry, sobriety, rationality, self-control, initiative, and the "Protestant work ethic," what would be the fate of the farm? It would be disaster. The owner-occupier farm could not exist without these character traits.

Everything translates into one omnipresent and omnipotent fact: the owner-occupier farm ran on human labor. It was the intense, excruciating amount of physical labor on the farm that produced the work ethic; without that ethic, the farm family perished. This is to say that cultural background, religious background, and even economic background were secondary influences in producing the obstreperous work ethic of Americans. They were not insignificant but simply paled before the demands put upon the laborer to make "yeoman" farming operate. Not surprisingly, yeoman farmers of the South and upland southerners in the Great Lakes had virtually the same work ethic as did northern Yankee farmers.[61] It is worth noting that neither feudalism nor the English estate system of tenant agriculture produced the intense Protestant work ethic, but it did arise among seventeenth-century English yeomen—that is, it was the unique ideological construct that originated with the family farm.

A second battle cry among the historians depicting a wholly capitalist farming North is the stress, often found in political rhetoric, on the slogan "the dignity of labor." Again, one has to draw out the implications of the work effort of northern farm families and add to it the code of gentility. What farmer would say that labor was undignified when he knew that not working meant the collapse of the family? Farmers were an important part of northern society precisely because they did labor, and their labor made and kept them small property-holders. It is valuable here to bring in the code of gentility and remind ourselves of the identity of those who called work undignified. They were the ones whose ownership of property enabled them to stop laboring and to live a life of leisure—they were the landlords of the soil, earning monopoly profits from rental agreements. The aristocracy and gentry of England were the ones who had declared manual labor was undignified for a man of substance. Why would not American yeomen, North and South, claim the dignity of labor against those whose lives purposely elevated leisure to the status of a virtue by depriving those who worked the fruits of their labor?

One also has to question the idea that northern farmers were driven by capitalist rationality and calculation. Perhaps somewhere on paper these family farmers tried to figure out whether they made profits or losses from a calendar year's effort, but they are rare indeed. There exists a considerable number of calculations at the end of diaries and ledgers, but they do not seem to be related to summarizing income and expenses to determine profits. Most of the time, these small farmers seemed to have been more inclined to have had a general knowledge of whether they had enough produce to pay debts, to repair machinery, or to purchase some needed item. They certainly kept track of what they owed others, but there is lacking an overall statement of the financial health of the enterprise.[62]

And then there are the obvious problems with nineteenth-century northern farming as a capitalist endeavor. Obviously these farmers believed in private property and individual choice in purchases and sales, so they met certain basic definitions of capitalism. But certain other features are simply abnormal from a capitalist perspective. The most visible oddity is that these farmers rejected growth. Average farm size decreased for nearly a century. Unlike the landlords of England, or the planters of the South, northern farmers were not adding acreage to their farms; aristocrats, gentry, and planters pursued growth of farm acreage with a shameful avidity. If one of the signs of capitalism is the continuous reinvestment of profits into an enterprise or into new enterprises, American northern farmers were disgraceful capitalists. There is no indication of reinvestment for expansion; reinvestment existed to maintain the working status of the farm. If some accounts are right, the farmers were socking away money: they earned income, saved it, and did nothing with the savings. Most likely the savings tendency of mature farmers was to prepare for old age when they no longer could do the physical labor the farm demanded.[63] The gross statistics of northern farming—the number of small farms, their constant increase, the decline in acreage per farm—is not a tale of capitalist growth and consolidation.

Moreover, the family farm was not economically dynamic. In some ways, it was a drag on the national economy. Jeffrey G. Williamson and Peter Lindert argued that American wealth and income distribution in the nineteenth century worsened because people stayed on the farm (where the annual rate of income growth over the century was 00.0 percent) instead of moving into manufacturing (where the annual rate of income growth was 2.0 percent). Jeremy Atack and Fred Bateman also noted the oddity of low returns to farmers in the nineteenth century and wondered about their economic rationality.[64]

And then we come to the ethos that, by all historical accounts, did exist in nineteenth-century northern farming: competency, a concept that among historians became important in northern agricultural history due to the findings of Virginia DeJohn Anderson and Daniel Vickers.[65] The reigning ethic of northern farming was to occupy enough land in fee simple to maintain a family in good health and material condition over the years and pass on land to sons and more rarely to daughters. Competency does not exhibit the acquisitiveness supposedly lodged within the system of capitalism, does not have the growth aspect of capitalism, and indicates that, once the producers achieved a level of material welfare, all that was to be sought was to stay at that level. Satisfaction with life was achieved at rather low levels of comfort.

In antebellum times, writers employed the word "competency" continually in discussions about American farming, especially the individuals writing travelogues and guides to North America. These individuals had their axes to grind, but they made two striking claims. First, they enticed people to come to western lands to procure a "pleasant and independent home, with comfortable competence"; with work and a little stock "a farmer makes himself comfortable, nay, independent, for life"; in older areas in the United States one was "delightfully impressed with the comfort and competence, though not luxury and wealth, which are everywhere enjoyed."[66] The second idea was that those coming to the United States could rationally expect competency through hard work but not a pecuniary fortune. One piece in a New England farm journal—however much credence one should give to such expostulations—warned farmers about dreams of wealth and urged them to be content with a small farm: "money, beyond that point which secures independence, is not an element of happiness or progress. When sought for, for its own sake, it is ever a means of evil."[67]

To understand the worship of the yeoman farmer, the ethos of competency needs to be employed. Since Jefferson, if not before, orators sang all sorts of praises to the yeoman farmer. What they glorified was neither his subsistence nor commercial style of agriculture but the fact that in North America the yeoman owned his own land and had "independence."[68]

It is possible to understand this stress on independence and landownership in a more modern idiom. Robert Leslie Jones in his recent study of Ohio agriculture decided that "most farmers really liked farming. They relished 'being their own men.'" On occasion one runs across these exclamations of self-ownership. Daniel Thomas wrote in 1854 in his diary that "I did not work as much as I was my own employer and engaged in digging my potatoes." Another diarist noted the times that a hired worker "*works for himself*" and not for the employer. The Indianan Fielding Beeler jotted down the problem he had with the independence of northern men when he worked on a road with others: they were "the hardest set of cases I ever had to deal with, every man being his own supervisor, and working on his own hook."[69]

If one looked at Europe's past, then the common European social and political experience was to be ruled by a master: an aristocrat over his serfs, a master craftsman over a journeyman and an apprentice, a father over wife, children, and servants. It was a political and economic system of mastery, of one person ordering others. Independence in the post-Revolutionary United States was a cultural reaction against the continual presence of mastery by superiors. They called the condition of living under mastery "dependence," and the condition of escaping the mastery of others "independence." Throughout the world, a system of mastery was at work.[70]

In the nineteenth-century North, the desired goal was to escape the mastery of others, and the way to reach that goal was by owning a small farm. Given that these people lived in a world in which mastery was the most obvious characteristic of social and political life, the desire to be independent formed a conviction so strong that it led numerous people away from a market mentality and from a desire for ever-increasing hoards of wealth. In economic terms, the preferences for independence and escape from mastery were so powerful that, after a person obtained farm property and a competence, the microeconomists' marginal utility theory and its indifference curve apparatus collapsed into a singularity.[71] Of course, these farmers lived in a nation where the existence of mastery was an unmistakable reality. And therein lay a danger to their prized possession of competence through family farms.

6

The Plantation Assault on the Southern Yeoman

Historians of the antebellum South have concentrated on the institution of racial slavery in the distinct areas of the slave's life, the slave-master's control and ideology, the nonslaveholder's condition and interaction with slavemasters and slaves, and how the region fared in comparison to developments in the North. Within their enormous and sophisticated body of work there are not many gaps, but one of them is the availability of land. For well over a century, everyone has known that the planter class (ownership of twenty or more slaves) occupied the high point on the southern social pyramid. Their wealth was beyond dispute, but they were largely understood in terms of their ownership of slaves, not ownership of acres of land.[1] The vitality of the question of who possessed the land inheres in the fate of the majority of southerners, the nonslaveholding common folk. Standard interpretations argue that the common folk were landholders (some 70 percent of the adult white male population), were primarily noncommercial yeoman farmers, had strong kin attachments, and were fiercely independent.[2] This view has been challenged in local studies, as a number of historians have found tenancy to be far more widespread than previously thought, perhaps numbering up to 40 percent of the farming population. Others emphasize growing impoverishment among the nonslaveholders inhabiting the mountain regions.[3]

Nonslaveholders depended on access to land to maintain their yeoman culture and independence, and that was precisely what the plantation was absorbing. In the antebellum years, the plantation was a prosperous, efficient, and growing enterprise. The question of slave labor or free labor may not have had much to do with the efficiency and profitability of the plantation; the plantation earned its economic preeminence because it was big. And that was the element that warred on the prospects of the common folk. Wealth was obtained by using cheap labor on the land; rich southerners obtained their cheap labor by the institution of racial slavery, the racial aspect being the manner in which they obtained a social consensus on its legitimacy. But slavery required vast quantities of land to operate successfully, so the plantation needed to be big. Then came the demographic reality: slavery in the United States was unique among the world's plantation economies because only in the southern states did the slaves have a positive population growth rate. That meant, as the number of enslaved grew, more land had to be obtained to make slavery profitable so long as slavery remained tied to agriculture. Ultimately, the nonslaveholding portion of the community either had to sell out, migrate, or become poor whites obtaining occasional wage work.[4]

Geography and Basic Numbers

In 1860, the slaveholding states contained 891,000 square miles (the North had 548,000) and a total population of 12,240,000, of whom 3,951,000 were slaves. (The nonslaveholding states excluding Oregon and California contained a total population of 18,119,000.) Population densities in the slaveholding states were amazingly low; for the entire area it was 13.7 persons per square mile. These numbers, however, are skewed because of the presence of Texas; subtracting it from the totals of population and square miles yields a density of 18.9. If England and Wales had approximately 1.2 million people in 1066, its population density was 20.3. After two and one-half centuries of English expansion into North America, the southern states still had fewer people per square mile than England and Wales had at the time of the Norman conquest, nearly eight centuries earlier.

These numbers merely underscore the obvious: the South in the nineteenth century was a rural society. Hamlets and villages existed, but few cities of any size, and most of them were on the periphery: St. Louis, New Orleans, Baltimore, Charleston, Mobile. There was a secondary tier of cities that was growing, and some believe that southern urban development was on the rise when the Civil War intervened: cities like Louisville, Knoxville, Nashville, Memphis, Vicksburg, Raleigh, Richmond, and Norfolk.[5] Nonetheless, at least 80 percent of the population was engaged in agriculture, and it is generally accepted that the South was 90 percent rural.

Cotton, sugar, rice, hemp, and tobacco were the semitropical crops that flourished in the South and gave the region its prosperity, especially cotton in international trade. Grain crops did not fare so well, especially wheat and oats due to the nature of southern soil. While corn grew well enough, the northern staples, so common in England as well, did not. Moreover, the grasses were different. Southerners could not raise sheep and cattle as easily as could the North (although Virginia, Kentucky, Missouri, and Maryland could), and the manure from the animals was not as regenerative of the soil as was the sheep manure of England. Robert Russell noted in 1857 that the South had "no good perennial grasses native to these soils," resulting in poor nutrition for animals and an inadequate means to maintain soil fertility. This, he wrote, was the reason for southern soil erosion.[6]

Given the type of soil and lack of basic mineral resources, it was not possible for the South to mimic the North's economy. The region had to differ in the types of agricultural products it cultivated, in the types of industry it pursued. The South may well have been doomed to have been underpopulated compared to other regions. The heat and humidity would have turned immigrants away. Given the commentary of Britons about the heat of New York, Massachusetts, and Ohio, cotton South states such as Alabama, Georgia, and Florida could only have inspired terror. Population growth in the South, especially urban growth, would ultimately depend on twentieth-century technology—air conditioning.[7]

Geography and the climate, however, did not dictate landholding patterns; that was a man-made determination. And the question just sits and festers. Why would a region with a population density of only 18 persons per square mile have poor, landless folk? This means not subsistence farmers, who may have chosen a lifestyle, but the propertyless poor over age thirty. Given this density of population, accommodation of all families in landownership should have been easily achieved. The answer was, and could only be, that the plantation system ruined the chances of small farmers to obtain land.

Plantations and Acres: An Accounting

Historians have recorded some of the acreages possessed by the largest slaveholders, the information most often being found in state and county histories. Table 6.1 offers a glimpse of what can be culled from these sources. Historians have generally found that the large planters needed to own 500 improved acres. While these numbers are scattered and episodically produced, they do emphasize that, for their time, southern plantations were huge. In the context of Europe, with agriculturists debating the merits of the peasant size of farm (20–50 acres) versus the large English farm (250 to 1,000 acres), southern plantations in their improved acreage stand out as some of the largest agricultural entities compared to those in Western Europe.

However, these figures are scattered data from various works and do not aggregate the amount of land the slaveholders held in each state. The vital concern in understanding the social dynamics of the slaveholding South is to determine how much land was available to nonslaveholding farmers and

TABLE 6.1 Plantation Acreage from Historical Works

Place	*Size of Plantation*
1. George Foster Pierce, Ga.	9,900 acres
2. Lowdowick Meriwhether Hill, Ga.	7,344 acres; grows to 8,229 plus 2,655 acres scattered
3. Jeremiah H. Brown, Ala.	13,566 acres
4. Joel E. Matthews, Ala.	3,500 acres improved, 2,605 unimproved
5. Charles A. Poellnitz, Ala.	2,500 acres improved, 5,000 unimproved
6. Francis M. Gilmer, Ala.	3,550 acres improved, 7,240 unimproved (5 plantations)
7. Samuel Townshend, Ala.	7,650 acres (8 plantations)
8. James Wright, Ala.	5,250 acres improved, 3,000 unimproved
9. Nathaniel Heyward, S.C.	14 rice plantations, 4,390 acres
10. Francis Surgett, Miss.	15,150 acres (plus 50,000 in Arkansas)
11. Samuel Davis, Miss.	15,000 acres
12. James A. Everett, Ga.	14,000 acres
13. Lodowick M. Hall, Ga.	8,229 acres
14. Joseph Bond, Ga.	19,388 acres
15. Col. Farish Carter, Ga.	14,690 acres
16. John Graham, Ga.	27,578 acres
17. James Rucker, Ga.	13,000 acres

SOURCES: No. 1, Bonner, "Profile of a Late Ante-Bellum Community," 671. No. 2, Flanders, *Plantation Slavery in Georgia,* 97. Nos. 3–7, Sellers, *Slavery in Alabama,* 21–39. No. 8, Phillips, *Life and Labor of the Old South,* 284–304. Nos. 9–10, Weaver, *Mississippi Farmers,* 102–3, 108. No. 11, Moore, *Emergence of the Cotton Kingdom in the Old Southwest,* 108, 117–23. Nos. 12–16, Flanders, *Plantation Slavery in Georgia,* 43–48, 87–92, 97–120. No. 17, Bonner, *History of Georgia Agriculture,* 87–88, 100–101, 189. For a list of North Carolina's planters in 1830, see Censer, *North Carolina Planters and Their Children,* appendix. Some of the sizes of the South Carolina rice plantations are given in Joyner, *Down by the Riverside,* chap. 1.

whether that amount was sufficient to maintain a "subsistence-plus" standard of living. To figure this out requires an estimate of how much land the slaveholders owned. That amount of land subtracted from the state totals of improved and unimproved acreage can give the amount available for nonslaveholding farmers.

Estimating the amount of land possessed by slaveholders requires some risky assumptions and rubbery numbers. One of the few studies that looked at the largest plantation masters in an entire state was the 1964 work of Joseph Karl Menn on the large slaveholders of Louisiana.[8] His book, based on manuscript census records, is a compilation of the large sugar, cotton, and rice plantation-owners of Louisiana who owned 50 or more slaves. In his appendix are the slaveholding individuals or companies (he looks at slave-owning rather than merely agriculturists), the number of slaves owned, the crop output, the number of improved and unimproved acres, the number of work animals (mules, horses, oxen), and the value of farm machinery, all broken down by parish. It is on Menn's findings that I make my assumptions about landownership in the southern states. He found a considerable concentration of landownership: the landownership of those possessing 50 or more slaves (a total of 1,628) represented 42.7 percent of the improved acreage of the state and 34.5 percent of the unimproved.[9]

From Menn's work, I recorded only the cotton producers (easily identified) and then averaged their slave ownership, acres improved and unimproved, and their total work animals. This elite group of cotton lords, 738 entities—if considered as people (not quite true) and ignoring multiple farm ownerships—constituted 1.1 percent of Louisiana families that contained more than one member. On average the great cotton planters of Louisiana owned 840 improved acres, 1,620 unimproved acres, 94.5 slaves, and 54.5 work animals. I assumed that owners of slaves in the planter class (20–49 slaves) owned acres at the same rate as the larger planters of Louisiana (50 and more slaves); that is, owners of 20 and more slaves held per slave 8.84 acres of improved land and 17.05 acres of unimproved land. For the categories of owners of 1–9 slaves and 10–19 slaves, I used the number of slaveholders in each category (from the census of 1860) and multiplied that times an estimate of the number of acres a slaveholder would likely own. The number of owners of slaves in the 1–9 category can be calculated per state straight from the 1860 census tables. However, it has to be adjusted for urban slaves. The number of urban slaves (but not the number of slaveholders) was given in the volume on population of the 1860 census near the end of each state's population profile. I assumed one owner for each urban slave. Subtracting the urban-slave category from the total number of slaves gave an estimate of the number of rural slaves. I took the proportion of rural slaves to all slaves in the 1–9 category for each state and multiplied it times the number of slaveholders to obtain the number of rural slaveholders in the 1–9 slaveholding category.[10]

For the acreage possessed by those having fewer than 20 slaves, I multiplied the number of holders in the 1–9 group by 100 acres improved and 200 acres unimproved to get a total acreage. For the slaveholding category of 10–19 slaves, the number of slaveholders was multiplied by 200 improved acres and 400 unimproved. (Details of the calculations are given in appendix A.) To calculate the number of nonslaveholding farmers, the occupation numbers of each state for planters, farmers, and farmworkers were added. (The occupation figures are found at the end of each state's population statistics in the 1860 census volume on population.) Although many authors believe that small slaveholders were essentially yeomen (some arguing for ownership up to 8 slaves), I finally decided that any slave ownership was a clear demarcation between family-first/subsistence farming and a commercial orientation to agriculture.[11] Moreover, slaveownership virtually demanded landownership in rural areas. It theoretically might have been possible for a tenant to own slaves to work a medium-sized farm, but the reality of such a situation seems not to have materialized.[12]

First, the prevalence of slavery in southern farming operations needs to be recognized. After subtracting out the urban slaveholders and determining the number of rural nonslaveholders, the percentage of rural slaveholders in the cotton states approached a majority position (see table 6.2). For the Border South, the percentage dropped considerably.

The results of this exercise to find how much land was available to nonslaveholders are given in table 6.3. Throughout the South, the average number of improved acres left for the nonslaveholder is decidedly lower than was available in northern family farming and by itself would have put southern nonslaveholders in the classification of peasants by European standards. Given the low amounts of average improved land for nonslaveholders, the extent of landlessness and tenancy must have been considerable—the estimates of 40 percent make sense.

TABLE 6.2 Estimate of Total Slaveholders and Farming Population in the Old South, 1860

	Total Slaveholders Rural	*Total Farmers, Farm Laborers, and Planters*	*Percent Holding Slaves (%)*	*Percent Non-slaveholder (%)*
Alabama	28,216	67,743	41.7	58.3
Georgia	31,649	90,083	35.1	64.9
Mississippi	28,361	57,318	49.5	50.5
Louisiana	14,322	26,952	53.1	46.9
Florida	3,286	12,831	25.6	74.4
South Carolina	22,424	43,970	51.0	49.0
Texas	18,222	58,365	31.2	68.8
Virginia	41,764	125,511	33.7	66.7
Maryland	12,360	41,234	30.0	70.0
North Carolina	29,018	104,438	27.8	72.2
Kentucky	32,350	147,571	21.9	78.1
Tennessee	34,432	129,903	26.5	73.5
Arkansas	9,967	57,263	17.4	82.6
Missouri	20,049	164,389	12.2	87.8

SOURCE: Computed from tables of occupations in *Census of 1860* 1: *Population.*

The existence of unimproved land for the yeoman, however, muddies the picture. On the one hand, one surmises that the great planters took the most fertile land, and so the remaining undeveloped area was largely waste—pine barrens, sandy soil, swamps, and mountainous areas. On the other hand, these lands may have teemed with animal life and permitted nonslaveholders to supplement their farming with considerable food supplies and clothing from hunting. Such a possibility accords with much of the literature on the common folk.

Nonetheless, the implication seems inescapable. If nonslaveholders wanted to be subsistence or family farmers, they had to migrate to the West—the huge amounts of available land in Arkansas and Texas could only have enticed them. This conclusion, however, points to a different origin of southern expansion than the one of positing planter lust for fertile lands to replace those exhausted by cotton cultivation. Rather, the plantation system, which by its engrossment of land that pushed white nonslave-

Table 6.3 Estimate of Improved and Unimproved Acreage for Nonslaveholders, Southern States, 1860

	Total Acres Held by Slaveholders (millions)	*Total Acres by Census of 1860 (millions)*	*Acres Left to Non-slaveholders (millions)*	*Number Non-slaveholders (rural) (thousands)*	*Average Acres for Non-slaveholders*	*Total Nonslaveholder Farm*
Alabama						
Improved	5.147	6.386	1.239	39.5	31.35	
Unimproved	10.628	12.719	2.091		52.90	84.25
Georgia						
Improved	5.462	8.063	2.601	58.4	44.51	
Unimproved	10.757	18.588	7.83		134.01	178.52
Mississippi						
Improved	5.098	5.066	-0.032	29.0	0	
Unimproved	10.031	10.774	0.743		25.66	25.66
Louisiana						
Improved	3.184	2.707	-0.477	12.6	0	
Unimproved	6.238	6.501	0.263		20.82	20.82
Florida						
Improved	.645	.564	0.009	9.5	01.00	
Unimproved	1.269	2.266	0.997		104.45	105.45
South Carolina						
Improved	4.390	4.572	0.182	21.5	08.45	
Unimproved	8.626	11.624	2.998		139.14	147.59
Texas						
Improved	2.649	2.651	0.002	40.1	00.05	
Unimproved	5.247	22.693	17.446		434.65	434.70
Virginia						
Improved	6.241	11.679	5.438	83.7	64.93	
Unimproved	12.357	19.679	7.322		87.43	152.36

Maryland						
Improved	1.379	3.002	1.623	28.9	56.20	
Unimproved	3.135	1.833	-1.302		00	56.20
North Carolina						
Improved	4.432	6.517	2.085	75.4	27.65	
Unimproved	8.769	17.246	8.427		112.40	140.05
Kentucky						
Improved	3.970	7.694	3.724	115.2	32.32	
Unimproved	7.915	11.519	3.604		31.28	63.60
Tennessee						
Improved	4.505	6.795	2.290	95.5	23.99	
Unimproved	8.947	13.874	4.927		51.61	75.60
Arkansas						
Improved	0.124	1.983	1.859	56.3	33.02	
Unimproved	0.249	7.590	7.341		130.40	163.42
Missouri						
Improved	2.329	6.241	3.92	144.3	27.10	
Unimproved	4.650	13.738	9.088		62.96	90.06

Average of Southern Nonslaveholders Compared to Northern Farms and the English Standards

Southern State	*Northern State*	*English Farm Standard*
Alabama: 31.35 (52.90)	Connecticut: 78 (42)	<3 acres, cottager
Georgia: 44.51 (134.01)	New Hampshire: 90 (44)	5–30 acres, peasant
Mississippi: 00.0 (25.66)	New Jersey: 51 (18)	30–50 acres, subsistence
Louisiana: 00.0 (20.82)	New York: 88 (28)	50–99 acres, family

(*continued*)

Table 6.3 Estimate of Improved and Unimproved Acreage for Nonslaveholders, Southern States, 1860 (*continued*)

Average of Southern Nonslaveholders Compared to Northern Farms and the English Standards		
Southern State	*Northern State*	*English Farm Standard*
Florida: 1.00 (104.45)	Pennsylvania: 67 (46)	>100 acres, capitalist
South Carolina: 08.45 (139.14)	Vermont: 78 (32)	>249 acres, large farm
Texas: 00.05 (434.60)	Illinois: 95 (53)	
Virginia: 64.93 (87.43)	Indiana 02 (61)	
Maryland: 56.20 (00.0)	Iowa: 55 (95)	
North Carolina: 27.65 (112.40)	Michigan: 64 (33)	
Kentucky: 32.32 (31.28)	Minnesota: 58 (29)	
Tennessee: 23.99 (51.61)	Wisconsin: 84 (52)	
Arkansas: 33.02 (130.40)	All North: 70 (59)	

NOTE: Acreage is improved; acreage in parenthesis is unimproved; English acreage is improved.

SOURCES: Computed from data in *Census of 1860* 1: *Population*; *Census of 1860* 2: *Agriculture;* see appendix. England typology from chap. 2. Size of northern farms from Atack and Bateman, *To Their Own Soil,* table 7.2, p. 112.

holding farmers to the West, was the mainspring of southern expansion. The conditions facing the nonslaveholder help to explain why southern proslavery politicians stressed race in their antebellum struggles with the North rather than the insult of letting northerners prohibit slavery; for nonslaveholders, prohibiting slavery in the West could be an appealing policy that enabled them to escape plantation land engrossment, but the emphasis on race would draw nonslaveholders politically to the side of the planters.[13]

Beyond 1860, the implications of table 6.3 are fearsome. The size of the southern family in 1860 averaged about 5.5 members. That figure indicated that the nonslaveholding population would double in about twenty-five years. Thus on average the available acreage for the nonslaveholder in 1885 would have been halved—in essence, a surge in the number of the landless if no other outlet were available. Because the engrossment tendency of the slave plantation would only have grown stronger because of the positive growth rate of the slave population, the southern slave states were heading into a social cataclysm. Landlessness and poverty among the nonslaveholders in the old slave states would have grown to frightening proportions while only slaves had employment on the plantations. The Civil War, of course, voided that potential outcome, but the roots of the Populist outbreak and agrarian discontent were actually visible in the landholding patterns of the antebellum South.[14]

Although the numbers in table 6.3 are undoubtedly inexact, they demonstrate the tendency of the plantation over time. The plantation was growing in geography and leaving little for the nonslaveholder; the yeoman needed geography in order to continue to exist. In 1860, the friction between the slaveholders and nonslaveholders was not readily apparent. Few clashes between them marred the political and social landscape. The tendency, however, was noted; where the plantation system took root, yeoman farming withered. The *Albany Cultivator* observed when discussing the rise of large grazing areas in Kentucky that the small farmers were selling out to the wealthy graziers "who are monopolizing the lands in the vicinity."[15]

This problem needs to be seen in light of the British historical experience. Through one means or another, landownership in Britain became concentrated in the hands of the gentry and the aristocracy. The gentry eventually eliminated peasants and yeomen by waiting for them to become mired in financial distress due to weather and bad crops; then the small landholders could be bought out. In the American South, planters could act similarly if small farmers held land suitable for the great staple crops of sugar, rice, and cotton. In this scenario not every yeoman was destined to disappear—just as in England, some small farmers managed to endure. With urban development, some small farmers could profitably supply city dwellers with fruits and vegetables. Nonetheless, as a part of the southern agricultural community, the proportion of independent nonslaveholding farmers simply had to shrink, while the proportion of landless rural dwellers had to increase.[16]

The condition of southern yeomen in 1860 was precarious, though little overt friction was visible. The danger lay in the future because of population growth. Because yeoman farmers had families numbering nearly six members, population would double. The children of antebellum yeomen were going to be unable to obtain land due to the engrossment of the plantation. This process was visible in the Appalachia region in 1860. But plantation owners did not need the cheap labor of the white nonslaveholders; they had the slaves. And the slave population was growing as well, giving an extra impetus to increasing the size of the plantation, which simultaneously decreased chances of landownership for yeomen.[17] So the final option for free youth if they stayed in the South was to become some sort of itinerant wage-worker. The crush that was to befall the yeomen in the future boded a clash of some sort. One portion of the working class was firmly embedded in the institution of slavery while the other portion was developing through the inability of yeomen to obtain land. A future conflict could have erupted; the only question was who would have been the enemy that the yeoman-to-be-wage-worker would have aimed at, the plantation lord or the slave.

The Efficiency of the Plantation

Inseparably intertwined with the question of the plantation's threat to the small southern farm is the reason for its inevitable victory. That translates into the question of why the plantation master using slave labor was more efficient in using resources to earn income than the yeoman farmer, which is essentially a comparison between extensive agriculture and intensive agriculture. In this area, the experience of the large farm in England will prove a useful guide.

Many scholars have investigated the slave plantation of the Old South, and their opinions have produced a battleground that only in the last two decades has become a resting place and memorial for its warriors. The earliest scholarly depiction of the plantation was made by Ulrich Bonnell Phillips, who gave interpretations that went unassailed for nearly forty years. He determined that slavery could only be profitably used for certain semitropical staples, that plantations were losing money and doomed to failure by rising slave prices and soil erosion, and that slavery deflected investment from modern activities like railroads and industry. Though the insightful agricultural historian Lewis Cecil Gray contested many of Phillips's conclusions, Eugene D. Genovese in the 1960s repackaged many of Phillips's ideas (minus the racism), wrapped them up in a Marxist framework of hegemony, and then repeated Phillips's pronouncement of doom upon the plantation. For the plantation was not of this modern world but dwelt in a twilight zone between medievalism and capitalism. Planters exhibited few capitalist traits and ran on precapitalist ideas of honor and paternalism. Kenneth M. Stampp had several years earlier attacked the idea of the unprofitability of slavery and emphasized how the planters extracted profit from slave labor by employing physical violence. Then came the economists' assault. First Alfred Conrad and John Meyer in 1958 tried to determine the profitability of slavery by using a "capital asset pricing" model, and found slavery to be as good an investment as any alternative. This procedure of using econometric models became the new standard, the most famous being the work of Stanley Engerman and Robert W. Fogel in their 1974 study *Time on the Cross*. Then came the deluge. The practitioners of cliometrics engaged in jousts over the questions of the extent of southern industrialization, the efficiency of slave labor compared to free labor, the rate of exploitation, the appropriate ratios of planting corn to cotton, the ability of the South to feed itself, the capacity of the southern economy to develop new activities rather than simply grow its old ones, and the mentality of the planters (capitalist or premodern cavaliers). Traditional historians stood aghast.[18]

Many of the questions raised in the exchange over the economic impact of the slave plantation remain areas of controversy, but some matters about the plantation have achieved a consensus. The plantation was profitable; it was likely to be profitable for another twenty years, possibly a half-century. The South could feed itself; it was not in antebellum times so devoted to cotton that it failed to raise food crops, forced to import them from the Old Northwest. The planter mentality remains something of a divide, many social historians doggedly holding on to the patriarchal model suggested by

Genovese, while political and economic historians tilt the balance toward capitalist attitudes. At the least, however, most have decided that the planters dominated their societies; perhaps not totally, but politically and socially their desires could not easily be thwarted.[19]

One controversy remains. While it is agreed that the plantation was profitable, the source of its profitability remains contested. Conrad and Meyer decided it came from the western states and the interstate slave trade; Ransom and Sutch found it in the physical violence used on slaves to extract excessive amounts of labor and from the tendency of plantations to move to more fertile lands in the West. Gavin Wright has forcefully shown how cotton stood at the center of plantation profitability; demand for cotton grew at a rate of 5 percent a year from 1830 to 1860, and burgeoning demand for cotton kept prices unusually high and served as the backbone for the planter profits. These arguments largely exist because economists are unhappy to admit that terror—i.e., slavery, in which people are terrorized onto an economic path—may be a more effective form of labor organization than free choice. The two economists who represent the efficiency of labor organization under slavery are Fogel and Engerman. They first deemphasize the coercion and instead stress the positive economic incentives for proper agricultural behavior, and then determine that the slave plantation method for cotton production obtained economies of scale by using the gang system of labor organization.[20]

Whether the employment of gang labor made slave labor more productive than free labor I leave to the technical expertise of the cliometricians and their search for better data and more revealing models. Rather, it is time to take a cue from British historians as to where the efficiencies of large farms came from—they arose from the use of capital over a broader area. Removing the complications that always result when slaves are treated as capital stock, a difficulty not encountered in the British example, economies of scale arose simply because the big farm was big. A plow that could cut furrows on a thirty-acre farm could do so with no less depreciation or waste of time than on a farm of three hundred acres. Large agricultural farms obtained economies of scale; they drove down the cost of production per acre. It was not that the large farm increased crop yield by better application of capital to the soil; it was that the large farm lowered production costs per acre.[21]

The English landed magnate made about 3–4 percent return on his investment; the southern planter made 6–8 percent, and perhaps 10 percent. There are reasons for the superior performance of southern planters over British aristocrats, most of them involving the difference between using slave overseers rather than tenants. The British gentry remained locked in a battle between living the leisured life and yet paying enough attention to business affairs to make sure they earned enough to afford a luxurious lifestyle. Because of the desire to display ostentation, the British landed magnate hired a steward, accountants, and eventually bank officials in London. These hired folk kept the books and oversaw operations, while the owner only provided enough attention to ensure no underling engaged in financial knavery. But this was in most cases an unnecessary management expense. Southern planters kept the books themselves, oversaw their slave overseers, arranged their own finances and merchandising; the cotton factors—the agents who arranged finances and disposal of the cotton crop for the planter—hardly represented a management superstructure. Moreover, the size of the farms on the British estates remains a statistical mystery. The large farms were there, but seemingly the estates included a number of small tenancies as well. Thus the British system might have captured economies of scale on some farms, but not on all of them.

Finally, the British mode of farming broke the estate into multiple farms run by tenants; the southern plantation did not break the farm into smaller fragments. Certainly tenants grew surly about the rent they had to pay, but the other side of the coin was that the income the landlord earned was reduced by letting the tenant keep a portion of the income for himself. Southern planters only had to let tenants have part of the crop when they resorted to leasing land; otherwise, their only supervisory expense was paying the overseer (anywhere from $200 to $500 annually). On the large plantations, the slaveowner might have several overseers, but in no instance does one ever hear of an overseer making the equivalent of £1,000 (roughly $5,000). The southern planter was willing to perform supervisory tasks and thus escaped the management expenses that mounted up in the British system of estate agriculture.

Two other areas produced economies of scale (or size) for the plantation vis-à-vis the northern small farm. The first was work animals. The source of power for the farm was animals; in Illinois and Indiana, those animals were principally horses. In the rest of the North, it was a combination of oxen and horses—very few mules found agricultural service in the northern states. Southerners used mules instead of horses for agricultural work generally; the mule withstood the heat and environment of the South better than the horse. The cost of a mule is assumed to be equivalent to that of an ox or the low end of the price range for a horse.[22]

The effect of size of farm on the use of animals is readily visible. Table 6.4 is based on the published census returns for agricultural counties in the North and large slaveholding counties in the South. By tabulating the improved acreage and dividing it by the number of work animals, one sees a considerable difference in the amount of acreage farmed per work animal.[23] This means that the cost of farming per acre using work animals was smaller for the plantation than for the family farm.

Another area in which economies of scale were significant was the use of wagons and carts. Farms engaged in a constant transportation of crops, equipment, and manure, not to mention trips to marketplaces. Wagons were expensive; a double wagon might cost $50 to $100, an ox cart perhaps $20.

TABLE 6.4 Comparison of Improved Acres per Work Farm Animal, 1860, North and South

Region	*Ratio of Improved Acres to Farm Animals (Horses, Mules, and Oxen)*	
Great Lakes	18.9	(Selected counties from Ill., Ind., Mich., Ohio, and Wis.)
New England	22.7	(Selected counties from Vt., N.H., and Maine)
Mid-Atlantic	23.3	(Selected counties from N.Y., N.J., and Pa.)
Cotton South	29.3	(Selected counties from Ala., Ga., Miss., and S.C.)
Kentucky and Missouri	12.3	
Virginia	31.7	
Nonslaveholding Counties		
Cotton South	15.8	
Border South	16.9	

NOTE: Figure for Kentucky and Missouri skewed because these states supplied mules and horses to the Deep South.

SOURCE: Computed from improved acres and animal columns of the selected counties in *Census of 1860 2: Agriculture.*

TABLE 6.5 Capital Costs per Acre, Plantation vs. Family Farm

Plantation: Animals and Equipment		*Family Farm: Animals and Equipment*	
15 plows and 8 harrows	$480	2 plows and 2 harrows	$80
4 horses	$160	1 span horses	$80
2 yoke oxen	$80	1 yoke oxen	$40
3 wagons	$240	1 double wagon	$80
6 carts	$120	1 cart	$20
3 harnesses (ox and horse)	$45	2 harnesses	$30
15 mules	$600		
15 mule harnesses	$225		
TOTAL COSTS	$1,750, call it $1,800	TOTAL COSTS	$330

Cost per acre when the improved acreage was:

500 acres	840 acres	1,000 acres	Family farm of 60 improved acres
$3.60	$2.14	$1.80	$5.50

NOTE: Price of mules put equivalent to price of horse ($40); cost of mule harness guessed at one-half cost of a horse harness. Other prices contained in narrative and in notes.

Solon Robinson indicated a harness was priced at $16 to $20. Plows and harrows together might cost $44, while a plow by itself ranged between $10 and $12.[24] The typical plantation and northern family farm do not exist, but to demonstrate how size of farm reduced capital costs per acre the rendering in table 6.5 is provided. I have used plantation sizes of 500, 840, and 1,000 acres to demonstrate the impact that farm size had on costs. The implements for a plantation were taken by guessing an approximate number from available inventories.[25] These estimates do not include the other obvious instruments used on a farm—axes, hoes, shovels, carpentry tools, various other mechanical devices—but the impact would have been about the same.

There were, of course, costs specific to certain crops, such as cotton gins and tobacco barns, but that does not undermine the basic point: plantations were efficient economic organizations because they were big and captured economies of scale that small farms could not. Slavery figured into this situation not because slave labor was particularly efficient (though it could have been) but because it was cheap: such a source of cheap labor allowed the plantation to take form, grow, and capture economies of scale. But if the price of labor had been high, then the labor cost would have overwhelmed the economies of scale and would have pushed landowners into selling their property—as happened in England after the Plague and as was happening in the northern states. So long as slavery existed, the incentive to grow the plantation larger was positive and attractive. And that meant the southern yeoman was under attack if he or she owned land susceptible of cultivating the great southern staples. The economies of scale were obtained not because of a specific crop but because of the size of the farm. As long as slave labor was available, the farm could grow in size, and the same economies of scale could be obtained no matter if the crop were cotton, sugar, wheat, or corn.

Besides economies of scale, or perhaps as a companion to them, the plantation could produce specialization of labor. Farm accounts bore witness to this in some rudimentary ways. On the Rosedew Plantation in Georgia, the entries during late September show consistently that one person was assigned to care of the mules, while in Virginia entries show that William Payne relied on two slaves, Pleasant and John, to do the bulk of the plowing; some slaves elsewhere were evidently entrusted with hauling material in wagons and carts; probably one or two were assigned to be cooks. In wheat production, certain slaves became the cradlers, and it was possible that some slaves were assigned hog butchering.[26] Overseers had a special advantage by having a large workforce during the seasons when everyone had to perform certain tasks, such as harvesting. What particular advantages specialization of labor had over the family farm is difficult to determine; most of the chores were well known in both instances, although the ability to allocate labor in amounts most needed at specific times must have given some edge to the plantation. Likewise, it is not certain whether having a larger workforce meant a better yield per acre than relying on family members and one or two paid workers.

Assuming family workers and slave workers were equivalently diligent—a major bone of controversy—cultivation should have been better on the plantation than on the family farm. The plow was used to cut away weeds during the growing season, and when the weeds were too close to the plant, the hoe was employed. In northern farm diaries, one never reads of more than one plow being used; in southern plantation journals, entries like "5 plows working" and "4 plows laying by corn" occur with frequency.[27]

As on northern farms, machines, wagons, tools, and plows broke and had to be repaired. The same leather craftsmen, boot and shoe makers, blacksmiths, and carpenters needed to sustain the family farm were required to sustain the plantation. Here, the plantation differed enormously from the northern family farm. Northern farmers, except for simple repairs and construction of buildings for animals, required mechanics and craftsmen to repair their equipment. The expense was probably limited because the times a farmer had to employ them were few. Rural mechanics and craftsmen eked out a living because there was a plentitude of small farms and thus potentially a considerable demand for their services.[28]

Plantations evolved a different way. Planters sought to create their own craftsmen and mechanic workforce out of slaves, to replace the free (usually white) laborer with a slave laborer. As an example of the process, William Wickham inked in his diary on July 3, 1829, that he hired four reapers (probably cradlers) to harvest his wheat and paid them $38.50. Four and five years later in his diary, he mentioned four reapers but no payment. Usually when planters hired people for a job, they noted the full names and the amount they were paid; when slaves did the job, the planters merely noted the job.[29] In a list that many would repeat, James B. Sellers said that the plantations of James Asbury Tait of Alabama "were all equipped with gins, bailing presses, grist mills, saw mills, [and] blacksmith shops." U. B. Phillips would add to that list the overseer's house, stables, corn cribs, wagon sheds, threshing and pounding mills, ice pit, smoke house, carriage house, poultry house, and slave cabins.[30] In actual practice, the planters made certain the crop was the first priority of the slave workforce, and so the planter employed itinerant and occasional free white labor. Over time, however, as the plantation grew and as the workforce expanded by natural increase, the planter could replace skilled white labor with slave labor. This was a process continually eating away at the vitals of the white skilled-labor community.

The greatest visible difference between northern and southern agriculture was the presence or absence of a village. Because of the multitude of small farmers, northern farming drew forth villages to service their needs. The evolution of small plantations into great ones voided the necessity of an accompanying village largely because the "slave community"—the row of huts the slaves lived in—provided most of the skilled labor the plantation lord needed. The plantation master through his connection with a "factor" took care of finance and marketing; the slave community yielded the skilled labor necessary to keep the plantation functioning. In the comparison between a slave-based, plantation agriculture and a family-farm agriculture, the stark and gaping difference between the two systems was not so much in the fields as it was in the villages the systems created: the slave community represented in southern agriculture the northern rural village. Whereas the northern village thrived, the slave community was a stagnant pool of poverty.

Replacing free skilled labor with slave labor has traditionally been explained as the planter's search for total self-sufficiency; it was also totally economically rational to do so when the slave population grew to a sufficient size.[31] In table 6.6 are averages derived from the published volumes of the 1860 census for rural skilled labor in each section of the country. The figures evoke some interesting questions about whether slave labor depressed white wages or not, and the quality of the reporting certainly comes into question as the wages for Virginians seemed unusually low while those for Alabama and Georgia seemed suspiciously high.[32] Various estimates put the yearly subsistence expense of slaves at $48, $32, and $61.[33] By giving a reward of up to $100 per year, the planter could give the slave an incentive to perform in one of the skilled trades and still have cut the cost of hiring free labor by one-third to one-half. The only matter left for the planter to determine was whether the slave was needed more in agricultural or in artisanal activity.

The census takers only recorded the activities of free workers—and actually, one is not even sure if they made returns for free blacks in manufacturing—and slaves were not defined economically as anything other than slaves. Hence the number of slave artisans is a conjecture. From his data, Fogel estimates, on large plantations perhaps 20 percent of the male slave population managed to obtain craft or management positions.[34] Comparing numbers of craftsmen in northern rural counties with slaveholding counties leads to the conclusion that the census returns show the South had a deficit in skilled craftspeople. (This assumes that the northern counties exhibited an appropriate ratio of artisans to the county population, an assumption that may be erroneous.) The results are given in tables 6.7 and 6.8.

The total number of employees in the areas of agricultural implements, blacksmithing, boots and shoes, carpentry, carriage-making, cooperage, flour and meal milling, cabinet furniture making, leather, sawed lumber, saddlery and harness making, and wagon construction within the slaveholding states (Missouri was excluded) was 59,000. If the slaveholding states had had craftsmen employees in the same proportion to the population as in northern rural counties, the total should have been 90,000, meaning that 31,000 might have been slave artisans. If one uses the total manufacturing employees instead of only skilled craftsmen employees per county, then the potential number of slave artisans jumps to 60,000.[35] In either case, the likelihood is that slave artisans represented 34 to 50 percent of all craftspeople in the slaveholding rural counties. Thus, slavery was eroding the position of the free whites of the South by substituting slave craftsmen for free craftsmen.

Beyond the absorption of land and the development of artisanal slaves, the plantation engaged in many activities reminiscent of the northern family farm. Clearing land and chopping lumber were a constant everywhere. During winter in the border states, workers cut blocks of ice out of rivers and ponds to use in ice houses. Gardens and orchards needed to be tended. It seems that at times planters exchanged laborers among themselves the same way northern farmers assisted each other. Repairing fences, laboring on roads, hoeing out the weeds from fences, ditching, and other work continued throughout the year. Taking care of the animals and removing dung from stables and pens was no minor activity for plantations that contained scores of horses and mules. Hog butchering was a prominent part of plantation life, the major difference between the plantation and the family farm being the number of pigs slaughtered. The main crop may have been wheat, tobacco, rice, cotton, or a combination of them, but they did not consume all the time of the plantation; there were plenty of other chores to perform.[36]

Several matters concerning the plantation in the antebellum decades deserve special mention. First is the claim made by U. B. Phillips that the plantation was limited in size to a one-mile radius (two miles maximum) because it took so much time (one-half an hour) for a gang to trudge out to fields. English farmers faced the same problem when they used gangs, and there is no reason to believe that the American planters would not have used the same solution had the plantation kept on growing: move the laborers out by wagons. The British extended the reach of the laborer out to five miles and sometimes ten.[37] The second claim has to do with manuring. A common charge against the slaveholders was that they did not take care of the land, that they grew their money crop till the land gave out and then either moved west or took the route of the long fallow—the reason they had about one-half to one-third of their plantation in unimproved land.

TABLE 6.6 Average Wages for Selected Rural Occupations in Rural Counties, 1860

	Agricultural Implement Employees	*Blacksmiths*	*Boot and Shoemakers*	*Carpenters*	*Carriage Makers*	*Flour and Meal Millers*
Slaveholding Counties						
Deep South	$336	$284	$330	$338	$384	$224
Border South	$280	$263	$239	$384	$305	$217
Nonslaveholding Counties						
Deep South	$240	$248	$301		$291	$212
Border South	$222	$266			$269	$337
Northern Counties	$337	$318	$306	$211	$323	$341

	Furniture Cabinet Makers	*Leather Tanners*	*Lumbermen Sawyers*	*Saddlers and Harness Makers*	*Wagon Makers*	*Coopers*
Slaveholding Counties						
Deep South	$342	$284	$224	$424	$273	
Border South	$360	$204	$223	$311	$257	
Nonslaveholding Counties						
Deep South	$198	$252	$180			$218
Border South	$251	$232	$249	$276	$269	$336
Northern Counties	$327	$296	$263	$325	$317	$315

NOTE: Figures are 1860 dollars. Blank areas indicate too few entries to yield pertinent calculation. Calculations based on returns of the following counties: For slaveholding, Alabama—Autauqua, Dallas, Greene, Monroe; Georgia—Baker, Burke, Crawford, Jasper, Monroe, Greene, Houston; Mississippi—Amite, Claiborne, Hinds, Lowndes, Noxubes, Yazoo; South Carolina—Beaufort, Chester, Richland, Williamsburg, Abbeville, Greenville; Virginia—Amelia, Brunswick, Caroline, Goochland, King and Queen, Louisa; Kentucky—Boyle, Clark, Garrard, Lincoln, Oldham, Scott, Woodford. For nonslaveholding counties: Alabama—Blount, Calhoun, Cherokee, Coffee, DeKalb, Randolph; Kentucky—Campbell, Pendleton, Pulaski; North Carolina—Catawba, Randolph; Virginia—Grayson, Marion, Marshall, Mongolia.

For North: Illinois—Bureau, Fulton, Kane, LaSalle, Sangamon, Vermillion, Will; New York—Allegheny, Chautauqua, Franklin, Ontario, Schoharie; Vermont—Addison, Franklin, Orleans, and Windham. For each occupation in each county the number of workers and the total amount of wages were given; the number above was the average found by dividing the number of workers into the total wages. Women were included.

SOURCE: *Census of 1860* 3: *Manufacturing.*

TABLE 6.7 Ratios of Craftsmen to Total Population in Rural Counties

Region	*Ratio to Total County Population*
Great Lakes	00.952
Mid-Atlantic	01.108
New England	01.281
Deep South	00.571
Border South	00.976
All North	01.078
All South	00.709

Ratio of all North to all South = 1.52

TABLE 6.8 Selection Basis for Northern Rural Counties and Southern Slaveholding Rural Counties, Plus Some Northern Industrial Counties

	North	
County and State	*Number of Farms*	*Manufacturing Employees*
Bureau, Ill.	2,424	92
Kane, Ill.	1,858	438
Sangamon, Ill.	2,219	519
Brown, Ohio	2,322	489
Knox, Ohio	2,437	579
Monroe, Ohio	2,743	111
Cattaraugus, N.Y.	4,500	1,022
Cortland, N.Y.	2,795	621
Addison, Vt.	2,418	486
Franklin, Vt.	2,405	382
Fulton, Ill.	2,764	474
LaSalle, Ill.	3,859	841
Vermillion, Ill.	2,149	137
Huron, Ohio	2,987	1,011
Lorain, Ohio	2,964	266
Allegheny, N.Y.	4,448	830
Chautauqua, N.Y.	5,890	1,610
Franklin, N.Y.	3,101	601
Caledonia, Vt.	2,360	602
Orange, Vt.	3,021	259

TABLE 6.8 *(continued)*

	South	
County	*Slave Population*	*White Population*
Autauga, Ala.	9,607	7,118
Greene, Ala.	23,598	7,251
Burke, Ga.	12,052	5,013
Monroe, Ga.	10,177	5,753
Claiborne, Miss.	12,296	3,339
Lowndes, Miss.	16,730	6,891
Beaufort, S.C.	32,530	6,714
Colleton, S.C.	32,907	9,255
Amelia, Va.	7,655	2,897
Caroline, Va.	10,672	6,948
Dallas, Ala.	25,760	7,785
Lowndes, Ala.	19,340	8,362
Jasper, Ga.	6,954	3,771
Houston, Ga.	10,241	4,828
Hinds, Miss.	22,363	8,940
Noxubes, Miss.	15,496	5,171
Chester, S.C.	10,868	7,098
Fairfield, S.C.	15,524	6,373
Brunswick, Va.	9,146	4,992
Goochland, Va.	6,132	3,814

SOURCE: Compiled from *Census of 1860* 2: *Agriculture,* and vol. 3: *Manufacturing.*

Eugene Genovese thought it was technically impossible for southerners to use enough manure, while Roger G. Kennedy posited that the planters were too money-mad to conserve soil fertility. And certain contemporaries would have agreed with them. One southern writer in 1846, in the midst of northerners talking about agricultural reform and regeneration by fertilizers, declared it was useless to tell southerners to manure; northerners could because they "can well afford to fertilize their little spots of ten or a dozen acres," but southerners had five hundred to six hundred acres to fertilize, and sufficient piles of manure did not exist.[38]

The records left by contemporaries have qualities that make them suspect for broad generalization, and the journals and accounts maintained by planters may exist because they were reforming agriculturists trying to keep abreast of modern developments; the general planter may have had a much more blasé attitude. Nonetheless, extensive manuring appears in the entries, and fertilizing may have been on the increase in the antebellum decades. For example, George Kollock in his plantation journal records extensive hauling and spreading of manure during March, assigning eight of his workforce of twenty-nine to the chore. One account book contained published instructions about plantation management and included detailed instructions for manuring.[39] There is also reason to suspect that, if expansion were no longer an option, planters would have successfully turned to fertilizing. In comparing American to British agriculture, Arthur Cunynghame in 1851 noted that Americans did not have the costs English farmers faced: two-thirds of the cost in English agriculture was in manuring. In the North, it was sowing and harvesting. The English got around the problem of exhausted soil by using exhausted bodies; cheap labor in British agriculture made manuring possible. Slavery was another form of cheap labor, and planters could have followed the English example.[40]

Supervision of labor was an area that clearly distinguished the northern family farmworkers from plantation workers, and gave plenty of fodder for northern antislavery agitators to trumpet the benefits of free labor. Free labor on the farm was expected to do chores without great supervision. Sometimes the farmer was there pacing the work, sometimes elsewhere doing other chores. What made the situation favorable to the northern farmer, in the midst of the eternal grumping going on, was that often his workers were relatives and had an interest in the farm doing well. The slave had no such interest. The slave's mental horizon for the future had to be dark and forbidding. Being a slave was pretty much all he or she could expect. There was no reward for being attentive. When supervised, slaves probably performed adequately and perhaps matched the free laborer's efforts in the North. But when unsupervised, the slave had few incentives to do anything. William Payne put in his diary an informative comment about the slaves' behavior when he returned from a business trip to a local center: "Hands doing little or nothing." An overseer who could not manage slaves properly might produce inertia among the slaves and bring ruin to the plantation.[41]

Yeoman's Paradise Lost: A Tale of Slaveholding and Yeoman Counties

Most historians of the Old South resist the idea that yeoman culture was being strangled by the plantation. Instead, the planter and the yeoman, with some friction, managed to find common ground on such things as patriarchy, sanctity of the domicile, the inferiority and therefore necessary subjugation of Africans, and the necessity of eviscerated government so individual liberty could thrive. Nonetheless there was an undercurrent at work in the slaveholding states that was eliminating the yeoman culture: the aggrandizing quality of the plantation was irresistibly shrinking the land available for small farmers and destroying the village environment that the rural artisan required.[42]

A way to demonstrate this process and its results is to compare the fates of small farmers in the regions where the large slaveholders achieved dominance to those in which they did not. Migration out of Virginia, North Carolina, Maryland, Kentucky, and Tennessee occurred between 1790 and 1820 and resulted in the yeoman farmers of those states settling the southern portions of Ohio, Indiana, and Illinois. Beneath the town of Mt. Vernon (Jefferson County), Illinois, the land in southern Illinois was known as "Little Egypt" and populated with upland southerners. In southern Indiana, Richard Nation relates, the counties were filled with the yeomen of the old South; they kept to their southern folkways, abhorred Yankees, and pursued family-first farming. Yet both regions developed a geographic profile that mirrored almost exactly the patterns occurring in the northern portions of their states. It was not the cultural background of the settlers that made a free-labor environment—it was the distribution of land.[43]

Table 6.9 exhibits the size of farms in the counties of southern Illinois and southwestern Indiana in 1860 and compares them to the farm distribution pattern in southern counties that contained numerous plantations. The counties depicted here reflect congressional districts, so the population totals are roughly equivalent. The Indiana and Illinois counties represent what could have been the South had plantation slavery not existed: a land of small farms with villages to service them. The small town development is evident in these two northern states even though the settlers supposedly held anti-urban attitudes. Town development was not due to culture; it was

Table 6.9 Comparison of Farm Size and Town Formation between Counties in Plantation States and Southern Counties in Illinois and Indiana

Illinois and Indiana					*Slaveholding Counties*				
County	*Num. Farms 3–99 Acres*	*Num. Farms 100–499 Acres*	*Num. Farms >500 Acres*	*Num. Towns*	*County*	*Num. Farms 3–99 Acres*	*Num. Farms 100–499 Acres*	*Num. Farms >500 Acres*	*Num. Towns*
ILLINOIS					ALABAMA				
Alexander	213	17	0	1	Baldwin	157	30	1	0
Gallatin	495	94	4	4	Choctaw	530	267	24	0
Jackson	757	84	3	No Return	Clarke	702	253	36	0
Jefferson	1,109	266	5	1	Conecuh	320	167	38	0
Johnson	1,154	49	0	No Return	Dallas	187	331	193	0
Hardin	345	54	0	No Return	Marengo	290	322	166	1
Massac	679	33	0	1	Monroe	470	255	39	0
Monroe	1,488	120	1	2	Washington	274	334	117	0
Perry	721	196	3	No Return	Wilcox	237	40	8	0
Pope	740	36	0	1	MISSISSIPPI				
Pulaski	283	6	0	1	Adams	38	77	99	1
Randolph	1,684	191	0	2	Amite	274	208	47	0
Saline	1,029	79	0	No Return	Claiborne	73	135	97	1
Union	1,173	85	0	2	Copiah	458	289	39	0
White	1,148	177	2	2	Covington	260	86	1	0
Williamson	1,818	86	0	No Return	Franklin	313	142	25	0

INDIANA				
Clark	1,095	283	3	12
Crawford	744	108	0	8
Davies	1,363	260	4	10
DuBois	1,430	76	0	6
Harrison	1,477	258	2	13
Knox	874	250	2	10
Martin	689	107	0	9
Perry	950	57	0	9
Posey	1,186	212	1	12
Scott	764	102	1	3
Spencer	1,347	145	0	10
Vanderburg	927	67	1	9
Warrick	1,348	182	0	11
Washington	1,200	545	4	14

Greene	110	17	1	0
Harrison	55	4	3	0
Jefferson	63	161	95	2
Lawrence	446	180	10	2
Marion	239	81	3	0
Perry	189	29	0	0
Pike	482	197	10	0
Wayne	103	40	3	0
SOUTH CAROLINA				
Abbeville	469	604	100	1
Edgefield	762	763	171	1
Laurens	340	790	141	2
Lexington	1,048	272	9	0
Newberry	381	426	34	0
Chester	270	427	94	0

SOURCES: The 1860 congressional districts from which I generally took the counties are in Illinois, district 9; Indiana, district 1 but somewhat scattered; Alabama, district 1; Mississippi, district 5; South Carolina, districts 4 and 6. Farms in the categories above can be calculated from *Census of 1860* 2: *Agriculture;* towns are enumerated toward the end of the state population numbers of whites, slaves, and foreign-born in *Census of 1860* 1: *Population.*

due to the existence of small farms. The comparison with plantation counties is staggering: the number of small farms collapses, and there is next to no town development.

The paucity of town life—really, village life—in the South is startling and is difficult to convey. However, one way to portray it is to inspect the pages in the 1860 census that enumerate the towns in the North versus those in the South. It must be kept in mind that numbers from the census are suspect; northern agents may have been overly zealous, while southern agents may have been unusually careless. Yet even factoring in sloppiness, the contrast is startling. The 1860 census, in its population volume listed at the end of state population, totals the towns and cities in the state's counties. Comparing northern counties to southern counties reveals the chasm between the sections. For example, one of the census pages for Ohio enumerates the towns in Ashtabula and Athens counties plus partial accounting of the towns in Ashland and Auglaize counties; on this single page there are 69 towns, only 2 having more than 2,500 people and most ranging from 500 to 1,500. And Ohio had 88 counties. The *total* number of towns listed for the states of Mississippi, Tennessee, and South Carolina was 48, which included Charleston, Memphis, and Nashville.[44] This condition was emphatically the result of plantations gobbling up the land.

Both the plantation and the British estate system of agriculture have undeserved reputations for paternalism, a term used to signify human relationships that were governed by personal interaction and human sympathies and not by a marketplace motivation of the individual to acquire higher income regardless of the consequences for others. The reality was that both the plantation and the British estate system were engines of wealth and income accumulation so the owner could live a life of gentility—a life of leisure unsullied by manual labor. Cheap labor was the fuel for the engine, and the result of the machine's operation was the spread of economic misery throughout the countryside. In the case of the British, the miserable were free agricultural laborers; in the case of the antebellum South, the miserable were slaves and, slowly but relentlessly, small nonslaveholding farmers.

Lack of overt class conflict between nonslaveholding yeomen and the planters has long been noted by historians, and researchers have often pondered why yeomen fought for a slaveholding republic. Perhaps part of the answer lies in the comparison between the upland southerners of the lower counties of Illinois, Indiana, and Ohio and the yeomen of the South. Once those people moved out of a slaveholding environment, they no longer felt the need to support the great beneficiaries of slavery. Those yeomen who stayed in the slaveholding environment, however, could not break free of the imagined obligations they had to a slaveholding community. This demonstrates, perhaps, the operation on southern nonslaveholders of Eugene Genovese's theory of hegemony.

Southern yeomen understood in a general way what was happening to them. That was made clear in the battles over slavery in both early Indiana and Illinois. The yeomen knew exactly what they wanted to avoid: the plantation system. By excluding the plantation, southern yeomen in Illinois and Indiana created the communities they would have preferred to have had in Virginia and Kentucky, ones that bore a striking resemblance to the rest of the rural North. However, in the northern states, upland southern yeomen retained one element that produced among them confusion as to who was their economic enemy: their racism toward African Americans. In the battles over slavery in the antebellum years, it would often be the racism that would be most observable in upland southerner hostility to northern abolitionists and to emancipation. At the same time, however, the yeomen did not lose recognition of how plantations monopolized the land. It was merely a matter of time, especially for those in southern Illinois and Indiana, to determine that the plantation masters, not their slaves, were the enemies of competence and independence.[45]

7

Inspiring Fear

THE PLANTATION, THE BRITISH GENTRY, AND MASTERY OF OTHERS

To understand northern fear of the South and its peculiar institution, one needs to obtain a solid grasp of the very real differences in scale of business between northern family farming and the plantation. Certainly slaveholding had a number of features that could worry nonslaveholders and invoke questions of morality, but essential to understanding the distrust existing between slaveholders and northerners was the matter of size. People who had so much land and mastery over others seemed magnified in their personal power. Their size dwarfed commoners and demeaned their manner of living; they seemed to be the Goliaths while the free farmers of the North looked like the Davids. In the historiography of the period, this is not the picture of the situation usually given. Instead, it is northern capitalists—New York bankers and the Boston Associates—who are styled the Goliaths, while the planters, struggling in their paternalistic way to make ends meet, are the Davids. The planters may have been taken advantage of in merchandising and financial arrangements, but the picture at least needs a major readjustment to bring in the masses of northerners, the family farmers.

What follows is a comparison of selected rural counties in the North to slaveholding rural counties in the South. This is not by random sampling. To choose northern rural areas, a comparison was made between the number of farms in a county and the number of manufacturing employees; when the number of farms was a multiple by two over the number of manufacturing employees, and the total number of the latter was normally less than two thousand, the county became a potential candidate for further inspection. For southern counties, I simply looked for counties in which the number of slaves was two times the number of whites. Table 7.1 shows the results of the procedure for some of the counties.

TABLE 7.1 Comparison of Farm Sizes in Selected Northern and Southern Rural Counties

	North				*South*		
	Number Farms	*Number Improved Acres (Average)*	*Number Unimproved Acres (Average)*		*Number Farms*	*Number Improved Acres (Average)*	*Number Unimproved Acres (Average)*
Bureau, Ill.	2,424	116.9	30.5	Autauga, Ala.	730	180.4	421.1
Fulton, Ill.	2,764	80.8	47.8	Dallas, Ala.	793	367.5	402.7
Kane, Ill.	1,858	119.8	36.9	Greene, Ala.	793	349.9	356.5
LaSalle, Ill.	3,859	62.3	4.8	Lowndes, Ala.	943	245.2	289.7
Sangamon, Ill.	2,219	141.6	29.4	Burke, Ga.	675	371.6	458.5
Brown, Ohio	2,118	78.2	51.9	Houston, Ga.	487	378.0	383.0
Huron, Ohio	2,987	63.8	28.5	Monroe, Ga.	662	293.2	181.9
Knox, Ohio	2,437	81.8	44.9	Claiborne, Miss.	296	430.1	517.9
Lorain, Ohio	2,964	60.7	27.6	Hinds, Miss.	595	320.3	353.4
Allegheny, N.Y.	4,448	64.8	49.5	Noxubes, Miss.	595	273.6	300.3
Cattaraugus, N.Y.	4,500	66.0	57.2	Beaufort, S.C.	917	298.8	673.1
Chautauqua, N.Y.	5,890	65.9	35.7	Chester, S.C.	791	231.5	223.3
Franklin, N.Y.	3,101	57.0	46.6	Fairfield, S.C.	683	341.6	416.4
Addison, Vt.	2,418	108.9	39.5	Amelia, Va.	397	275.3	248.6
Caledonia, Vt.	2,360	86.7	45.3	Brunswick, Va.	642	250.6	277.4
Franklin, Vt.	2,405	94.6	49.2	Caroline, Va.	733	250.8	167.9

SOURCE: Computed from *Census of 1860 2: Agriculture.*

The first point of comparison is the number of farms in the selected rural counties and the average number of acres per farm, both improved and unimproved. Table 7.1 merely gives numerical form to that which is already known but which needs to be put in more memorable form. The North was populated with a plentitude of small farms while the slaveholding regions held few. Moreover, on the basis of twenty to sixty acres as the standard for subsistence farming, most northern farms barely eked over the borderline, while southern plantations were not just commercial but massively so in terms of English agriculture. Table 7.2 gives the number of manufacturing employees and mechanics in these counties. The rationale for the table is that the paucity of white rural artisans in the South is not wholly conveyed

TABLE 7.2 Manufacturing Employees and Mechanics in Rural Counties

	North			*South*	
	Number of All Manufacturing Workers	*Number of Mechanics*		*Number of All Manufacturing Workers*	*Number of Mechanics*
Bureau, Ill.	92	53	Autauga, Ala.	364	321
Fulton, Ill.	474	278	Dallas, Ala.	287	240
Kane, Ill.	438	266	Greene, Ala.	123	109
LaSalle, Ill.	841	210	Lowndes, Ala.	200	194
Sangamon, Ill.	519	262	Burke, Ga.	89	87
Huron, Ohio	1,011	594	Monroe, Ga.	53	51
Knox, Ohio	579	392	Claiborne, Miss.	118	91
Lorain, Ohio	266	183	Hinds, Miss.	269	259
Monroe, Ohio	111	96	Lowndes, Miss.	324	209
Cattaraugus, N.Y.	1,022	823	Beaufort, S.C.	97	93
Chautauqua, N.Y.	1,610	764	Chester, S.C.	114	97
Franklin, N.Y.	601	401	Fairfield, S.C.	0	0
Addison, Vt.	486	311	Amelia, Va.	71	46
Caledonia, Vt.	602	255	Brunswick, Va.	65	61
Franklin, Vt.	382	265	Caroline, Va.	85	79

NOTE: Mechanics defined as the total of the following occupations: blacksmiths, carpenters, coopers, joiners, mechanics, flour and meal millers, saddlers and harness makers, lumbermen sawyers, boot and shoemakers, leather tanners, wagon and carriage makers, agricultural implement employees, and machinery makers.

SOURCE: *Census of 1860* 1: *Population;* vol. 3: *Manufactures.*

by percentages; the absolute number of people involved brings a fuller realization of the presence of slave artisans.

Finally, table 7.3 gives the distribution of farm sizes in the selected counties and aggregates them by states. (The full listing of counties used is given in appendix B.) As shown here, the plantation states skewed the distribution to the large, commercial farms, and the small farms were becoming disappearing appendages. Meanwhile, in the North, the bulk of the farms were in the 3-to-99-acre category, a subsistence-plus-minor-surplus category for most of Western Europe at this time. For comparative purposes, I also looked for a smattering of southern counties that exhibited a predominantly free population and tallied the distribution of farms into these three size categories. The distribution of farms in the southern nonslaveholding counties resembled closely the distribution of northern farms. To state the obvious: without slavery, the size distribution of southern farms would have matched closely the size distribution of northern farms. Soil and climate determined some of the crop choices and altered the methods of soil preservation, but they did not dictate the existence of the plantation. That was a human decision, and it was guided by the desire for leisured wealth in the fashion of the British gentry.

Comparisons of the income of northern farmers and southern plantation owners is difficult because almost no one has published the earnings of the plantation lords; all the attention has been on ownership of slaves. However, as guesswork, we can assume a yield of two bales of cotton per acre, a bale equaling 400 pounds, thereby totaling 800 pounds of seed cotton per acre. Using the formula that, for every 3 pounds of seed cotton, 1 pound of lint cotton was obtained, the output from 1 acre translates into 266 pounds of lint cotton. That total times a price of 11 cents per pound equals about $29 per acre. Thus the income from 100 acres in cotton equals $2,900; for 300 acres, $8,700; for 500 acres, $14,500. In English pounds, these dollar amounts would be roughly equivalent (at $5 equal £1) to £580, £1,740, and £2,900 respectively. The latter amount put the large southern planter in the category of the English gentry.[1]

One direct way to compare northern farms with southern plantations is the wheat harvest. Northerners up to 1860 continued to grow wheat. In the Great Lakes, corn-hog agriculture was just getting started. Northerners were at best growing 20 to 30 acres of wheat and getting anywhere from 7 to 15 bushels per acre. For a middling amount, consider the output to

TABLE 7.3 Farm Size Distribution by State Using Selected Rural Counties' Percentages

	Farm Size Category			
	% 0–99 Acres	*% 100–499 Acres*	*% 500+ Acres*	*Number of Counties Used (Total Number of Farms in Selected Counties)*
NORTH				
Illinois	58.1	40.6	01.3	7 (17,453)
Indiana	80.7	19.0	00.3	5 (9,677)
Michigan	76.9	23.0	00.1	5 (15,979)
Ohio	80.6	19.3	00.1	8 (22,246)
Wisconsin	81.8	18.0	00.2	5 (18,256)
New Hampshire	68.7	34.3	00.1	4 (11,694)
Maine	87.7	12.2	00.0	6 (21,457)
Vermont	64.0	35.7	00.3	7 (18,303)
New Jersey	65.6	34.3	00.1	5 (9,581)
New York	77.5	22.4	00.1	10 (39,427)
Pennsylvania	85.2	14.8	00.0	7 (27,321)
SLAVEHOLDING SOUTH				
Alabama	60.4	28.7	10.8	5 (5,757)
Georgia	30.6	48.4	21.0	8 (3,556)
Mississippi	33.3	46.7	20.0	6 (3,264)
South Carolina	49.9	41.1	09.1	8 (6,793)
Kentucky	39.5	55.2	05.3	7 (4,619)
Missouri	60.2	37.5	02.3	7 (6,298)
Virginia	30.5	59.8	09.7	11 (6,538)
SOUTHERN COUNTIES WITH MAJORITY FREE PEOPLE				
Alabama	82.0	17.5	00.6	6 (7,610)
Georgia	78.8	20.6	00.6	4 (3,206)
Kentucky	84.1	15.8	00.1	4 (5,178)
North Carolina	76.3	23.4	00.3	3 (3,671)
Virginia	68.0	31.6	00.4	4 (3,916)

SOURCE: *Census of 1860 2: Agriculture.*

have been 280 bushels of wheat; at about $1.25 per bushel, that amounted to about $350. Of course, northern farms engaged in many activities other than wheat growing—cattle, hay for animals in the cities, dairying, and garden and fruit produce—but wheat provides a basis for comparison. And much of the time, northern farmers practically dribbled their harvest to market, selling two or three bushels at one time, four at another, two later on. Wheat farming on slave plantations in Virginia and Maryland had been established by the 1820s, and the owners were definitely not dribbling their harvest to market in piddling amounts.

Table 7.4 lists from Virginia diaries and account books some of the wheat output that individual plantations produced in Virginia in the antebellum decades.[2] From the vantage point of the northern farm, these are immense amounts. The famous agricultural reformer Edmund Ruffin published his harvest totals in the *Southern Planter* in an effort to goad Virginia's stubborn planters to use advanced methods of farming and fertilizing. He obtained a huge amount of crop. On a farm of 750 improved acres, 227 unimproved—a *big* farm!—using 22 slaves, 8 mules, 4 oxen, and machines for threshing and shelling, he produced in 1845, 1846, 1847, and 1848 an output of wheat totaling 1,977 bushels, 2,432 bushels, 3,511 bushels, and 5,127 bushels respectively. Richard Eppes, who owned the Bermuda Hundred in the Piedmont, was principally a wheat and corn planter; he had 100 slaves. He made enough from his operations to take a European honeymoon with his new wife lasting several months, which cost him $2,570. Moreover, the planters in Virginia were raising corn for sale, not simply for animal feed or domestic consumption. Eppes sold on September 8, 1853, some 4,347 bushels of corn for $2,391. Eppes's sale of corn by itself was two times the average income of a northern family farmer.[3] Eppes was not alone. Ruffin also reported crops of corn between 1845 and 1848 of 2,830 bushels, 1,600 bushels, 3,600 bushels, and 4,500 bushels. Evidently Virginia farmers had found a market for their corn and corn meal in the West Indies. A number of sources indicated that Virginia planters were raising corn principally for sale, not for cattle feed or domestic use.[4]

The alteration in Virginia agriculture between 1840 and 1860 has been somewhat unappreciated.[5] Wheat suddenly became prominent again as a staple crop in Virginia. The James River area had been ravaged by poor agricultural practices when wheat was attempted in the 1750s as an alternative to tobacco. That substitution ruined the soil, produced abandoned

TABLE 7.4 Examples of Wheat Output by Virginia Plantations

Richard Eppes Account Book
- September 10, 1852, 6,802 bushels
- August 18, 1853, 6,593 bushels
- October 17, 1854, 4,096 bushels
- November 27, 1855, 2,622 bushels
- September 1, 1856, 4,520 bushels
- October 9, 1857, sale of wheat $10,751.70
- October 7, 1859, 5,882 bushels

William F. Wickham Diaries
- Dec. 25, 1828, 2,800 bushels
- Sept. 11, 1829, 3,480 bushels
- Sept. 7, 1830, 5,340 bushels
- Aug. 28,1831, 3,462 bushels
- November 23, 1832, 4,719 bushels (sold)
- Dec. 31, 1844, 4,700 bushels
- August 31, 1859, 5,800 bushels

John H. Cocke Journals
- September 19, 1851, 3,510 bushels
- September 9, 1852, 2,350 bushels

Purchases of Wheat, Waring-Baylor Papers
- August 31, 1840, account of Ro. P. Waring, 1,011 bushels
- October 4, 1839, account of Guynfield with Waring, 1,822 bushels
- Aug. 20, 1840, account of Rennolds with Waring, 655 bushels
- September 20, 1839, Henry Waring account with Ro. Waring, 678 bushels

G. B. Tyler and R. H. Tyler Farm Account
- Summary, 1852, 921 bushels
- Summary, 1855, Shelter farm, 636 bushels

land, and sent farmers and planters scurrying out of the area. But proper methods of cultivation and the application of fertilizers began a rejuvenation of wheat agriculture by the 1840s, or so it was advertised. By 1850 Virginia had become the third largest wheat producer in the nation. The resurgence of Virginia agriculture from its doldrums was the proud theme song of an agricultural reformer, Willoughby Newton. In 1854 he declared that "ten years have wrought a mighty revolution in the fortunes of Virginia." Virginian planters expected to obtain 25 to 40 bushels of wheat to the acre and to match the output of the international leaders of wheat cultivation, Great Britain and Flanders. Reports in the press flourished about one wheat planter producing 21,000 bushels of wheat, another 12,000 bushels. These numbers compare well with the output of some of the largest British farms.[6]

This surge in wheat production was not coming from family farms; it was coming from slave plantations. Newton made it explicit. Slavery was eminently practicable in wheat cultivation. It had not been slavery that brought desolation to the James River area between 1770 and 1830 but awful husbandry. Now that planters knew how to take care of the land, the earlier problems had disappeared and wheat production by slave plantations was eminently profitable. The attack on slave labor had been totally erroneous. Slavery was "tractable, efficient and profitable," he said, and with slaves "we may defy the competition of the world."[7]

Wheat cultivation in Virginia showed none of the traits that contemporaries and historians believed slavery induced. One fault in particular was proven mistaken—the mating of slavery with machinery. The wheat planters of Virginia talked about harvesters all the time, as well as drills and steam engines. The planter John H. Cocke recognized that machinery was not only labor saving but also yield enhancing. He saw a result from use of the reaper: "Made trial of McCormicks Reaping Machine," he wrote in 1843; "It cut 3/4 of acre of wheat in 35 minutes leaving less wheat standing than the best cradling I ever saw—cost $100." Harvesting by cradle, sickle, scythe, and other hand implements probably resulted in a significant portion of the crop being left on the ground. The new machines captured all the harvest and thereby increased the yield per acre that was actually obtained.[8]

What had altered the soil situation in Virginia was the form of plowing (deep versus shallow), proper crop rotation (arguments over the three- versus four-field rotation system), and fertilizing. Guano was certainly mentioned and often advocated, but it was not the only method discussed. Some planters reported spectacular—also suspect—results from using native plants. Among the fertilizers—"manures"—mentioned were clover, lime, field peas, clover seed, gypsum, marl, and plaster, as well as stable manures.[9]

The effectiveness of the manuring campaign has been questioned, but the Virginians were experimenting, seemingly with some success.[10] The slave plantation was escaping the supposedly rigid boundaries of semitropical staples. Virginia planters grew wheat and corn for market; those were the two staples that supposedly were the heart of the northern farm economy. In fact, the wheat plantation using slavery looked far more profitable than the family farm in the Great Lakes area. One can conjecture that the North was early saved a ferocious battle over slavery in the Old Northwest because of fortuitous timing. Because the early English settlers of the Tidewater were so consumed with planting a crop that earned an immediate profit, they turned to tobacco. Concentration on tobacco deflected them from considering other crops, the primary one being wheat. It took time to learn how to cultivate wheat without ruining the soil, and evidently by the 1840s the proper techniques were being discovered. If planters had established successful wheat plantations between 1730 and 1770, one wonders if the Northwest Ordinance could have ever been passed banning slavery in the Great Lakes area. If the profitability of slave-grown wheat had been proven by 1780, Virginia planters would have made a strenuous battle to enable expansion of slavery into Ohio and Indiana.

The Virginia experience, however, raises the question of what factor in the plantation was most crucial: the scale of the organization, permitted by cheap labor, or the cultivation time taken by the crops. The general interpretation is that the short growing season of wheat favored the small free farmer while the long growing season of cotton required longer labor services and made the slave-labor plantation feasible. The Virginia plunge into large-scale wheat production upsets this interpretation. There was no agricultural endeavor free farmers did that the large plantation could not do better because of specialization of labor, economies of scale, and ability to allocate labor as needed. Planters chose the tropical staples because they offered immediate profits that could not be assailed competitively, but if the option of those crops disappeared for one reason or another, the plantation seemed sufficiently adaptable that it could have engaged in any kind of crop production done throughout North America. There was, in short, no reason the slave plantation could not have successfully gone above the Mason-Dixon line.

The southern plantation dwarfed the northern family farm. Such extremes in size had to have induced an apprehension that such giants could misuse their powers and injure the smaller parties. If nothing else, such men as commanded these enterprises could not have had much respect for those operating on a small scale nor have had much concern for their "rights." After all, wrote one Georgia editor in a foul humor, "Free Society, we sicken at the name, what is it but a conglomeration of greasy mechanics, petty operators, small-fisted farmers, and moon-struck theorists? All the Northern, and especially the New England states, are devoid of society fitted for a well-bred gentleman." It takes no effort to see the leisured class's aristocratic disdain toward manual labor in this passage.[11]

Confronting the Size of the Plantation

For the United States, the plantation was the economic behemoth of its day. Authors and visitors recognized its size and recorded the number of acres the great planters possessed. Reports on the size of farms in North Alabama put the common size at 800 to 900 acres. A certain Dr. Russell published his observation in the *London Times* on the sugar plantation of one planter. He was impressed that one could gaze upon "6,000 acres of the finest land in one field, unbroken by hedge or boundary."[12]

But the agricultural press shied away from elaboration on the contrasts between the family farm and the plantation. Editors and contributors seemingly were more interested in the feed of mules, horses, hogs, the timing of sowing and reaping, the types of buildings to erect on farms—in the details of farm life. Few articles appeared pitting slave labor against free labor (although there were some).[13] This was not the case of British visitors to the United States. The only problem was that they had so many axes to grind when they tramped through the nation that it sometimes makes their comments less susceptible to clear interpretation.

For those visitors hostile to the United States and its republican institutions, slavery fit their purposes in obvious ways. Entering the slave states, Francis Hall noted in 1818, one went from the "smiling villages" of the central and New England states to the "splendid equipages of a few planters and a wretched negro population." Both in the free states and the slave states, a European confronted a racial bias that was strong beyond comprehension—it was nearly impossible for the European "to appreciate the intensity of this feeling." Americans looked upon Africans as "a distinct species from the rest of the human race."[14] Various travelers discovered the shocking torpor and desolation that had overrun Virginia, deplored the lack of town life, found the heat insufferable, and complained of the lack of liberty in the South.[15]

As to the economic actors in slavery, numerous British writers bemoaned the material condition and physical treatment of the slave. Slaves were terrorized by physical violence and had no possessions. They were sold from one master to another, and slave marriage was nonexistent. The auction block shocked British observers no matter their opinion on the system of slavery. Robert Russell made one interesting comparison and comment after another about southern slavery and was open to considering a number of its positive economic attributes. But when he went to a slave sale of a pregnant woman holding a child being separated from her husband, he was aghast: "Slavery, I must confess, was never brought home to me in a form so repulsive as this exhibition."[16] One of the ways English sojourners measured the welfare of American slaves was to look at their housing; for those harboring antislavery opinions, slave huts were "pig-sties." An Englishwoman, Marianne Finch, in 1853 claimed that slave housing was "nothing like the clean pretty cottages in which some of our English peasants live," but was better than an Irish cabin or even some of the housing of the English "labouring poor." Geologist Charles Lyell made this observation a little more intelligible: "The negro houses are as neat as the greater part of the cottages in Scotland (no flattering compliment it must be confessed)."[17]

These expressions evoked a famous comparison that was commonly made regarding American slavery: were American slaves worse off than the lower class in Europe and Great Britain? Most wayfarers gave the surprising answer of "no." Matilda Houstoun had been in the South for one year and decided "*domestic* slaves are the least unhappy *menials* in the world," and she went on to describe the slave's material life as comfortable *for a lower-class people.* Others wrote that slaves seemed happy and contented, while Alexander MacKay admitted that the material condition of a slave's life was better than that of European "free" laborers.[18] Within this interchange between British visitors and southern planters was an interesting comparison that took the Britons by surprise: contrasting English agricultural workers with American slaves. When Lyell was in Georgia, a planter used the publications of Henry Colman against the antislavery animus of Lyell: he used "Mr. Colman on English agriculture, in which the poverty, ignorance, and stationary condition of the British peasantry are painted in most vivid colours." Some admitted that southern slaves were better off than "the poor peasantry of my own country."[19]

Views on southern nonslaveholders reflected views of what the observer believed that the system of slavery had wrought. Among the multiple crimes listed against slavery was the teaching of indolence among all whites, because only blacks did labor in the South. This was a general indictment, and all kinds of Britishers—not to mention abolitionist northerners—found or theorized that slavery produced among whites a distaste for work. Thus economic torpor suffused the entire region.[20] A few went further. Plantations were huge operations that pushed out small farmers. One writer told Europeans not to go into the South if they wanted agricultural work: "Whites cannot obtain—neither should they seek employment, on plantations." A more intriguing interpretation was given by William H. Blane, who said that the plantation system ruined life for nonslaveholders. Slavery prevented towns from developing, which limited possible artisanal employment. The reason was that each plantation owner had "his own shoemaker, tailor, carpenter, &c. on his own estate—all slaves." The only thing a nonslaveholder could do was to migrate elsewhere.[21]

Britons made a formidable attack on slavery based on economics to explain why the institution was retrograde and inappropriate for the new age of steam and innovation. Most of their comments come under the category of the virtues of nonslave labor—that is, free labor—versus the vices of slave labor. There were multiple arguments, but the basic ones come under the heading of inculcating the wrong habits, a degraded morality. First it taught whites to obtain income by physical violence rather than hard work; slavery undermined the dignity of labor and earning a competence by the sweat of one's brow. For the slave, the system taught men and women that hard work earned no reward, that work was to be shirked and avoided, that dissipation and indolence were greater values than attentiveness and diligence. Many held that slavery purposely kept people uneducated and ignorant.[22]

Robert Russell toured the United States in 1854–55 to investigate American agricultural practices, and his observations varied considerably from those who disliked democracy and who held antislavery sentiments. Russell's willingness to be objective in his assessment of slavery, however, led Frederick Law Olmsted to challenge him later in his famous book, *The Cotton Kingdom*.[23] Russell had no fondness for slavery but saw its strengths and weaknesses. He believed that in certain crops—tobacco, sugar, and cotton—slave labor was clearly superior to free labor, largely because the planters got economies of scale and exercised strict supervision over the labor force. They also obtained a division of labor. Like modern economists, he saw that slave women and children were used in cultivation, thereby helping its profitability. And he had a different view of the effects of slave labor on whites: there was some degradation "but not to the extent perhaps that is commonly supposed." He witnessed slaves and free whites working together, and "freemen in the Slave States think it no more degrading to cultivate their small properties with their own hands than they do in the Free States." The economic backwardness of the South came from the plantation because the planter did not aid local development and the slave population was a "non-consuming class." He noted the size of plantations and saw them as constituting a large farm system. In discussing economic matters with a Mississippi planter, he heard that the South did have trouble feeding itself—corn grew less exuberantly in the South than in the North—but he was told that the return on investment if a planter achieved self-sufficiency was 24 percent. He saw slavery's end when slavery no longer had fertile soil to remove to, and the increasing slave population would eventually depress slave prices so much that ownership would hold no attraction. However, over the next decade or two, he saw little sign that southern slavery would end.[24]

When visiting Britons considered the large planters and their lifestyles, they started making the connection between the English landed estate and the southern plantation. They produced some hostile commentary about extravagance, manners, and dissipation, but often they complimented the gentlemanly manners, education, and refinement they found among the planters—and among no other Americans. Many Britons were impressed with the homes of the planters; some "approach[ed] to splendor," some were like "our old English mansions," and in 1856 William Ferguson said that some of the plantations had "more of the home-look of English scenery than is usually met with."[25]

Comparing the southern planters to the English gentry became a common theme and easily lapsed into the subject of aristocracy. James Stirling wrote in 1857 that the planters were "to all intents and purposes an Aristocracy, nay, an Oligarchy." Tocqueville denied that the great southern slaveholders were an aristocracy by European standards, but most commentators noted that in the United States the large planters were "truly [a] leisure class," the only real one in the republic. And they also were willing to call the great slaveholders aristocrats because they were "living on their estates by the sweat of their slaves, having no taste for work." They were "the only class who live on the labour of others, and without their own personal exertions."[26]

The British travel literature on North America has an oddity to it. While the sojourners approached the topic of comparing the slaveholding planter to the British landowning aristocrat, they stopped short of a full investigation. Perhaps part of the reason was the existence of African slaves. The slave community in the southern states was so exotic and out of the British traveler's plane of existence that it became a subject unto itself, and they simply could not make a leap to equate the slaves on the American plantations to the agricultural workers on large English farms.

Societies of British Estate Agriculture and American Plantation Agriculture

There are unsettling commonalities between the southern plantation and the large British farm. They both exhibited from the start a concentration of landownership in a few hands. The planters could not compete with the British aristocracy in this area, but nonetheless the lopsided wealth distribution had similar effects on both. Neither group cultivated land with their own hands; the extent of landownership dictated that they had to manage others who labored, or in the British case manage the managers of labor.

Of all the comparisons to be made between the southern plantation and the British large farm, the most provocative one is between the slaves and the British farmworkers. This contrast has been almost totally ignored, as the most common comparison has been between southern slaves and European and North American industrial workers. The reason for that exercise partially has to do with the nature of the proslavery argument, in which southern authors loved to take the details of English working-class life from the Chartists, Carlyle, Dickens, and parliamentary investigations and measure them against the material standards of slave life. This obsession also derives from the establishment of a new movement among historians starting in the 1960s. Practitioners of the New Social History employed various forms of class analysis that befitted urban environments, found the enemy of mankind in industrial capitalism, and wanted to compare the exploitation in slavery with the exploitation of the factory system. They cared little about the life of the British agricultural worker.

Nonetheless, comparing the two agricultural labor groups needs to be done because it goes to the heart of a number of interpretations about antebellum American society. Both the slave plantation and British estate agriculture were sufficiently capitalistic to put aside the question of patriarchy and leftover medieval economic practices. Here the operating definition of capitalism would be profit-seeking owners (the strict maximization postulate can be ignored) who kept reinvesting their earnings and sought to employ labor at the cheapest price possible.

From the descriptions and analyses offered of the material condition of the slave and the "free" British agricultural laborer, one comes to the conclusion that slaves had a better life. If the reports were true that British agricultural workers only had cheese, milk, and bread for supper, the diet of the slave (pork, bacon, corn meal, plus some fruits and vegetables) may have been superior. Robert W. Fogel estimates southern slaves on average had diets containing over 4,000 calories; Mark Overton has written that the nineteenth-century English agricultural laborer only obtained 2,000.[27] Cliometricians have spent some effort on height and weight of persons as being markers of their well-being; one is impressed by the commentary on British farm laborers that emphasized their diminutive size and slowness of gait. Slaves operated under threat of coercion, family disruption, and sale; slaves were immobilized—had no capacity for movement by voluntary choice. Supposedly British farmworkers escaped these conditions, but it seemed to have little impact on their lives. Deference was so extreme that mere facial movements from the tenant farmers and gentry may have been enough to elicit the behavior desired. The promise of parish relief may have had something to do with worker immobility, but perhaps so did the tradition of attachment to place. While the antebellum literature is full of the charming English village, the writings on the homes of British agricultural workers describe filthy hovels, wholly equivalent to the quarters for slaves.

Equally fascinating is the hint that domination by landlord or by slavemaster produced about the same set of observations concerning working-class character. British agricultural workers were described as stupid, slow, mentally dull, unable to learn, resistant to change, given to excessive drunkenness, and prone to sexual license. These are about the same adjectives given to slaves, except for the racist depiction of slaves detailing their enjoyment of servitude, singing, and child-like attitudes (the "Sambo" portrayal). For the British workers, except for sex and drink, there was no joy in living—at least if the descriptions of them are accurate. Olmsted was disturbed that British agricultural workers had no pride of being Englishmen or of belonging to the English nation; for all religion they had a profound contempt. The extreme immiseration of the group had wiped out all sense of community attachment and cross-class ties. For the English agricultural laborer, paternalism was a sick joke. Slaves also probably doubted the paternalism of their masters and had a powerful resentment toward their domination. The comparison of southern slaves to British agricultural workers finds few rays of light; it is a dismal story for both groups that lasted for centuries.[28]

Plantation Ideology and Gentry Ideology

A writer justifying the Confederacy in 1864 opined, "Southern society has most of the virtues of an aristocracy, increased in zest by the democratic form of government and the freedom of discussion on all topics fostered by it," and had "a landed gentry, [which constituted] its founding families." This association of the southern plantation states with English aristocracy and gentry was not uncommon. *De Bow's Review* had an article on MacKay's travels in America, and the writer approvingly quoted MacKay as saying "land is possessed, as is with us [England], in larger quantities; the owners, as in England, generally living on their estates." Thus the South has a "nearer approach to English country life" than anywhere else in the United States. During the Civil War, an American writer glommed onto the southern worship of British society by remarking that American slaveholders "have attempted an exaggerated imitation [of the British aristocracy] in that they aimed at the ownership of labor as well as of land."[29]

Antebellum southern society had its tensions and odd contradictions, one of the larger ones having to do with equality. On the matter of African inferiority there existed throughout the United States in the antebellum decades a stultifying consensus. The British, however, did not base their idea of inequality on race; inequality existed regardless of race. The inequality practiced by the British aristocracy and gentry was aimed at their own ethnic stock. Most southerners did not have this disposition, and certainly in politics the advocacy of equality could be brazenly affirmed by slaveholding editors and political candidates.[30]

The British aristocracy and gentry had traits well known in the Western world. Its members were the great landowners of the realm, they had privileges involving control of government, they performed no manual labor and lived a life of leisure. Moreover, its members had a similar philosophy of social life and government. The gross imbalance in the distribution of landed wealth required a justification, although the basic ideas were imbedded in English society centuries before as rule by a monarch and an aristocratic elite commenced with the Norman invasion. The main authors for the defense of inequality were Robert Filmer in the 1650s and then Edmund Burke in the 1790s. They were the philosophers of the natural order of inequality. Society needed order and stability; that was provided by a father figure who exemplified the principle of paternalism. The leaders possessed an intelligence and capacity for achievement and military ardor that the mass of mankind did not; to exercise those gifts, the leaders required wealth without labor so they could devote their time to social order and the affairs of state. They ruled and preserved property rights, the basis of civilization. Although the common people ("the laborers") were appreciated, and a few that had ability could rise, they were to remain in their place and show deference to the leaders. The English put a heavy emphasis on the importance of landed property and family maintenance over time. For that reason inheritance was a crucial subject, and out of this concern for keeping the estate in family hands legislators enacted the laws of entail, strict settlement, and primogeniture. Simply put, the upper class controlled government directly to maintain property rights, which for keeping estates in the gentry's possession was essential, used the national church as a pacifier of rebellious sentiments, and considered the sufferings of commoners the will of God. On the latter point, the aristocracy and gentry simply did not much care about the maladies of the common laborer; the poor were so fated and that was to be their unmitigated lot.[31]

Of the philosophy of inequality that the British elite espoused, certain distinct doctrines stood foremost. First was their utter disdain for democracy or rule by the masses; the masses had neither the leisure nor the intelligence to understand complicated matters. Democracy threatened property rights with agrarian laws. Also, the British elite stood foursquare on the principle of individual liberty, and they recoiled in horror at the idea of the masses writing laws telling individuals how to behave or individuals censoring themselves out of fear of public opinion. Finally, the philosophers of inequality and their adherents made it plain: natural equality and natural rights did not exist and could not exist. Religion and Nature called for inequality, not equality, to provide the necessary order to enable human communities to survive with minimal conflict.[32] William E. Gladstone exclaimed that "inequality [in England] lies imbedded in the very base of the social structure," and Mathew Arnold agreed that Englishmen looked upon inequality almost as a religious principle.[33] This subject of equality versus inequality put the United States and England at intellectual loggerheads for most of the nineteenth century.

As a republic, the United States had no formal nobility and therefore differed markedly from the political structures of Western Europe. Discussions of the class structure of American life by foreigners led to an emphasis on the middle. William Brown, a British traveler, defined the middle in 1849: the United States was composed of the "middle classes of society—farmers, mechanics, and merchants." The definition, often repeated, is worth underlining: the farm-village economy of family farmers and village mechanics was seen to be middle class. What travelers noted was the fierce resistance Americans had to anything smacking of aristocracy and monopoly, both being the evils of the European system of inequality.[34]

The great exception was the slaveholding plantation South. The two Hungarian travelers with Louis Kossuth in 1851, Frances and Theresa Pulszky, testified that planters resembled "the aristocrat of north-eastern Europe" and had the outward earmarks of aristocracy: "of large estates, of elegant houses, of carriages." Northerners and European visitors seized upon two elements to determine the existence of aristocracy: wrote Chevalier, "slavery necessarily requires great estates, which in fact, form aristocracy." In the republican United States, the charge that the planters represented an aristocracy was politically damaging. Nonetheless, based on the English ideal of gentility, there was some truth to the charge.[35]

However, for a number of reasons, the planters were not typical British aristocrats. They were not handing down family estates the way the British did (and many commentators ascribed this to the lack of the laws of primogeniture and strict settlement), the planters had too much in common with a business class to be a true leisure class, and their wealth had not reached atmospheric heights. One indication of the lower gentry status of the American planter was his failure to keep a stable of hounds to go hunting. F. M. L. Thompson reported that the Duke of Cleveland paid £500 to £2,000 annually to maintain his kennel of 125 hounds.[36] But the comparison did not spring out of thin air. Americans were well aware of the status, prerogatives, and wealth of the British aristocrats, and they testified to the parallels between them and southern planters.[37]

One special parallel deserves investigation. Just as the aristocracy needed an explicit defense for their elevation above all others, the slaveholders required a defense of their embrace of slavery—hence the proslavery argument. It had as its central idea the inferiority of Africans, who were fitted only to perform manual labor guided by European intelligence. Of great interest to many historians has been the proslavery advocates' criticism of free labor and manufacturing, showing how capitalists exploited factory workers. Proslavery writers also used biblical justifications of slavery; they employed climatic explanations to show that the cultivation of semitropical staples could only be achieved by using darker-skinned races. Some advocates expounded on how slavery kept a society conservative: god-fearing, patriarchal, female-submissive, family-oriented, militaristic, and ordered.[38]

Yet the extreme proslavery argument so beloved and belabored by historians of the Old South is essentially an undisguised imitation of the Burkean defense of aristocracy and inequality. Within proslavery advocacy was the crucial aristocratic defense of inequality—that societies required a leisured, propertied class to do the governing, preserve and extend the arts, and maintain intellectual standards—that is, to keep the standards of high civilization unsullied. That mission could not be accomplished if the talented—the superiors—were dragged down by "drudgery and brutalizing effects of continuous toil." After the war, Edward A. Pollard defended slavery because it permitted some to escape "the demands of physical labour." Abel Upshur unhesitatingly affirmed that it was necessary "that one portion of mankind shall live upon the labors of another portion." These statements easily folded into the common proslavery belief that all of the great civilizations of past ages relied upon a ruling class that did not labor.[39] The subject of manual labor revealed a considerable division in the antebellum South. Proslavery advocates did not claim that manual labor was degrading or unneeded—in fact, they all agreed that it was essential to the prosperity of a society. They did insist that an inferior type of person had to perform the work, while a more enlightened type of person had to be relieved from it. Elsewhere in the South, however, one heard hosannas to manual labor similar to such exhortations found in Yankee land.[40]

Two other areas of convergence between the Burkean justification of aristocracy and southern proslavery were democracy and liberty. In the decades before the Civil War, many southerners had a barely concealed contempt for the idea of democracy. A North Carolina planter confided in his diary his fear of democracy. It allowed the unpropertied the right to legislate about his wealth: "to take away, or destroy the value of our Slave property, and the demon democracy by its leveling ing [*sic*] principles, universal suffrage and numerous popular elections, homestead laws, and bribery are sapping the foundation of the rights of property in every thing." This was essentially John C. Calhoun's opinion, clothed as it was with mystical references to the concurrent majority; the majority had no right by nature or by any power to rule over others. The rejection of majority rule and democracy was tempered by only one consideration: in order to draw the nonslaveholder to acceptance of slavery, slaveholders had to admit equality of race. Known as Herrenvolk democracy, it is the doctrine that all people of one race are equal but people from other races are not.[41]

Closely connected to the fear of democracy was the exaltation of individual liberty. But the liberty the planters and their allies claimed was the European aristocratic version of liberty. There are unspoken words in this version: for whom is the claim of liberty made, and for whom does the claim of liberty not apply? The British aristocratic version was straightforward: liberty belonged only to those people who could responsibly exercise it, and the judges of that condition were the aristocrats themselves. In the years of the French Revolution and until the 1820s, common people espousing republican ideas were deemed unfit for liberty and tossed into jail for mouthing such notions. What was missing was a key word and a phrase that made liberty democratic: *equal* liberty *for all*. Aristocrats never said that. The clearest example comes from one of the famous phrase-makers of the nineteenth century, Lord Acton, whose phrase "power corrupts, and absolute power corrupts absolutely" appears unendingly in scholarly works and in the popular media. Yet Lord Acton was an aristocratic conservative who supported the Confederacy. After the Civil War he wrote an open letter to Robert E. Lee thanking him for fighting "the battles of our liberty." Elsewhere he had stated that the "democratic system poisons everything it touches." After raising the torch of liberty to such heights and denouncing democratic processes, he went on to defend slavery as potentially a Christian institution, although in 1866 he recanted that opinion. Nevertheless, one has to wonder how a person who touted liberty with such ferocity could initially find no grievous fault in chattel slavery.[42]

Understanding "liberty" and its uses in political rhetoric is crucial to understanding the position of slaveholders and the northern fear of them. They did not argue liberty for others, only for themselves. This is why the seemingly incongruous marriage of slavery and liberty can be rather easily explained; the southern apostles of liberty never meant liberty for everyone—only for the select few, for the aristocracy. British aristocrats had no need for slavery, so arguing for it or against it was nonessential; but stopping the majority from impinging on the freedom of action of the elite was important. To sum the matter up, the catchword "liberty" in the nineteenth century was an elite rhetorical device of class domination. It never meant "equal liberty for all."[43]

There is confusion among some writers about the "libertarian" southern antebellum impulse. It was not libertarian, but it was hostile to the federal government because the federal government could impinge upon the liberty of slaveholders. If one takes seriously the writings of libertarian proponents, one cannot make antebellum slaveholders libertarians. Writes David Boaz: "Libertarianism is the view that each person has the right to live his life in any way he chooses so long as he respects the equal rights of others." Because slavery is an act of coercion, libertarians must either condemn the peculiar institution or admit they are not libertarians. The attempt to understand how Americans rationalized and justified the simultaneous existence of freedom and slavery has been a subject pursued for many years, the most forceful delineation given by Edmund S. Morgan in *American Freedom American Slavery*. More recently, François Furstenberg has argued that the two could be joined by arguing that those who fought for freedom deserved it, but those who did not merited slavery, thus providing a new justification for the coexistence of two opposite labor systems. However, I think it could be successfully argued that the British heritage in America included the aristocratic notion that liberty was always reserved for the few. The dilemma is resolved by aristocratic liberty; coercion is perfectly allowable for some, liberty for others. Southern spokesmen shouting liberty had no libertarian ideal of liberty, they had an aristocratic one.[44]

When and where did southern political leaders ever argue for liberty for others? Except for the revolutionary generation, southern political leaders talked of liberty only for themselves. And when one takes equality out of the phrase "equal liberty for all" and substitutes "liberty only for a few," one has turned the word "liberty" into a synonym for "tyranny."[45]

The concurrence between southern planter and British aristocratic political ideals brings into question some of the reigning depictions of southern antebellum political values. According to historians like William J. Cooper Jr., Lacy Ford, and Kenneth Greenberg, southerners held to the values of old republicanism: a rural society of propertied citizens who had economic independence and who could not be manipulated by the wealthy or the socially privileged. Republicans scorned economic dependence, which was characterized by a wage-earning society such as was evolving in northern cities.[46]

This interpretation might be confirmed by consulting the rhetoric of southern political leaders, but in the agricultural structure of the southern states it makes no sense at all. The plantation was gobbling up the land, without which the majority of the population simply could not become rural property owners. While the situation in 1860 was not desperate for the nonslaveholder, population growth insured that it would become so in the not distant future. The South was following the same path as agricultural Britain. Since 1660, the British estate system of agriculture consolidated land in the hands of the gentry and eliminated the peasantry and the yeomen. The great plantation in the South was doing the same thing to the small southern yeoman farmer. Republicanism required a society composed of a majority of citizens who owned property and exhibited independence; the plantation made that kind of society impossible. Many have noted proslavery fears of the nonslaveholders in southern society; that fear had a material basis.[47]

The structural concordance between the southern plantation and the great British estate made for a common political ideology. Both bred the same rural inequality, and both gave birth to, for lack of a better expression, an ideal of aristocratic liberty and independence, an ideal that required property and independence only for the few, not for the many. The South was moving distinctly toward aristocratic liberty. Southern leaders had the British gentry idea of liberty and politics; liberty for themselves, not for others; independence of action for themselves, not for others; and political power for themselves, not for others. Their ideas of states' rights was akin to the British gentry ideal of a landed estate being a small kingdom into which no other power—specifically, a central government—was to intrude.

But this sense of liberty only for themselves ties into the personality trait of the planters, and of the British aristocracy, that frightened northerners: mastery over others. While some of the ideal of mastery might have had cultural origins abroad (such as English medieval paternalism and Scottish clan practices), it was directly linked in a practical fashion to slaveholding. In slaveholding, the advice was clear: establish mastery over one's slaves. "Remember that a slave in a state of insubordination is an enemy," cautioned one editor in 1829, but "that in a state of perfect subjection, he is a kind, willing, good-humoured, and useful friend." One writer for *De Bow's Review* advised in 1851, "Firmness of manner and promptness to enforce obedience will save much trouble, and be the means of avoiding the necessity for much whipping. The negro should feel that his master is his lawgiver and judge, and yet is his protector and friend, but so far above him, as never to be approached save in the most respectful manner." Mastery explained why slavery was the most efficient form of production. In Edmund Ruffin's phrasing, slavery allowed manual labor to be directed by "one directing mind, and one controlling will."[48] Slaveholding as a daily routine taught slaveholders mastery of others. It could hardly have been more opposite of social relations on the family farm, between farmers and other farmers and between farmers and farm laborers. From the vantage point of the treatment of others, the two systems were totally alien to each other.

How much the mastery exercised over African slaves affected the slaveholders' treatment of other whites is a matter of some debate, but in one area the desire for mastery was explicit: mastery over the policies of the federal government.[49] So long as state governments had all the authority and slaveholders ruled the states, policies that could damage slavery, impinge upon the property definition of slavery, or reduce its profitability were unlikely to become laws. The federal government was different. There nonslaveholders came to the national legislature having no interest in the welfare of the peculiar institution. Therefore the nonslaveholders in Congress, the majority, had to be neutralized. The southern congressional delegations were on a quest to make sure only their laws passed Congress, and with as little compromise of their program as possible. Northern antislavery groups did not see this as the operations of a normal interest group, but rather as an outward manifestation of the southern drive to mastery. And from it they constructed the slave power conspiracy—that the southern legislative agenda was to expand slavery into the territories, to erect more slave states, to extend slavery into the northern states, and establish a slaveholding aristocracy to rule the United States.[50]

PART III

Agriculture and Sectional Antagonisms

The structural chasm that divided the family farm and the plantation produced a contrast of values. The family farm promoted equality, independence, local community, and self-mastery. The plantation evoked the ideas of inequality, a hierarchical order to society, and mastery over others—a repetition of British aristocratic ideals. However, the two agricultural systems were separated by geography. Should the geographical separation between them disappear, then the contrasts would emerge politically with a fearful force. Territorial expansion would be the agent obliterating the geographical separation.

8

The Agricultural Basis of the Free Labor Ideology in the United States

Historians have given a multitude of interpretations about sectional animosities in the United States from the Revolution to the Civil War, but one central thread appears in them all, the free labor ideology. The obvious dividing line between the South and the North was the presence of slavery. So sectionalism became the slave states versus the free states. For most researchers the meaning and content of slavery was obvious, but the meaning and content of freedom was not. Ever since the 1960s, writers have worked to offer the definition of freedom that nineteenth-century northerners intended, and the answer came in the concept known as the "free labor ideology," a catch phrase for a congeries of values that northerners, antislavery people and Republicans, held inviolable. Over time, the work on the ideology moved away from an understanding of freedom being the absence of someone's domination over another to an assertion that mid-nineteenth-century northerners were really espousing capitalist values: freedom to participate in the consumer marketplace, to choose one's employer, and to rise in social position. These capitalist values were then construed to show that northerners were actually sanctifying wage labor, and wage labor in those days meant life at the lowest wage possible, subsistence labor. The end result of these studies is that free labor advocates were using the word "freedom" to disguise the new type of poverty being imposed on the country by industrial capitalism, which for workers meant wage slavery. The moving force in the United States was the creation of a capitalist economy; the free labor ideology was the means of making the system legitimate in the eyes of the populace, and the force of the change to industrial capitalism demanded the South sacrifice its premodern ways and adopt capitalist practices. The South refused and the result was, after a bewildering array of political movements, civil war.

In the first years of the twenty-first century, the reigning interpretation among antebellum historians is now the capitalist takeover of all things in North America. The free labor ideology occupies center stage in this understanding of the evolution of the United States. The interpretation is largely incorrect. The only area that comes close to being represented by this view of American development is urban America, and even there major qualifications need to be made. The reason for the interpretation's erroneous assertions stems from the sources and interests of the investigators of the free labor ideology: they are hypnotized by the rise of urban America and industrialization. The analysts of the ideology failed to include rural America, where nearly three-quarters of northerners lived. When rural America is included, it becomes obvious that free labor ideals sprang not from manufacturing development or urban areas but distinctly from rural areas, and the goal was never subsistence wages but competency.[1]

The Free Labor Ideology: Its Development and Growth

Two developments in the 1960s set the stage for the arrival of the free labor ideology. The first was the work of Eugene D. Genovese on southern society. In *The Political Economy of Slavery,* he hypothesized a South that was precapitalist, doomed to economic stagnation because of the backwardness of slavery, and expansionist because of soil exhaustion. Ultimately, the reason for the conflict with the North was the aggressive nature of the North's capitalist society and its ravenous appetite for profits and control of the federal government.[2] The second was the arrival of the New Social History, which took as its method of operation investigation of class relations and particularly the lives of the lower class—and the lower class came to mean specifically the urban working class. From this perspective followed a renewed interest in the ideology of a people, or more broadly in the set of values and ideas a people express to show their understanding of the world and the proper moral order.[3]

Eric Foner then laid down the foundation for analyzing antebellum political rhetoric. In *Free Soil, Free Labor, and Free Men,* a study of the values of Republican Party members in the 1850s, Foner devoted his first chapter to defining the free labor ideology as the dignity and worth of labor, the need for each laborer to be assured of obtaining the full fruits of his labor, the ascent of men from wage-labor to property ownership, and a final advance into self-employment. Such a system promoted economic growth and social happiness because men responded to incentives and, importantly, could not be economically bullied or manipulated. Republicans did fear monopoly and wanted equality before the law for all citizens. In general, Republicans envisioned a society of small property-holders and social mobility for the young; they did not have a set of values that justified what would become the giant corporation in the late nineteenth and early twentieth centuries. However, they believed morally fit persons advanced to self-employment; if some people stayed wage-earners, then either some misfortune had befallen them or they had severe character flaws. In his second chapter, Foner showed how the free labor ideals of the Republicans clashed dramatically with the practices of southern slavery. Slaves were trapped in a caste for all their lives, denied the fruits of their labor, and had no social mobility. Because slavery deprived workers of an incentive for enhanced productive effort, slavery produced an economy mired down in backwardness.[4]

The floodgates were opened, and out poured a deluge of articles and books using the free labor paradigm. Its most prominent use was interpreting the abolitionists. David Brion Davis in 1975 employed the ideology to explain that the overall influence of British abolitionists was to justify wage labor used in factories because the greatest evil in the world was slave labor.[5] Shortly thereafter, historians explored the known problems abolitionists had with the labor movement in the United States and found them centered in the free labor ideals that justified capitalism. However, many students of the antislavery radicals temper the class implications considerably because the abolitionists were the only group who agitated for racial equality.[6]

Elsewhere in slavery, cultural, ideological, and labor studies, historians increasingly turned to various forms of class analysis to establish a framework for evaluating the pronouncements of the leaders of a society and their relation to the society's social structure. The brunt of these studies was that all the words used in free labor exhortations were meaningless to the developing urban working class in the United States. Within these studies one particular argument became prominent: when proponents talked about the glory of free labor, they really meant the magnificence of wage labor. And these historians found no magnificence in wage labor, most adopting the position that wage labor meant a life at the lowest level possible.[7]

John Ashworth uses this understanding of free labor advocacy to explain political realignment in terms of how the parties reacted to it, and writes that the antebellum story is "the rise to dominance of the northern labor system, with wage labor an indispensable part of it." Jonathan Glickstein has detailed the bottom rung of laboring America and argued that Republicans used social mobility themes to mask capitalist exploitation.[8] Looking at the difficulties of common labor in Baltimore in the early Republic, Seth Rockman dismissed free labor rhetoric as an attempt to disguise what was really transpiring: "Historians have described 'free labor' as an ideological project whose success as a shorthand for the social relations of the wage system owed much to crafty wordsmiths, clever lawmakers, and those far more likely to hire labor than to perform it."[9]

The essential interpretation that has emerged from free labor studies follows this pattern. The North became economically more powerful because of industrialization, and its articulate members justified its new economy as progressive and more morally sound because it did not employ physical violence. Meanwhile, the South was mired in an old patriarchal mode of production and either slowly developed or simply maintained the status quo. The economic growth of the North fed a self-righteousness that evoked a morality to condemn the South's economic sluggishness by focusing on its forced labor system—and northerners attacked the South in waves, first by the abolitionists, then by the Liberty Party, then the Free Soilers, and finally the Republicans. The change in economic structure promoted a belligerent intolerance of those who did not adopt the same economic principles. This was the situation that caused southern separation; the North could not contain its evangelical urge to dominate the nation's economic system and govern it according to one set of free labor principles.[10]

For the past fifty years, antebellum historians have built an interpretation of sectional antagonism that uses free labor as a device to show the aggressions of industrial capitalism. This effort has been almost wholly misguided. The free labor ideology in the United States was a family farm ideology intended for the countryside. It was awkwardly thrust on urban America, where it did not fit. The point is underscored by glancing at the person most identified with the free labor philosophy—Abraham Lincoln. What connection did Lincoln have with urban industrialization? Answer: none. Ray Ginger, a historian writing about 1890s Illinois, understood perfectly: “The Lincoln ideal was rural. The image was of Abe the rail-splitter, not of Abe the machinist.” The consequence of misreading the free labor ideology is that antebellum historians have misconstrued the reasons for sectional collision.[11]

Theory and the Labor Market

Whether or not one wants to engage in Marxist theory, it becomes a necessity because writing on antebellum free labor ideology is so thoroughly saturated with it. Certain ideas of Marxism are crucial to evaluating the theory's applicability to the United States in the nineteenth century and particularly to rural America.[12] In the various Marxist versions of the labor market, capitalism creates a condition that guarantees the workers the lowest wage possible. This hands to the capitalist the highest profit possible. However, wages can never go below subsistence because if they do the worker dies and the capitalist has no access to “labor power.” Therefore subsistence wages are the baseline, and at one point Marxist historians argued that capitalism drove wages to that level. That wages necessarily must be at subsistence is a point of some contention, but for the theory to work out its prognostications, subsistence wages are in fact required.[13]

Every writer on social relations has the eternally frustrating problem of defining “class,” and most of the time they use some notion of ownership of property (“ownership of the means of production”), which is basically a wealth variable, but in terms of social power so many other avenues exist to exercise influence over others that everyone who engages in this exercise in defining class ends in a state of exasperation. At the least, however, we can say what class does not permit: it does not permit massive movements of people out of one state of property ownership into a higher state of property ownership. Class as a social mechanism has to operate as a sort of prison for its members. Only then will the conditions of life begin to draw forth from people a consciousness about why the conditions of life exist as they do and initiate a search for what has caused those conditions and how to change them (or, in the case of capitalists, how to preserve and justify them).

For social historians, what determines how one fares in a society is ownership of the means of production. Owners of capital can always outwait workers, and thus workers in order to survive have to capitulate and accept whatever wages the capitalist wants to offer, which is as little as possible. As laborers realized in the nineteenth century, the labor market was always overstocked with people, and therefore wages always plunged toward subsistence. The competitive element in capitalism always ruined the chances of workers for high wages.[14]

Regardless of how one might evaluate this critical analysis of the labor market, there are some ramifications—really, the backbone of the way in which the capitalist expansion thesis is presented—that are vital for understanding the antebellum North. The first is the use of the word "permanent" to describe the wage-earning class.[15] Employments in which workers entered and exited rapidly and continuously did not satisfy this requirement. For purposes of antebellum history, permanent wage-earning among workers probably was occurring in textile factories and in iron-manufacturing establishments, although the continuation of the artisanal tradition diluted the sense of becoming a permanent wage-earner. This explains why social historians have had such a tussle with northern agriculture: there was no permanent wage-earning class, and one would not appear for over a century (except in California). If one defines capitalism as the emergence of the *permanent* wage-earner, then northern agriculture was not capitalist.

A second difficulty occurs with the analysis of wages. Part of the moral condemnation of capitalism is that it condemns a majority of the population to poverty through subsistence wages. For this to occur, wages have to be set at subsistence or an average near that level over time. If wages are not at subsistence, stark theoretical inconsistencies occur. First, wages moderately above subsistence might have a comfort effect which could dilute any class consciousness. Second, the real difficulty arises if wages increase enough to allow savings. While in the early nineteenth century, the opportunities for persons with small savings to invest their money in financial instruments like bonds or stocks were either miniscule or nonexistent, they did have the potential to allow people to buy property or implements—to set up shop on their own. Wages high enough to enable savings produce a contradiction in the Marxist analysis: savings enable a worker to be a worker and a capitalist at the same time. Savings, in short, undercut the analysis that posits subsistence wages. As it turns outs, savings play a large role in understanding what people *expected* from a free labor economy and the rate of wages that a free labor economy was expected to establish.

The Principles of Nineteenth-Century Free Labor

Free labor ideals had distinct elements, some primary, some derivative. First of all, free labor advocates celebrated the act of labor, both mental and manual. Second, labor was dignified because by it an individual contributed to the well-being of society. Third, labor deserved its reward, the fruits of labor; by the antebellum decades this largely meant the wage, but the term "wage" was usually a synonym for reward. Many historians to the contrary, "wage" did not imply belief in a permanent wage-earning class, and almost without exception it did not mean a subsistence wage. Fourth, the worker who received the full fruits of his or her labor had the ability to save. The whole antebellum idea of free labor hinged on wages being munificent enough to allow savings. Fifth, through savings a person could obtain property. And sixth, the goal of the system was to lead individuals to property ownership that enabled a person to be independent of others, that is, self-employment.

Certain generalizations about these primary qualities of free labor advocacy must be immediately recognized. This was not a set of values that would lead to a socialist conclusion. Some sort of free-market economic system is absolutely required for it to operate. It not only accepted the importance of property ownership, it elevated, justified, and glorified private property. Private property was the mainspring of free labor ideals. And it was individualistic. This was not a group philosophy at all; individuals earned the fruits of labor, not groups; individuals were rewarded, not groups; and the mobility was to self-employment, not to rising in some hierarchical enterprise. Finally, it posited fairly explicitly—some deviations allowed—a laissez-faire government. The society imagined by the free labor exhorters was one in which individuals could rise from poverty to property ownership to self-employment by saving money from wages. If government could tax heavily, then government was hamstringing the capacity of wage-earning people to save and obtain property. Thus a fairly unobtrusive government was required for free labor principles to work.[16]

The set of primary qualities in the free labor program could evoke claims about the effects of a free labor economy, but the claims were not indisputable. This distinction is made because the primary qualities could actually produce a number of different outcomes. It was assumed that allowing people to obtain the full fruits of their labor would induce industriousness and greater productivity. That assertion was an extrapolation from the primary qualities that might be false. Individuals upon obtaining a competence and independence might simply stop there and be satisfied with their economic position. There was nothing in the value system to sanction continual striving for higher levels of profitability and ever-increasing piles of wealth.

The free labor ideal educed from its core concepts that the resulting economy was egalitarian and thereby supported republicanism and democracy. Free labor practices reinforced political and social egalitarianism by creating a society of small property-owning producers. Because of the smallness of enterprise, property ownership was thought to be widespread; those without property would support the system of property rights because they had a rational belief that they would eventually become property owners. This outcome assumes a number of economic conditions that many in the nineteenth century never owned up to. First was the availability of resources—and in the concern for available land contemporaries did recognize this requirement. Just as important was the extent of economic competition and how it could eliminate self-employment among the majority, allowing only winners of the competition to survive. Also not taken into account was the potential for the appearance of large enterprises through a combination of financial manipulation and economies of scale. That their principles of free labor might lead to large enterprise would have seemed a paradox to antebellum free labor advocates.

Two generalizations not to be drawn from primary free labor principles are a division between mental and manual labor and the glorification of the pursuit of wealth. Free labor proponents celebrated labor and made no distinction about how the labor was performed. To be precise, farm labor was primarily manual, but it certainly obtained mountains of rhetorical respect. Moreover, the free labor formulation in the United States aimed originally and directly at creating an independent competency. The values of free labor were not about obtaining wealth. There was no ambiguity among orators and writers about making competency more significant than wealth. What made a "good" society "good" was a widespread ownership of property in which people felt they were contributing to society's welfare and to their own comfort. The "American dream" in the nineteenth-century North was a dream of independent competency, not of becoming wealthy. The rise of the "American dream" of becoming inordinately rich developed in the twentieth century and itself is a marker of the change in American practices, institutions, and values.

Finally, the free labor ideal simply does not work well with the factory system for an obvious reason. As David Montgomery pointed out sixty years ago, it was one thing for contemporaries to argue that a person could rise from farmwork to farm ownership, but it was entirely unreasonable to think that all fifty workers in a small factory could all eventually set up their own factories. The number of workers simply militated against factory workers rising to self-employment. The free labor ideal of the mid-nineteenth century had no reality for those engaged in factory employment. They needed a different kind of free labor justification, one that perhaps occurred in nineteenth-century protectionism or mid-twentieth-century progressivism. By having a high wage, a factory worker could accumulate savings, own his or her own house, and consume goods—in short, have a good material life and a type of competency compatible with urban living—but not independence. In truth, giving up the latter goal of independent competency was in the United States a long quest that would take most of the twentieth century.[17]

Agriculture and Free Labor Ideals

Abraham Lincoln provided the clearest, most precise elaboration of the meaning of free labor, and historians have referred to his pronouncements for the best examples of the genre. He enunciated his view at the Wisconsin State Fair in September 1859, repeated parts of it later in Cincinnati when he was giving political speeches, and inserted it prominently in his second annual presidential message to Congress when explaining the primary difference between the Confederacy and the Union: "The prudent, penniless beginner in the world labors for wages awhile, saves a surplus with which to buy tools or land for himself, then labors on his own account another while, and at length hires another new beginner to help him. This, say its advocates, is free labor—the just, and generous, and prosperous system, which opens the way for all, gives hope to all, and energy, and progress, and improvement of condition to all. If any continue through life in the condition of the hired laborer, it is not the fault of the system, but because of either a dependent nature which prefers it, or improvidence, folly, or singular misfortune."[18]

Lincoln was indeed a clever wordsmith but not an original one. His basic ideas about starting poor, earning wages, saving, accumulating property, rising to competency and self-employment, and then hiring others were indeed the primary qualities of the free labor argument, but he got them from his own experience in Indiana and Illinois and from the speeches and writings of others—they had lain about for decades in rural America.

In the early years of the American republic, the basic formulation of the free labor argument emerged, and it was strictly connected to the countryside, not the city or manufacturing. J. Hector De Crèvecoeur is known among historians for his early declaration that the Americans were a new people with a new political edifice, but his *Letters from an American Farmer* (1782) were also an early salvo of the free labor ideology. The English squire did not exist on American soil; poor immigrants arrived, worked for others and "receive[d] ample rewards for their labours; these accumulated rewards procure them lands; those lands confer on them the title of freemen."[19] That is, people start out poor, labor, earn high wages, save, and then become family farmers. In a pamphlet put together to inform would-be immigrants, Thomas Cooper noted that American cities were much the same as British cities, but in the countryside one could obtain a farm. If the immigrant did not bring a fortune, "he will find he must at first live low, and work hard." The promise was that his children would do well because North America did not have as did England "rich proprietors and great lords."[20] Just before the Revolution, an author detailing American agriculture noted how New England was divided into small farms and how the "lower classes" worked so arduously because they hoped to obtain land: "This great ease of gaining a farm, renders the lower class of people very industrious." The writer then explained how this came to pass by inference; looking at Old England and the laborers and small farmers there, he saw only misery because the wages of labor were so low that the lower class could not aspire to property ownership.[21]

Celebration of free labor principles began at the time of the Revolution, but the wars with the trans-Appalachian tribes between 1770 and 1815 displaced the recognition of social mobility going on in rural America. The conflict over western land instead elevated the activities of the backwoodsman, the pioneer, and the Indian fighter. After 1820, when the Anglo-Americans began attacking the trans-Appalachian forest, the free labor ideals reappeared and circulated widely in the commentary about the American rural community; it persisted and grew throughout the rest of the nineteenth century.[22]

From 1820 onward, discussions about the American family farm all emphasized hard labor. For emigrants to the West, the order of the day was "work!—Work!!—Work!!!" Solon Robinson wrote, "The West is no place for pride or laziness; we want industrious farmers and mechanics; we don't care how poor a man is, if he is industrious, he cannot remain poor." An English tradesman came to the United States and worked as a cradler for some months on a farm: "really it was very hard work," and Simon Ferrell in 1832 around Cincinnati observed, "Farming people in Ohio work hard."[23] In an 1851 compendium of tributes to American yeomen, one woman offered poems with lines such as "Labor is worship!—the robin is singing," "Labor is life!—'Tis the still water faileth," and "Labor is glory!—the flying cloud lightens." Manual labor on the farm was celebrated and appreciated, according to James Phillippo in 1857: "In America there is a proper estimation of the respectability and dignity of labour, and of field labour in general," and "all regard labour, especially agricultural labour, as one of the great elements of society."[24]

Visitors from abroad and writers at home all remarked on the existence of the farm laborer and his fate. First, from the beginning farmworkers obtained "wages"; there was no novelty in working for wages in colonial and early republican America. Second, observers were amazed at how high the wages of American workers were. Patrick Shirreff was one of the early travelers interested in agriculture, and he translated wages into bushels of wheat for comparison. Illinois workers obtained annually the equivalent of 580 bushels of wheat; in comparison, he figured English farmworkers obtained 70 bushels of wheat, Scottish workers 62 bushels, and the Irish 30 bushels. That meant the Illinois farmworker made eight times the amount of the English, almost ten times the Scottish, and nearly twenty times the Irish. Given the amount of English wages that were spent on shelter and food, which the American farmworker was given, Shirreff's calculation was close to the mark. This high pay meant few farmworkers remained farmworkers. "A servant [farmworker]," he said, "if industrious and saving, can lay by, in the course of two years, sufficient to purchase eighty acres."[25]

"At such a rate of remuneration as this," wrote John Regan about western wages at sixteen dollars per month, "the labourer cannot but save money, and so he is fairly on the road to enable him to purchase land." Here is the key that made the system of family farming dominant in northern agriculture for nearly two centuries: the ability of farmworkers to save. For modern readers, the alleged high wages of nineteenth-century farmworkers requires a nearly impossible leap of the imagination, but nearly everyone back then mentioned it, and the statistics of northern agriculture support the continuous creation of family farms throughout the century. By one route or another, agricultural workers were becoming family farmers, and presumably savings played an important role. This movement of masses of agricultural workers out of the category of hirelings into the category of property holders violated all the definitions of class. In 1879 an English Commission on Agricultural Depression came to the United States. Its report stated, "The farm laborer can hardly be said to exist as a distinct class in the United States, unless it be among the colored people [sharecroppers] in the Middle and Southern States."[26]

By the 1830s, observers of American agriculture had already written down in abbreviated form Abraham Lincoln's 1859 Wisconsin State Fair address. George Flower, part of the English settlement in Illinois in the 1820s, penned, "The process has been constantly going on ever since the commencement of the settlement. The labourers are continually advancing to the condition of independent farmers. Their place is filled up by others that arrive, who also become independent farmers, and in their turn hire the newly arrived labourers." Patrick Shirreff in 1835 affirmed the property mobility: "The abundance of land induced labourers to turn landholders, and reward others with high wages, who likewise became landholders. Thus there was a constant progression in society, by the prudent and industrious labourers rising into wealth, and receding from the first point of settlement on becoming landholders. These movements continue up to the present time with the existing cause—abundance and cheapness of land, to which many of the peculiarities of the country and its inhabitants may be traced." Within the agricultural literature of the nineteenth-century United States, these statements abound.[27]

High wages and cheap lands resulted in a constant addition of small farmers to the northern farming community. People rose from poverty to farm ownership and to two highly expressive states: competence and independence. In 1775, it was noted that New Englanders had "very little elegance among them" but they did possess a "decent living." By 1792, Jedediah Morse agreed that the United States did not have luxury, but it was "enough if affluence, independence, and happiness are insured to patience, industry and labour." Archibald Prentice in 1848 said about the same: "A decent competence is to be had where in the old country there would be poverty and wretchedness." From Fanny Wright to Harriet Martineau to the agricultural journals, all agreed that American farmers earned a modest, perhaps comfortable life, but certainly not a luxurious one.[28]

Isabella Bird toured the United States and Canada in the 1850s and saw the expanding middle in rural society. She also noted that, "though a competence may be always obtained, I never heard of any one becoming rich through agricultural pursuits." She had a point. The gentry and aristocracy in England nearly monopolized the land and obtained monstrous incomes, but in the United States and Canada landownership fractured into a multitude of small farms, and everyone recognized that great fortunes were not to be made from farming. One guidebook told would-be farming immigrants, "Let him be in no haste to be rich," while Henry Colman told a Massachusetts audience to accept a lifelong goal of decent housing, food and raiment, educated children, small savings, and no indebtedness; this was all "a reasonable farmer ought to expect or wish."[29] And in a fashion they were right. While the names of the creators of the Lowell mills and later of industrialists in oil and iron and steel are well known, who remembers the name of a titan in antebellum northern agriculture?

Mrs. Bird also said the obvious about sophistication in the United States as compared to the aristocratic nations of Europe: "It must never be forgotten that the noble, the learned, and the wealthy have shrunk from the United States; her broad lands have been peopled to a great extent by those whose stalwart arms have been their only possession." The free labor ideology as it emerged out of agricultural America was a slap in the face at aristocracy. Aristocracy and the landowning gentry of England had set the standard: to be a gentleman one had to be above labor—certainly above manual labor but also mental labor if one truly dwelled in the vaunted life of leisure. In economic terms, to be a gentleman meant one did not labor but lived off the labor of subordinates. The rural northern United States was a repudiation of the code of gentility, of all the alleged graces that a leisure class supposedly gave to a society. What was valued was labor. William Hancock, an English visitor in 1860, phrased it quite well: the "motto of the republic might very well be, 'Respect for labor.'" He might have noted that the motto applied more to the rural North than to the urban North.[30]

The unique quality of the northern United States was the preponderance of family farmers. Visitors did go to Lowell to see the Lowell manufacturing girls, and they were interested in American iron making and coal mining, intrigued by the American system of manufacturing (interchangeable parts), and they marveled at its agricultural machine innovations. But in 1860, no visitor to the country had any doubt where the power and numbers in the economy lay: the United States was overwhelmingly an agricultural nation populated with owner-occupiers. Neither commerce nor industry appeared to be the chief characteristic of the American North; agriculture did. Long after the Civil War, British visitors continued to write in wonder about the owner-occupier quality of northern farming. By the 1880s the region was still a land of "yeomen, and those on a magnificent scale."[31]

The reference point for free labor principles was only partly slavery; it was commonly the estate system of British agriculture.[32] This becomes apparent when visitors' comments were made about what was happening in Canadian agriculture. There the free labor ideology fully emerged, and the negative referent was the dismal prospects of independency in British farming. Most of the travelers' remarks, except those about French Canadians, strikingly repeated the upward-mobility theme they penned for the American republic. Wages were high, savings were possible, and land was cheap. Isabella Bird wrote that "Persons of sober, industrious habits, going out as workmen to Canada, speedily acquire comfort and independence"; they started out as day laborers and after eight years had "substantial homesteads." In Canada, one obtained "competency" and with it the loss of the old servile behavior that tenants and landlords demanded in rural England. There were different circumstances in Canada involving land distribution, governmental policies, and land companies, but within the foreign travelers' commentary was the same free labor ideology for Canadian agriculture that they had expressed for the United States.[33] Canada had no slavery. Free labor ideals, in short, originated not solely in comparing an agricultural system to slavery but in comparing North American farming to British estate agriculture.

The northern farm was the reality behind the rhetoric of free labor. All the component elements, the primary arguments, arose from the rural experience of America. Free labor values actually had a fairly good correlation to the reality.[34] Social mobility on the farm was no phantom. Abraham Lincoln as its spokesman had certain difficulties, as Lincoln's love of the farming life was at the least dubious, but he was raised in a rural community and rose within it. That experience was what he drew upon to illustrate his speeches.[35] But within the set of principles were some interesting twists that were not enunciated but were crucial to understanding what contemporaries, especially Republicans and abolitionists, intended when they glorified free labor.

Wages and Property Ownership in the Free Labor Ideology: America and Britain

Much of the scholarly interest in the free labor ideology is lodged in abolitionist and political ideology studies, especially in speeches of the activists and newspaper editorials during U.S. electoral jousts. On occasion, laborers went on strike, and that action provoked in the public realm a discussion of free labor principles. Generalizations drawn from this rich literature are generally accurate and valid; abolitionists and antislavery politicians did indeed approve of a market society and had an idea of a labor market. Nonetheless, the generalizations are incomplete because the scholarship has almost entirely been focused on urban, industrial society. When the topic is expanded to include rural society both in the United States and Great Britain, a different version of free labor arises.

Antebellum Americans were aware of the plight of the English farmworker from the works of Henry Colman, whose words were widely reproduced in the agricultural press. He wrote that the wages of farmworkers were so pathetically low that savings were impossible: "in no case, under the best exertions, can they make, from the wages of labor, any thing like a provision for their old age and decay." A life of destitution was all they had to look forward to, relieved only by sex and alcohol, and when the infirmity of old age came they clung onto life solely through the inadequate agency of parish charity. Moreover, their status was absolutely fixed, not at all like the American agricultural worker, who "is, or always may be, the owner of the house in which he lives, and of as much land as he chooses to cultivate. . . . The laborer here [in England] is doomed to remain in the condition in which he is born—he cannot rise above it." In short, because of low wages there was no social mobility to a middling state.[36]

What then followed were applications of his understanding of the free labor system. The rural workers' "lot seems to be sealed, and they must remain in this condition of servility and dependence. They cannot rise above it. They are not slaves; but they are not free. Liberty and independence, to them, are words without meaning." And in this one passage, Colman blasted away 90 percent of the writing of modern historians on free labor ideals. Free labor meant a real chance at escaping poverty—it was never about the middle-income class moving into the upper-income class; the mobility was out of a state of youthful poverty into the middling element of society. Free labor promised most people competence and independence; it is crucial that Colman recognized that material circumstances gave the words "liberty and independence" their meaning. Then in this passage comes the hard kick: these workers were not slaves—experiencing no physical coercion, no ownership of the body in British agricultural labor—but he added what amounts to be the modern view, they were not free. Anticipating the progressives of the mid-twentieth century, he stated that the burden of poverty was so great that it obliterated the possibility, and even the meaning, of competence and independence.[37]

Colman then offered the American political viewpoint about why the condition of the British agricultural worker was so wretched. "[A] republican," he said, "would say that the evil is fundamental, and grows out of a constitution of society establishing different ranks, the appropriation of the land in a few hands, the high price of land, the depressing sense of dependence, and the hopelessness of competition and of all attempts to acquire influence, respect, or wealth."[38] His observations allow one to conclude that for Colman the English workers' misery was institutionally caused and not a result decreed by nature.

A second instance of questioning the relationship between poverty and free labor came in 1853 from a Scottish visitor to the United States and Canada, William Chambers. Told he needed rest and a change of view, Chambers traveled to North America and stayed in Providence, Rhode Island. There he talked to a group about Scottish rural workers. It was a story of wretchedness summed up as "[the worker's] condition, in short, . . . is *without hope.*" One of the listeners blurted out that it sounded as if the workers were "unprivileged serfs." Chambers replied: "They are free, and, under fortunate circumstances, may rise from a humble to a high station." And the American replied, "But the freedom you impart is associated with such depressing influence, that the chance of rising is very slender." The Rhode Islander continued, "Your aristocracy, having insured the ignorance and incapacity of the peasantry, turn round and say they are unfitted to exercise any political privileges—a pretty kind of liberty that!"[39] Again, material circumstances determined whether a worker was free or not.

The class conditions of British agriculture were widely discussed in North America in the nineteenth century. In the agricultural press, popular journals, and newspapers, Americans detailed the lowly condition of the British farmworker. Almost all the oratory about the glory of American yeoman farming had as its explicit referent the landed monopoly of the aristocracy, the unfairness visited upon tenant farmers, and the abject poverty of English workers. The reference point was not slavery, though it could be—it was the British agricultural system. And the point that was reiterated endlessly was that all British workers—not just agricultural workers—received "scanty wages" and that economic improvement was nearly impossible. The effect of subsistence wages? "Hope is annihilated within him from the very first." Another writer in 1862 exposed the abject humility the British agricultural workers had to display, recoiled in horror at their low pay, and said freedom was mocked when wages were this low: "Yes, it is a glorious *freedom* which her [England's] poor peasantry enjoy, when they dare not utter a word of manly remonstrance to the proud master who thrives upon their half-paid toils, lest they should be turned away to starve!" By this comparison, free labor ideals when applied to the rural North make perfect sense; it explains the different fates of agricultural workers in the American North and those in Great Britain.[40]

These examples demonstrate that behind free labor rhetoric was an expectation of outcomes. The vast majority of people were expected to rise from a position of poverty to a position of property ownership and independence. It was the outcome, not only the principles, that justified the system. That outcome was dependent on high wages. That high wages were absolutely essential to make the free labor system operate as idealized is simple logic. This attribute was often couched in negative terms. Isabella Bird wrote in 1856 what many had been saying: high wages made for savings, which made for social mobility, but "In England, in most branches of trade, the low rate of wages renders it impossible for the operative to save any portion of his earnings"; but in America, a man "has every hope of securing a competence."[41]

Wages high enough to permit savings and social mobility were indeed part of nineteenth-century American political economy. Many recognized the evils of subsistence wages for a society where universal white manhood suffrage reigned. Though such individuals interjected in their discussion of wages all kinds of connections not germane to mobility on the farm, they usually praised high wages because they led to savings and to competence. Among themselves, they argued vigorously on how to obtain the blissful state of remunerative wages. The grand loophole in the American discussion was numerical: exactly what was a wage high enough so that the average wage-earner could be expected to save? For extremely conservative individuals, any wage could be enough to allow savings. However, as a generalization intimated from the sources rather than specified there, most American free labor advocates rejected the attitude that savings should require "heroic" acts of self-control. The ability to save a portion of earnings had to be within the capacity of the majority of citizens. In the rural portions of the North, this was evidently true; it manifestly was not in many urban employments.[42]

Wages also fit into an unusual discussion of incentive and property ownership. The linkage between work performance and wages in contemporary accounts had some complications. The real formula was that free labor was no better than slave labor when wages were low; the fear of the whip and the fear of starvation were about equivalent in drawing forth effort.[43] As wages rose, the response from the worker was indeed to increase effort. But in the nineteenth century, the ultimate incentive was to use wages to get to property ownership. Indeed, most nineteenth-century commentators assumed that full work effort only came when the worker fully owned his (or her) own farm or shop. Working for *any* wage implied that full value did not come to the worker but was leached off to someone else, and therefore a full effort was not made. Wage labor always carried a stigma of a person being exploited and implied that the laborer never gave a complete work effort. Only property ownership elicited the industriousness that political economists celebrated.

Thus on yeoman farms one got a prodigious work effort because all proceeds arising from every drop of sweat belonged to the farm family. An American farmer, unlike a British farmer, "must work, it is true, but he works for the first time on his own land."[44] Opportunity for obtaining a farm stimulated work effort "since every man, who has any proper feeling of independence beating at his heart, would rather toil for himself and his family, as an uncontrolled yeoman, than labour as a confined servant to a stranger."[45]

Also tied to the question of wages and to eventual farm ownership was a motivation not economic but social. Wages had two aspects. One was the amount of remuneration. The other, however, was the fact of supervision. All wage work implied that the worker was under the command of a superior; the wage worker had to obey the will of another or be discharged. A wage worker was always being judged by a superior. Independency escaped this supervision, and on a family farm independence meant self-direction, self-judgment, and no superior. In this sense slavery and wage work had a common ground: both were systems in which superiors ordered inferiors, the primary difference being the amount of coercion allowed.

The question of who was master of whom played a vital role in American sectional animosities. Henry Fearon was a British traveler to the republic in the early nineteenth century who vigorously criticized the United States in order to stop a feared out-migration of English farmers. He found nothing good about American farming, and any incentive to hearty work escaped his notice; many of the farmers were squatters and "truly wretched" except in one area: such a person was "his own master." About the same time, another writer talked about farms along the Hudson River: "The owner can with Truth say, 'this ground is my own.' He holds it upon no slavish condition."[46] The family farm in its quest for competence and independence had as a social goal the rejection of any authority over the family—the man knew only himself as master and none other. The debate between the sections had in its background the struggle between those used to self-mastery against those accustomed to mastery over others, the family farm versus the plantation.

The goal of the family farmer, and of free labor advocates as well, was not so much economic growth as it was independence. This was given particularly memorable form in the writings of Anthony Trollope, a novelist who came to the United States in 1861. He was the son of the famous Frances Trollope, who in the 1830s had scandalized Americans with a vituperative depiction of the republic's crude social habits. Her son was not so vicious. He went to Milwaukee, met a small farmer, and was invited to his home. "He is his own master, standing on his own threshold, and finds no need to assert his equality by rudeness." Common education had given people a penchant for reading newspapers, and the extent of reading by ordinary people took visitors by surprise as their experience with agricultural workers and tenants in England led them to surmise that workers and tenants knew nothing except the weather, crops, animals, and beer. This Milwaukee farmer asked about Great Britain, and "he astonishes you by the extent of his knowledge. . . . To me I confess that the manliness of such a man is very charming. He is dirty and perhaps squalid. His children are sick and life is without comforts. His wife is pale, and you think you see shortness of life written in the faces of all the family. But over and above it all there is an independence which sits gracefully on their shoulders, and teaches you at the first glance that the man has a right to assume himself to be your equal. It is for this position that the labourer works, bearing hard words and the indignity of tyranny."[47]

City vs. Country, Mastery vs. Self-Mastery

As the North moved into the middle of the nineteenth century, its countryside had acquired renown for the equality among people because of its relatively equal distribution of property among farmers. Certainly the orators for the yeoman farmer stressed a host of character traits that drew raves from observers: thrift, intelligence, independence, honesty, unbiased opinions, strong work habits, and a willingness to fight for the country. Often overlooked by historians in these exaggerated claims of virtuous character was the social trait of equality. But the orators did stress the social and political equality in the North that drew its sustenance from the family farm. "We glory in our political and religious freedom; all of us. Here, we are all equal, from the President down to the pauper." A New York speaker, Ebeneezer Mack, boasted that, "in our country, the tillers are the owners of the land, or may become so by industry and economy," and from that condition came "equality of rights" and "equal protection of the laws." Another New Yorker applauded the "*comparatively equal distribution of property among you* [farmers]," and said the United States was a beacon to all the serfs, tenants, and peasants of Europe, who suffer material degradation because "the corner stone of the Orders of European aristocracy is the monopoly of the soil."[48]

The omnipresence of the family farm in the northern states was seen precisely as the foundation for American democracy that gave meaning to the affirmation of the Declaration of Independence that "all men are created equal." Indeed, several foreign visitors caught the reversal of political views caused by the wide distribution of property in the American North versus the concentration of landed wealth in Europe. The countryside in England was "conservative" due to the landed interest; in the United States, the countryside was "democratic."[49] For the foreign observers, the oddity was that in Europe the cities were the centers of democratic agitation, but in the United States these travelers saw urban areas as the incubators of a rising aristocracy. One wrote that democracy in the nation worked well enough in the countryside, "but it is in the large towns and cities of the Union [that] Democratic institutions are on their trial." Cities not only bred wealthy merchants and manufacturers but also uncontrollable mobs. In a withering attack on slavery in 1857, James Stirling noted the class divisions in the city but added, "in the rural districts the old yeomanry still prevails, and is the backbone of American society." Eventually, though, the cities would outpopulate the rural areas, and aristocracy would establish itself in North America as it had everywhere else.[50]

A budding rural-urban argument was in the making several decades before the outbreak of the Civil War, and it went far beyond governmental policies and alleged mental sophistication or simplemindedness; it involved political philosophies and ideals of justice and fairness. Free labor ideals worked in rural America; they only partially applied to urban areas and in some cases did not apply at all. The values of independence, individualism, hard work, and self-control had one outcome in the countryside—the survival of the family farm. In cities, the same traits created class division. Not the least important factor in promoting these divisions was property distribution: the northern countryside achieved a roughly egalitarian property distribution, but certainly not the city.[51]

The same division between northern rural areas and northern urban areas can be seen in one other arena of national life, although much more pronounced: the southern countryside versus the northern countryside. Besides the differences in the size of operations and the income derived, another contrast appeared. An ideological gulf separated northern farming from southern farming, and it can be summed up in one word: mastery.

Free labor concepts proliferated among the southern common folk. Hard and continuous work, self-control, some temperance, and rational planning were as vital to the small freehold farm in the South as in the North. Actually, the work ethic of the small farm was much the same no matter its orientation to the market. To make family farms operate at all, a strong work ethic was simply essential. The ideal of competency was fairly much the same North and South, as was the ideal of social mobility: to obtain a farm of 80 to 120 acres. Southern yeoman farms had more trouble with commercial farming because of the lack of urban areas and transportation facilities, but when they appeared the southern yeoman farmer made the switch from subsistence to semi-commercial farming without great difficulty.[52]

Much of the material relating to southern ideology is absorbed in the proslavery argument in which a number of southern authors, most notably George Fitzhugh, trumpeted slavery as the best form of socialism for the worker while capitalism thrust workers into the worst state of material deprivation imaginable. The writers of these publications scoffed at the idea of free labor being fairly or even adequately rewarded.[53]

Nonetheless, there was one branch of the proslavery argument that united a northern idea of factory labor with slavery—supervision. To get a plantation to yield income, one had to labor at supervising people. The advice was for planters to "Work yourself and make everybody around you work." Edmund Ruffin decided the reason slavery was profitable was because people were directed in their work effort by the owner.[54] Supervisory capacities in the South became the route to social mobility. One contributor to a journal told young men to start out as overseers of a small farm: "this training [gave] him a knowledge of what could be done; it gave him strength of body and mind, and habits of industry." After "beginning with [a] small number of hands, three, four, five or more, he became qualified to manage too many to justify his laboring in person, his profits increase to a share and a half . . . till at length he was enabled to establish himself." The road to independence was paved by knowing how to master other men.[55]

That the planter class directed the efforts of other men led to the conclusion that the planters were similar to European nobility: they directed labor and did not perform it. And with that conclusion came the same philosophy toward others that the English aristocracy had: people by nature and God were destined to certain ranks in the world, and most were designated to be made to labor by superiors. William R. Russell, an English sympathizer with the Confederacy during the war, related a conversation he had on a steamship in March 1861. A southerner scoffed at equal rights: "Some were born to be slaves—some to be labourers in the lower strata above the slaves—others to follow useful mechanical arts—the rest were born to rule and to own their fellow-men."[56] Slavery produced a hierarchy and a defense of that hierarchy. In the North, urban areas were generating new hierarchies without slavery, and in their wake came as well a justification for inequality.

But the heart of the matter in the antebellum decades was that southern slavery starkly represented the mastery of some people over others. This is hardly a novel finding. One of the obvious points of friction that would ignite sectional fires was that the southern mastery over others was brought alongside of the "self-mastery" of the northern family farm. In 1869 an agricultural publisher put the desire for independence and competency into a memorable anti-mastery framework: "His farm is his little world, and over it he reigns supreme, calling no man master."[57] So long as the two sections kept their systems separate, the clash between self-mastery and mastery over others was muted. But if ever the perception grew that mastery over others became aggressive, the fears of the small-farm Davids would be activated about the purposes and designs of the great plantation Goliaths. The territorial issue activated those fears.

9

Davids versus Goliaths

THE SECTIONAL CONTROVERSY AS A FIGHT BETWEEN AGRICULTURAL REGIONS

The structure of the United States between 1840 and 1860 did not pit a capitalist section against a precapitalist one, or an industrial North versus an agrarian South. Instead there was a struggle between sections founded on the plantation and the family farm, a conflict about which economic organization of agriculture would guide the policy of the federal government in choosing which one would be allowed to expand out of its borders. In broader terms, the contest was between the hierarchy of the plantation and the egalitarianism of the family farm.

For most of the nineteenth century, the two land-use systems of family-farm and plantation agriculture coexisted without overt friction as both headed, with the exception of the nonslaveholder exodus out of Kentucky and Virginia into the Northwest Territory, in a westerly direction and did not intersect in a north-south direction. Nonetheless, the material differences between the two systems were stark. The plantation system relied entirely on slavery, and it consumed land in huge chunks and converted slaves into artisans. The family-farm system parceled land out in diminutive amounts, thereby requiring less territorial expansion to suit the needs of its inhabitants, and also incorporated in its economy independent artisans who lived in nearby villages. If brought together, the systems could clash over land use and the freedom of people who performed artisanal work.

While the economic demands of each system on federal policy did not necessarily conflict, they could. One aspect of economic development influencing national policy formulation was the economic binding of the Old Northwest to the Old Northeast and Mid-Atlantic states by the railroad network. This had the unexpected effect of making free-state residents look toward the domestic market for economic improvement (the farm West trading with the manufacturing East) instead of looking abroad to foreign markets. Southerners continued to cultivate the semitropical staples, especially cotton, and their economic vision remained fixated on overseas markets. What brought these points of conflict to the forefront of political battle was territorial expansion. Though historians have tended to see the North as the aggressor because of its demands that the territories exclude slavery, the agricultural North could only look upon the South as the aggressor, for the struggle seemed to northern farmers to resemble the plantation Goliath bullying the family-farmer David into yielding his western inheritance.

Historians and the Causes of Southern Secession

Explanations abound for the reasons behind the sectional controversy between 1820 and 1860. This plethora of theories has produced some deep insights into the different societies, their respective mentalities, economic processes, and social orderings, but little has been achieved in whittling down the mass of ideas to the primary influences causing secession. Much of the work on Civil War causation involves ideological perceptions; the North absorbing free labor ideals, the South retaining agrarianism and old republicanism. Investigation of the South has become a growth industry, and its practitioners have stressed racism in creating sectionalism: southerners believed that northern Free Soilers sought to destroy southern white equality by emancipating blacks. Many adhere to the ideas of Bertram Wyatt-Brown that the South was an honor society that simply could not stand the affront of northerners insulting them about the peculiar institution. Some authors point to the origin of North-South differences in ethnic composition: the South was a Celtic society while the North was Anglo-Saxon. The agrarian nature of the South made its leaders wary of cities and free labor developments, thus enabling them to view northern society with disdain. The maniac desire of the planters to protect their wealth in slaves has been offered as a way to explain southern behavior, and Michael P. Johnson and William W. Freehling have emphasized planter fear of southern nonslaveholders. In general, the basic thesis of Eugene Genovese still holds sway: the plantation made the South a premodern society at odds with capitalist development in the North.[1]

Many historians dispute the conclusions of these studies and argue that the problem was at heart political. Michael Holt's work exemplifies this position, as he finds the seeds of sectional antagonism in the rise of new issues of religion, ethnicity, and immigration, which he sees as the forces that tore apart the party system of Whigs and Democrats. Edward Ayers emphasizes commonalities between North and South and argues for "human agency" and contingency in producing the odd political choices that led to secession. Earlier historians stressed the inadequate power of the federal government to hold the nation together, and emphasized the crucial nature of the acquisition of western territories. One group, the revisionist historians of the 1930s, flatly stated that no honest divisive issue existed at all, that sectionalism originated in the minds of radical agitators over slavery. Others have insisted that southerners enjoyed real power in Congress: Leonard Richards, William E. Gienapp, and Don E. Fehrenbacher have argued that the "slave power" was no figment of an abolitionist imagination.[2]

Behind most of the writing on the causes of the sectional dispute lies one particular force: the changes wrought in the North by industrialization, or its surrogate names of modernization, the market revolution, and the transportation revolution. In the early twentieth century began the analysis that has become the springboard for much of the historiography on sectionalism. The progressive historians Charles A. Beard and Mary R. Beard discarded the older tradition of explaining American history in terms of ideas and morality, and replaced it with an emphasis on clashing economic interests. They offered the view that the Civil War was a fight between southern planters, northeastern industrialists, and western farmers over tariffs, internal improvements, western lands, and banking legislation. The industrialists and western farmers allied together and, rather than admit defeat, the planters took the South out of the Union. The Beards' provocative thesis has remained a staple of Civil War historiography to this day, although most historians doubt these policies possessed the weight to break the nation.[3]

The Beardian hypothesis opened the door to further musing about economic change and the sectional conflict, and such contemplations appeared prominently in historical writing after 1960. The Beards' interpretation had the strength of including rural America, a feature that got lost over time, but their interpretation was based upon making simple economic interests extraordinarily important. People were guided by their pocketbooks, not by ideas or classes or a zeitgeist. In post-1960 historical writing, these latter qualities were added one after another. The central idea holding all the literature together was that the North underwent an economic transformation while the South did not.[4]

Besides Eugene D. Genovese's early interpretation that North-South antagonism reflected the contrasting ideologies born out of the material differences between a precapitalist system and a capitalist one, historians became attached to the idea of modernization that was then reigning in the social sciences. While the theory of modernization includes the advance of such social traits as impersonal exchanges, stronger central government power, the rise of bureaucracy, urbanization, and need for social services, the key element became more and more the birth of industrial capitalism in the North. The North became modern because it became industrial; the South remained patriarchal and premodern because it stayed rural. Besides altering political ideology by giving birth to free labor ideals, the arrival of industrialism destroyed the old parties and produced a new one of Republicans (the modernizers) versus the backward-looking Democrats.[5] Explaining the nation in terms of those who adopted industrial capitalism versus those who did not is the reigning paradigm in nineteenth-century American history, and it is the main contender—but not the only one—for explaining the genesis of the Civil War. The explanation is seriously flawed.[6]

Why the industrial capitalist thesis is a misguided conceptualization for understanding the dynamics that led to southern secession involves realization of two vital economic elements of the nation. First, the agrarian quality of the North has been ignored, while scholarly attention has been lavished upon the structural differences between industrial capitalism and slavery, defined as the difference between factory wage labor and ownership of another person's body. Supposedly the wage-earning regime of capitalism evoked the individualism of the marketplace while the ownership of another person's body yielded paternalism.[7]

That may be one way to look at the structural difference between slavery and freedom, but it is not the only way, especially if one redefines the structure to mean the actual operational requirements of the large plantation and the large industrial factory (and here "large" means over twenty-five people). Both the plantation and the factory required heavy capital investment, and both relied on cheap labor—which one being the cheapest is left to other scholars. The point is that the real structural problem of both the plantation and the factory was the supervision of labor: how to get laborers to perform tasks so the income demands of the owners were met. In economic terms, this is the problem of shirking, and shirking invokes the subject of supervision. With the plantation the means of supervision was overseers and slave drivers using physical violence; with the factory, it was (and is) foremen and supervisors threatening dismissal and blacklisting. Both use a form of coercion to get what they want.[8] From this point of view, industrial factories and plantations seem more similar than different.

The second defect with the structural analysis of the industrial/capitalist interpretation of American sectionalism is that it ignores the major structural dichotomy in the nation's economy: the small farm versus the large plantation. No other aspect of the nation shows such massive divergence. Moreover, the difference between large plantation and small farm goes beyond a physical description. Small-farm agriculture elicited an ideology different from the plantation's: manual labor versus mental labor, self-mastery versus mastery over others, cooperation versus command, equality versus inequality, democracy versus aristocracy. While the plantation and the family farm were both agrarian, they evoked very different understandings of the world. Ironically, the plantation and the factory drew from their owners and directors a similar set of values (order, superiors ruling inferiors, obedience to authority, liberty only for the wealthy).[9]

On one particular federal policy, the family-farm North and the plantation South could easily slide into confrontation—the issue of territorial expansion. Expansion of the family farm would be a replication of the family farm elsewhere. It could be desired as an outlet to an overpopulated agricultural population east of the Mississippi River—not to put excess industrial workers on the land, but to have land available for the progeny of eastern farmers. Plantation owners sought new lands for any number of reasons; perhaps because they were exhausting the soil back east, perhaps because they wanted to settle their progeny on plantations where there was available land, perhaps because they needed a safety valve for nonslaveholders, or perhaps because they were expansive capitalists. But this antagonism is direct, simple, eminently visible, and unavoidable. Family farmers wanted land carved into small plots while plantation owners wanted—insisted—that they be able to procure large holdings. One or the other had to triumph; both could not. So if the question of slavery's expansion ever came into the political realm, there would be conflict. And that of course is precisely what happened.

Just to drive the point home, industrial capitalism was not needed at all to generate the sectional conflict; modernization was irrelevant to sectional discord. The obsession of historians with urbanization and industrialization has had a warping influence on the understanding of American history prior to 1880. All that was necessary to produce the American sectional conflict was population growth and a westward migration of people. So long as the plantation system and the family-farming system kept separate geographically, the Union could persevere; but when the plantation owners demanded extension into areas family farmers thought was theirs, the conflict started. And the person most renowned for initiating the conflict, David Wilmot, came from no industrial setting. He represented well the agricultural basis of his electoral district.

Northern Fears about the Plantation

The northern family-farm system emphasized self-mastery: self-discipline and hard labor led to competency and independence. The abolitionist minister Theodore Parker called this the natural urge to be free, for each man "to be his own master." This attitude led to an embrace of the Jeffersonian idea of liberty, of "self-ownership." An editor for the *National Era,* probably Gamaliel Bailey, then said southern leaders claimed Jefferson for themselves, but in reality Jefferson was for them a "gaseous abstraction." Jefferson's message of self-ownership, Bailey wrote, instead came to be embraced by the "great numbers of the free laborers of the Northern States—the mechanics and small farmers." (Bailey identified as "free laborers" the mechanics and small farmers, not factory workers.) The connection between self-ownership, mastery, and the northern farm came to the fore in a postwar address by John A. McClernand, antebellum Democrat and then Civil War general:

> In a republic, particularly, founded upon the idea of the virtue and intelligence of the people, the true policy of the State is to foster the independence of the citizen as a member of the State. This can be more effectually done by encouraging him to become a free-holder than in any other way. Lord of his own domicil[e]; reposing under his own vine and fig tree, and tilling his own field, his dignity and means of subsistence are placed under his own guardianship and providence.
>
> On the other hand, the exclusion of the many from the ownership of the soil inevitably tends to make them the Slaves or dependents of the few who monopolize it. . . .

> ... And, even, in our own times and country, the engrossment of the best portion of the cotton and sugar growing lands in the South by the local class of Slave-holders, gave a character to human bondage in that section which, at the same time feeding the pride and arrogance of that class and offending the dignity of the free laborers of the North, at length, forced the train of civil war between those sections.[10]

McClernand laid out the economic basis of the collision between North and South: it was a question of who was going to control land in the United States. The northern system was self-mastery; the southern system was mastery over others.

Southern expansion and northern expansion, however, did not cross paths. Both moved in a westerly direction, and until the 1840s southern profitability depended on the cultivation of crops different from those raised in the North. So long as both sections moved strictly in a westward direction, hostilities were aimed usually at local targets: for the family farmer, at merchants and bankers; for plantation owners, at slave insurrectionists and disloyal nonslaveholders, plus the ever-present and dreaded Yankee abolitionists. However, should the paths of northern and southern expansion cross, as they did in the settlement of the Old Northwest prior to 1830, a joint of conflict was created. A joint of conflict is most obvious when two different groups need or compete for the same resource. Both the northern and southern agricultural systems depended upon a particular pattern of land use; one consolidated land, the other dispersed it. The potential material conflict between them was blatantly obvious.

This obvious material conflict, however, has been ignored. Historians since Charles Beard and Mary Beard have sought the joint of conflict in the policies advocated in the halls of Congress.[11] Henry Clay in the 1820s and then the Whig Party in the late 1830s championed the "American System," a program intended to build up the internal economy of the United States by using federal dollars to construct internal improvements (river dredging, canals, roads, and railroads), a national banking system to stabilize the currency, the slow disposition of the public lands to encourage urban development and markets, and a high tariff to stimulate manufacturing by eliminating foreign competition. The American System of Henry Clay outraged southern agrarians because it enlarged the powers of the federal government, had potential favoritism lodged in its implementation, and allowed federal bureaucrats to override state bureaucrats on the control of money. A protective tariff (high taxes on imported goods) translated into potential trade wars with European nations, higher prices for British goods (which planters avidly purchased), and a redistribution of income from the South to the North because southerners would be buying northern wares at inflated prices.[12] For most historians of the progressive persuasion, Clay's American System was the obvious political byproduct of industrialization, and it clearly delineated an economic clash over federal policy between industrialists and southern agrarians.

Never fully implemented and virtually repudiated after 1844, the American System certainly generated its share of heated verbosity and sectional antagonisms, but whether these policies were strong enough to have any material impact on either the northern or southern economy remains questionable. States undertook the construction of internal improvements until 1840, and after that date construction of railroads proceeded rapidly without government assistance. The nation did not have a national banking system, but a certain stability in financial affairs was obtained. A protective tariff never lasted long enough to determine its effects. Moreover, all of these policies were eminently open to compromise. Northern advocates of the American System lost on the policy front, but the North did not suffer economically from the defeat. Rather, the North entered a period of modern economic growth. Most historians doubt that these policy concerns had the power to generate disunion.[13]

Northern farmers and the village economy had an odd relationship to the American System. All farmers north and south had to spend several days a year working on county roads, so the countryside throughout the nation recognized the urgent requirement of transportation mechanisms. Whether they wished internal improvements to be done by federal or state authorities was another question. Farmers probably were suspicious of bankers and merchants for a host of reasons—some of which boiled down to farmers having to do manual labor while financiers earned huge sums without sweating. The tariff was strange territory for northern farmers. Agriculturalists above the Mason-Dixon line did not rely on foreign trade like southern planters did. Because of the canal and railroad system built between 1815 and 1860, farmers in the Old Northwest sent their crops directly to the East Coast by the iron horse or Great Lakes shipping and were abandoning the Mississippi River route. This tied the Old Northwest more securely with the economic interests of the Mid-Atlantic and New England states. Moreover, it was an *internal* market that mattered: a food deficit had arisen in New England because of urban development and poor soil. In some ways, the tariff, by stimulating eastern manufacturing and urban growth, aided western farmers by enlarging their market. Yet western farmers also were highly sensitive to European demand for American cereal crops. Thus, northern farm attitudes to the tariff were mixed.

In 1850, the plantation regime was locked out of the nonslaveholding states, and no obvious conflict between the plantation and the family farm existed. Nonetheless, by 1850 most northerners had a very good idea of how the plantation could ruin northern farm communities if ever it found a way to burst out of its boundaries. Northerners had plenty of information from southerners themselves that the plantation crowded out the small farmer, established monopoly of the soil, instigated a damaging competition between free labor and slaves, reduced employment for whites by training slaves to be artisans and mechanics, and degraded whites to ignorant poverty-stricken wretches if they could not migrate elsewhere. These results obtained wherever the plantation system became entrenched.

The origins of Ohio, Indiana, and Illinois were southern. Southerners left Kentucky, Virginia, North Carolina, and Maryland to populate the Old Northwest. They were generally antislavery, and in the first several decades of existence in these states, the southern settlers there beat back attempts to legalize chattel slavery. Some of these residents had religious reasons for doing so, but many were simply upset at the plantation society that slavery produced and how little room was left for yeoman farmers.[14] In 1832, Virginia legislators, disturbed by Nat Turner's rebellion and the outflow of their people into other states, debated the possibility of gradual emancipation. Some saw slavery as the cause of Virginia's decline. Thomas Marshall, the son of Chief Justice John Marshall, insisted that slavery ruined the yeoman farmer: slavery "banishes the yeomanry of the country—deprives the spinner, the weaver, the smith, the shoemaker, the carpenter, of employment and support." This observation was seconded by Philip N. Bolling; the peculiar institution "drives from us the laboring man—the honest, industrious poor. . . . The small freeholders are driven off also." Others complained that where the plantation presided, education was absent; the planters would not support public education.[15]

The plantation, so observers wrote, warred on the small farmer. In South Carolina, Francis Lieber said one person he knew spoke of "the incompatibility of large plantations with neighboring farms, and their uniform tendency to destroy the yeoman." Even so prominent a person as Senator Clement Comer Clay saw the process of southern agriculture as being the rich planter "pushing off the many who are merely independent," and ultimately there would simply be the "master [who] grasps the whole domain that once furnished happy homes for a dozen white families." In 1856 one Wisconsin newspaper published an editorial on "How Slavery Eats Up Small Farmers."[16]

Americans were well acquainted with aristocratic engrossment of lands throughout Europe and the results of such practice for common people. The operative phrase they used was "land monopoly," the operation of which on a society resulted in mass misery and slavery. The Milwaukee *Weekly Wisconsin* published an article on "Land Monopoly in Europe," stating, "Millions of Germans, Irishmen, Englishmen and Frenchmen . . . are perishing for want of the necessaries of life. By Land Monopoly, they are precluded from being their own employers, and although able and willing to work, they are denied employment." At a Dane County Whig convention in Madison, Wisconsin, the fourth resolution demanded elimination of land monopoly by restricting the amount of land a person could own to 640 acres. Back in New York City, reformers created the Land Reform Association to agitate for a homestead act (free land in the West for actual settlers) to alleviate miserable working-class wages. At the same time, all types of writers made it explicit that plantation agriculture was a form of land monopoly, and land monopoly created an aristocracy and a mass of pauperized workers. For proof, all they had to do was reference the example of England.[17]

Within northern commentary appeared frequently the general occupation of "mechanic" and "artisan." The words could have possibly meant urban craftsmen, but most likely were representations of skilled workers in villages servicing the farm economy. The terms were used constantly by orators and agricultural writers, and often they named the occupations vital to the farm-village economy: carpenters, blacksmiths, boot makers, harness makers, and others.[18] These characterizations did not allude to the urban factory. Thus when Richard Yates gave his notable free labor speech in 1860, he praised the policies that could elevate free labor by creating a better "market for farmers and mechanics." In an 1864 pamphlet aimed at working-class voters under the title of "How Slavery Injures Free Workingmen," the author referred to the farm-village economy when he wrote that slavery gave "to slave mechanics, slave shoemakers, slave blacksmiths, slave carpenters, slave wheelwrights, the labor, and to their wealthy masters, the profits, which of right belong to the free workingmen." In short, much of the free labor commentary made about the impact of slavery upon the white population was referencing the rural northern economy, not industrialization.[19]

Out of the land monopoly fastened onto southern society by slavery arose the poor whites of the South. Often a misperception that missed the large number of small landowning farmers who possessed no slaves, this general charge argued that the slave plantation eliminated employment for white mechanics and monopolized the best land so nonslaveholding white farmers had no opportunity for worthwhile remuneration: all became a mass of poverty living off the scraps of the plantation. The two most famous exposés of this condition were Frederick Law Olmsted's journeys through the slaveholding states and North Carolinian Hinton Rowan Helper's *The Impending Crisis.* There were other publications as well, but the message was plain: the plantation created a society divided into rich and poor. Once it was legally sanctioned, the process of land engrossment led to masses living in poverty. The only escape from the system once it took root was to migrate to a state that outlawed slavery. The comparison was stark: the egalitarian rural society of independent yeoman farmers of the North versus the dependent and impoverished poor whites of the South. John M. Palmer recollected—honestly, one hopes—that his father had moved the family out of a slaveholding state to Illinois because "lands [were] so cheap that we could all be landholders, where men were all equal."[20]

The last charge critics made against the plantation system was the unwelcome personality traits of the plantation owner. Northern farming communities had their rich and poor, and in the villages merchants, bankers, and lawyers represented an upper crust, but the overall structure was the multitude of small farms. This structure promoted an ideal of equality and especially of equal treatment to all, no deference to any except if merited by achievement. The plantation owner mastered men; he, it was believed, demanded deference in the same fashion that the gentry of England did. During the battle over legalization of slavery in Illinois in the 1820s, Morris

Birkbeck, writing as Jonathan Freeman, made telling points against establishing slavery because it would bring the "slave-gentry" into the land of "plain farmers, working with their own free hands, or the hands of free workmen"; planters were great men who rode around on horses "when I and my boys are working perhaps bare-headed in the hot sun." Slave masters would not be good neighbors who helped to raise barns and houses. Hinton Helper gave the slave "oligarchy" the characteristics of "arrogance" and "conceit." Tennessee's Parson William G. Brownlow, who was proslavery in the antebellum years, reacted against the oratory of William L. Yancey, an Alabama fire-eater, that labeled washing clothes and blacking boots as jobs for slaves, not free men: "That sort of slur upon honest labor may do in South Alabama, among purse-proud aristocratic Democrats, but in East Tennessee, where two-thirds of all our families own no negroes, and do their own work, it will not [take]."[21] The personality traits that came with the plantation were ones that the northern farm-village community reviled.

The Slave Power and the Territories

At the same time northern antislavery elements made their free labor exhortations, they also continuously charged that southerners dominated the federal government, that the interests of the planters constituted a slave power. All legislation in the national congress had to be shaped in accord with the economic interests of slaveholders. Many went further than that: the ultimate interest of the slave power was to make slavery secure from attack from any quarter. That could be accomplished by letting slavery expand by war and conquest of Mexico and then Central America and the Caribbean islands, and then permitting slavery to go into them. Next, according to some of the more extravagant writers on the subject, these territories would become slave states (as free whites would not migrate to territories where they would face the plantation complex), the numerical majority of slave states would grow, and eventually laws would be passed, aided by the decisions of the Supreme Court, to legalize slavery throughout the country. People like Abraham Lincoln labeled this potential scenario the nationalization of slavery. Those adhering to the slave power understanding of the political goals of the planters often made an argument separate and distinct from that of the free labor ideology, but the two were so intimately tied as to be different sides of a single antislavery coin. The one side was that the planters constituted an aristocracy that used its power to control politics for its own financial betterment and permanency (the slave power), while the obverse of the coin was that aristocrats obtained their wealth by oppressing labor and taking away from the laborers their fruits of labor (the free labor ideology). The free labor ideology and the slave power thesis were both derivatives of the political economy of aristocracy.[22]

Southerners dominated the Supreme Court, held the chairs of the important committees in the House of Representatives, obtained a nearly equal representation with the North in the Senate (until 1851), controlled foreign policy, largely determined the distribution of federal patronage, and usually denied northerners the presidency. Many believed the undue weight of southerners in national councils stemmed from the three-fifths compromise by which planters obtained additional representation for themselves because slaves were counted as three-fifths of a person. Adding to that power were northerners who bowed to southern demands—the "doughfaces" of the North, who largely resided in the Democratic Party.[23] However, who dominated federal policy is not clear. The Border South was far more amenable to Henry Clay's American System than historians have allowed, and the rural parts of the North had a possible material interest in low tariffs. Internal improvements were an important issue in the 1850s but not one that was going to produce national division. The Pennsylvania iron interests were outraged over the low-tariff advocacy of plantation-state Democrats, but Massachusetts textile manufacturers were satisfied with the low tariff of 1857.[24]

The reality about the federal government in the antebellum decades was that it was not particularly powerful. By certain policies it could only marginally aid or injure various economic activities. The weakness of the federal government has thrown some wrenches into the gears of historical interpretation. If the power of the federal government was so limited, then how does one explain the northern dread of the slave power—except to say that other motives moved northerners, such as cultural imperialism or a purely political obsession with having their own people obtain office and patronage. Combined with the impotence of federal power was a strong belief that the plantation was not suited for the prairies and arid regions of the West. Slavery, many believed, needed a semitropical climate to raise plantation staples; the small northern farm was supreme, it was said, in the raising of wheat, corn, oats, and hay. These were the arguments of conservatives and Democrats in the antebellum decades, later picked up by historians in the 1930s, and now currently enjoying a renaissance. Southerners were acting defensively against northern attacks on the peculiar institution, seeking to stop northerners from imposing their cultural values and American System policies on them, and had no intention whatsoever of pushing slavery into the North.[25] Behind this situation lay one central assumption: slavery, through its manifestation in the plantation, posed no material danger to northerners. This is an incorrect assumption.

In only two areas did the central government exhibit strength, distribution of public lands and foreign policy, and in one area it had a potential to exert power, control of the definition of property on a national basis. The unrealized power of property definition was starting to be acknowledged as it was the basis for Roger B. Taney's decision in the Dred Scott case that Congress could not prohibit slavery in the territories because the Constitution recognized slaves as property. The other two powers were indeed potent. The war-making power of the federal government was upsetting to all kinds of northerners, particularly when it was being used, it seemed, to find new lands for slavery's growth. The Mexican War may have had more sectional impact than historians have accorded it as the conflict gave strength to the apprehension that a slave power existed and was willing to use the federal government as a tool to satisfy its material desires by violence.[26]

The second strong power of the federal government was its ability to determine how to distribute public lands. The federal government owned the public lands, surveyed them, and auctioned them off once territorial enabling legislation passed Congress. Congress could impose taxes and appoint territorial officials; it set the price of land, although except for the homestead legislation passed during the Civil War it did not limit the size of purchase. The door to interminable controversy leading to national separation was whether Congress had the power to exclude slavery. In the midst of the Mexican War, David Wilmot and disgruntled Democrats broached the subject in the Wilmot Proviso and demanded the exclusion of slavery from any territory acquired from Mexico.

How important landed expansion was to the social and economic well-being of either the free labor North or the slave labor South has been a moot question for over a century. In the southern case, the argument has been that slaves were so sloppy and careless in their work habits that they exhausted the soil and were incapable of regenerating it by fertilizers;

therefore, for the plantation to continue to exist, it needed new land. Others have argued that the South had hardly exhausted the soil it did have, and Texas was only beginning to be settled.[27] Economically, the matter cannot be cleared up. So long as migration was easy and cheap, slaveholders would rather move than stay rooted and go through the expense of careful cultivation. However, if the British estate system of agriculture was an appropriate comparison, then the southern plantation did not require additional land. The one condition that remains shrouded in mystery was whether the plantation could have adapted to wheat culture and the northern staples. Virginia's experience in the 1850s would argue yes, but it was not conclusive. In any event, northern contemporaries seemingly had little awareness of the extent of slave-grown wheat in the Old Dominion.

An agitation for making small amounts of western land free to actual settlers gained momentum in the antebellum decades. The proposal, the homestead law, became a lightning rod for sectional controversy, especially in 1860 when President James Buchanan vetoed a measure that had passed Congress. The homestead ideal brought out the normal commentary about the glorious results accompanying the family-farm/village model compared to the dilapidated condition of society wherever the slave-based plantation entrenched itself. By itself, however, the homestead proposal did not produce a significant political agitation (unlike the tariff in the iron districts of Pennsylvania), but rather was wrapped in the more generalized fears about the slave power and the general apprehension that it was encroaching on northern society on all fronts.[28]

The North actually did not require the West economically at all. The states that composed the North in 1860 (excluding California and Oregon) were going to dominate the American economy until World War II—the trade pattern between the East Coast and the Old Northwest was the mighty engine that drove the American economy to productive heights for nearly a century. The population growth of the areas west of Iowa to the Pacific Ocean was miniscule for decades. The North in 1860 had plenty of land to make more farms; the number of farms grew from one million in these states in 1860 to two million by 1900.

Whatever the actual economic needs of the North and the South, many northerners reacted to southern expansionism with shock as it made the slaveholding states look "aggressive," an adjective endlessly repeated among antislavery writers and orators. After the Mexican War, southern actions, individualist and disparate though they were, did nothing to soothe any fears northerners had about using the powers of the federal government to acquire more territory. The incidents over Cuba (the Lopez expedition), the antics of filibusters like William Walker, the border strife between Mexico and the United States, and the Ostend Manifesto all seemed to be part of a larger pattern of slave plantation expansion at any cost.[29] What was happening, though, as apprehensions grew about an expansionist, aggressive South, was the surge in free soil commentary about the type of society that the plantation produced: land monopoly, impoverished white farmers, depressed wage rates, slave mechanics, no schools, and few villages.

Southern expansion had been below the nation's southern border and had little effect on northerners who thought of expansion for themselves only in a westward direction. All that the turmoil of 1844 to 1851 produced was a vague unease among many northerners that the plantation states were acting aggressively and their belligerence might take them into realms where they should not go. As for voting and arranging compromises, the political system of Whigs versus Democrats weathered the storm quite well. The issue, however, started a realignment in the southern states that produced in the Gulf states a monstrous Democratic majority.[30] The only possible compromise would have been an extension of the Missouri Compromise line to the Pacific Ocean. Northerners wishing to continue a migration strictly in a westerly direction should have been amenable to this extension, with the condition that the federal government would restrain from adventurism in foreign affairs.

The back-breaker for sectional peace came in the passage of the Kansas-Nebraska Act and its repeal of the Missouri Compromise line, thereby potentially enabling the plantation to move above the 36° 30′ line and establish slavery where it had been forbidden for thirty-four years. For the rural North, the "aggression" became palpable. They were now going to have to confront plantation society in a place that had been reserved for the North's family-farm/village society. The plantation and the family farm were antagonistic precisely because they differed in their land use, and the difference was irreconcilable. From the land use pattern came the differences in society: a yeomanry in a village setting of rough equality versus a slaveholding elite presiding over an impoverished mass of slaves and whites.[31] The swelling anger of the North was fed by the belief that the South had pledged itself in the past to observe the Missouri Compromise and now broke its promise.[32]

From this premise, it was a small step for northerners to think no region in the United States was safe from the desires of southern leaders to expand. Given the property doctrine in the Dred Scott decision, the fear that slavery would soon be a national institution was a logical, if highly unlikely, possibility. Moreover, the differences in cultural behavior became clear in "Bleeding Kansas," where, despite hyperbolic newspaper reporting, southern migrants—the "Border Ruffians"—openly relied upon threats of violence to win elections and give themselves political power. Perhaps, as some have claimed, Missourians in Kansas thought they were only seeking to preserve their honor and to acquire land that by geography belonged to them, but their actions showed how little they intended to observe the traditions of the northern village community.

Nicole Etcheson has mapped out precisely how the struggle between proslavery and antislavery people in Kansas between 1854 and 1860 was not about slaves themselves but about the freedoms of white settlers. The political effect of the Kansas conflict also requires comment. Kansas demonstrated what happened to northern Democratic loyalties when confronted directly with proslavery politics; they were tossed aside. Kansas Territory was a relentless machine that turned northern Democrats into Republicans. Kansas became an overwhelmingly Republican state by 1862.[33] In the cadence of events in the 1850s, the key one was the passage of the Kansas-Nebraska Act. That legislation convinced the rural North that the South was invading its exclusive domain.

Rural America and Congressional Districts

The nation's conflict between the economies of the northern family farm and the southern plantation can be illustrated by looking at the economic characteristics of congressional districts in the 1850s. What follows should not be looked upon as a replacement for existing voting studies. The many analyses of the realignment of the 1850s have established without question the importance of ethnic voting blocs, nativism, temperance, and religious divisions. What is presented here is an additional perspective illustrating an economic division that should be considered as part of the political upheaval.

First, a look at the northern congressional districts reveals the influence of the conflict between farm-village economy and industrialization. The published census of 1860 gives the number of farms for each county, the farm sizes by acreage category, and the manufacturing employees for each county. This information has some problems. The categories for small farms are numerous, but the category of 100 to 499 acres is unfortunate. If one takes from British studies the definition of a large farm being 250 acres or more, then the category of 100 to 499 mixes together different kinds of farms—but it is the number that has to be used. Second, the occupations listed for manufacturing include a number that are primary processing and artisanal, not at all industrial or factory production. Finally, it must be mentioned that the censuses of 1850 and 1860 never provided county totals for merchants, innkeepers, and transportation workers (state totals exist in the first volume on population). This transportation/service part of both the farm economy and the urban economy is not accounted for and thus is not represented in the analysis, even though this part of the economy was more important than manufacturing. With these flaws in mind, the counties can be used to depict the characteristics of congressional districts because most of the districts were defined by counties. This was not true for New England. For that reason, the New England states, except for Vermont, are dealt with separately and speculatively. Finally, California and Oregon are excluded because of their distance from the rest of the nation.

Northern congressional districts were divided into the categories of rural, balanced between agriculture and manufacturing, and industrial depending on the ratio of the number of farms to the number of manufacturing employees (combined total of male and female). When the number of farms was two times larger than the number of manufacturing employees, the congressional district was deemed rural; when neither the farms nor the manufacturing employees was twice the size of the other, the district was labeled balanced. The final category was industrial, in which the number of manufacturing workers was overwhelming. The results of compiling all the information in the counties that composed the congressional districts for the North are given in table 9.1.

The total number of free state congressional districts was 148; excluding California (2), Oregon (1), and New England (28) except for Vermont, there were 117 districts that could be characterized by using county information. In table 9.1, 75 of them were substantially rural. The numbers given are the averages of the number of farms in the congressional districts per state; for the rural category, the farming quality of these districts is overwhelming, as is the small size of the farm (3 to 99 acres). Most of these districts

TABLE 9.1 Northern Congressional Districts and Number of Farms and Manufacturing Employees: Rural, Balanced, and Industrial

PART A. RURAL CONGRESSIONAL DISTRICTS

State	*No. of Cong. Districts*	*Av. No. Farms 3–99 Acres*	*Av. No. Farms 100–499 Acres*	*Av. No. Farms 500+ Acres*	*Average No. Total Farms*	*Average Manufacturing No. (Male and Female)*
Ill.	9	10,643	5,002	129	15,774	2,486
Ind.	11	9,440	2,036	34	11,510	1,886
Iowa	2	21,266	4,904	26	26,196	2,838
Mich.	3	14,958	2,219	12	17,188	4,889
Minn.	2	8,948	332	1	9,282	1,048
Ohio	18	6,877	2,158	32	9,068	2,238
Wis.	3	17,293	2,455	24	19,772	4,850
N.Y.	15	6,855	2,191	11	9,057	2,989
Pa.	9	6,986	2,158	5	9,149	3,112
Vt.	3	6,423	3,828	34	10,285	3,499
Totals of Farms and Employees (not averages):						
Total	75	659,530	194,532	2,568	856,630	201,471

PART B. CONGRESSIONAL DISTRICTS NEARLY BALANCED BETWEEN FARMING AND MANUFACTURING

State	*No. of Cong. Districts*	*Av. No. Farms 3–99 Acres*	*Av. No. Farms 100–499 Acres*	*Av. No. Farms 500+ Acres*	*Average No. Total Farms*	*Average Manufacturing No. (Male and Female)*
Mich.	1	5,495	2,103	4	7,602	5,531
Ohio	1	4,676	1,392	17	6,085	4,911

N.J.	4	4,450	2,351	6	6,220	6,848
N.Y.	6	4,221	1,669	6	5,901	8,752
Pa.	7	4,744	1,624	2	6,371	5,428
Totals of Farms and Employees (not averages):						
Totals	19	86,509	31,934	118	118,565	119,729

PART C. INDUSTRIAL CONGRESSIONAL DISTRICTS

State	*No. of Congressional Districts*	*Total Manufacturing Employees (Male + Female)*
Ohio	2	30,268
N.J.	1	28,633
N.Y.	11	132,339
Pa.	9	151,304
Totals	23	342,458

NOTE: Av = Average of congressional districts.

SOURCES: Congressional districts taken from *Tribune Almanac* reports, 1853–60; supplemented by Martis, *Historical Atlas of United States Congressional Districts*. Totals for farms and manufacturing employees taken from *Census of 1860* 2: *Agriculture*, and *Census of 1860* 3: *Manufacturing*.

have three to four times as many farms as manufacturing employees. Moreover, if one uses one manufacturing person for every ten farms as a rule for the minimum number of village-support occupations (manufacturing) that farming required, then most of the manufacturing in these districts must have been local. The ratio of farmers and farm laborers to manufacturing people in fact bears a striking resemblance to David Hey's figure for seventeenth-century Myddle, England; for every eight farmers there was onc craftsman.[34] That ratio makes nearly two-thirds of the North's congressional districts resemble the pattern of English agriculture two and one-half centuries earlier. In short, these manufacturing occupations reflected the village economy, not modernizing industry.

There were 19 congressional districts that had a balance between farming and manufacturing. However, applying the one manufacturing employee for every ten farms rule to this group, the characteristic of this group tends to be rural. Lastly, there were 23 districts that were distinctly industrial (or, more likely, commercial-industrial), and they tended to be urban centers: Philadelphia, New York, Albany, Pittsburgh, and Cincinnati. Farm information was nearly nonexistent for these areas, and so none is given in table 9.1. The line yielding totals in the three divisions of table 9.1 contains the number of farms in each category and the total number of manufacturing occupations; they are not averages.

New England's division into rural, balanced, and industrial categories was estimated by looking at country totals, given in table 9.2. By making some guesses, I divided the congressional districts into: Maine, 4 rural, 2 balanced; New Hampshire, 2 rural, 1 balanced, 2 industrial; Connecticut, 1 balanced, 3 industrial; Rhode Island, 2 industrial; and Massachusetts, 1 balanced, 10 industrial.

Adding the speculative division of New England congressional districts in the economic categories with the division of the other northern states yields a total of 81 rural districts, 24 balanced, and 40 industrial. The districts that were balanced between rural and industrial would probably tilt more toward the farm-village economy than to the industrial economy, and so the numbers would finally imply 105 farm-village-economy congressional districts and 40 industrial ones—some 72 percent of northern congressmen came from the farm-village system. This merely reemphasizes how much rural life characterized the North and shaped federal policy. Individuals in Congress from these districts may have differed on the impact

TABLE 9.2 New England Counties Divided into Rural, Balanced, and Industrial

PART A. NEW ENGLAND RURAL COUNTIES

State	*Number of Counties*	*Total Number Farms*	*Total Manufacturing Employees (Male + Female)*
Maine	8	26,345	6,586
N.H.	4	11,636	3,484
Total	12	37,981	10,070

PART B. NEW ENGLAND COUNTIES BALANCED BETWEEN FARMS AND MANUFACTURING EMPLOYEES

State	*Number of Counties*	*Total Number Farms*	*Total Manufacturing Employees (Male + Female)*
Conn.	5	14,532	25,009
Maine	8	29,889	28,033
Mass.	3	6,061	6,832
R.I.	1	959	1,224
NH	4	12,820	9,755
Total	21	64,262	70,853

PART C. INDUSTRIAL NEW ENGLAND COUNTIES

State	*Number of Counties*	*Total Number Farms*	*Total Manufacturing Employees (Male + Female)*
Conn.	3	10,587	39,460
Mass.	11	29,495	210,589
N.H.	2	5,948	19,173
R.I.	4	4,315	30,980
Total	20	50,345	300,202

SOURCES: Same as table 9.1.

of internal improvements, national banks, western land policy, and the tariff on their constituents, but in the larger scheme of things the representatives of these districts spoke for the values of the family-farm/village economy. Their perspective on social interactions, economic undertakings, and political rights and duties had come from the family-farm/village economy. It could hardly be any wonder that they viewed plantation lords with vast suspicion and wanted no part of the plantation complex in their neighborhoods.

Southern congressional districts were divided into ones that seemed most to have the characteristics of plantation dominance, near plantation dominance, yeoman dominance, and industrial. They were further subdivided into the groupings of the Border South slave states (minus Delaware and Missouri) and the Deep South states (the original states of the Confederacy.) The rule I followed for making the separations was that, if 40 percent of the farms were in the combined categories of 100–499 acres and above 500 acres, then the district had plantation dominance. The likelihood was overwhelming that the acreage owned in plantations dwarfed that in yeoman farms (if the average plantation was about 2,000 acres, it was twenty to forty times larger than a yeoman farm of 50 to 100 acres.) The 100–499 acre category was added to the 500-plus acreage category because the total number of farms 500 acres or more was 21,501; the total number of slaveholders having twenty or more slaves was 37,662. Some of the planters had to be in the 100–499 acre category, and the assumption is that farms in this category tended to be large farms (greater than 250 acres) rather than yeoman-sized farms. If the combined total of the 100–499 and 500-plus acre category was greater than 30 percent of all farms but less than 40 percent, I put it in the division of near-planter dominance; and under 30 percent to the division of yeoman dominance. The industrial districts (and there were only 5 of them) were determined the same way as for the North; two or more times the number of manufacturing occupations than the number of farms.

The results are given in table 9.3 for the Deep South and table 9.4 for the Border South. There are several observations worth noting. First, the number of small farms was always depressed when there were a number of large plantations, which follows mathematically from the fact of a finite amount of acreage to be divided among farmers. In both the border and the Deep South, it is apparent that the large plantation was squeezing out the yeoman farmer. Moreover, one catches a glimpse at the process of consolidation by noting the nearly wholly yeoman character of the frontier states of Arkansas and Texas. Although plantations existed in these states, the overall characterization of them has to be a gigantic mass of small farms. Eventually, slavery would end up producing the large plantation and reducing the number of small freeholds; a comparison between the older states of Virginia, South Carolina, and Georgia with the frontier states of Texas and Arkansas provides a glimpse into the latter group's future had no emancipation intervened.

Second, the number of manufacturing occupations is small per congressional district compared to rural northern congressional districts, no matter if the southern district was planter or yeoman. Northern rural areas had more artisans by nearly a factor of two. Some of this disparity had to be due to slave mechanics, but some may have come from the lack of a village environment even among the yeomen. There were only 5 industrial districts; Baltimore had 2, New Orleans 1, Richmond 1, and Louisville 1.

Comparing North and South, the contrast becomes stark. The northern congressional districts are tilted to the small farm; the southern congressional districts toward the plantation. Overall in the South, given the rules established earlier, 35 congressional districts were plantation dominated, 18 nearly plantation dominated, and 23 yeoman dominated. Combining the plantation and near-plantation categories, the South had 53 congressional districts in which the plantation was the chief economic force, while the 23 yeoman districts were a clear minority (total of 81 districts, 53 had the plantation as the chief characteristic, or 65 percent of the total). In much the same proportions, the chief economic force in the North was the family farm and its attached village (105 congressional districts, 72 percent of the total). For the North, the industrial/mercantile/manufacturing congressional districts (40) were the economic minority. Thus battles in Congress can be seen as fights between the representatives of the family farm and those of the plantation, and the focus of the struggles was which economic form was going to rule the country.

The final piece of information to add to the political situation of the North was the determination of political party preference. To do this, an abbreviated table of the party winners in the congressional districts between 1842 and 1860 is given in table 9.5. The congressional districts are divided into the major categories of the rural districts, the balanced (between rural and manufacturing) districts, the industrial districts, and the New England

TABLE 9.3 Southern Farm Distribution by Congressional District and Total Manufacturing Employees

PART A. DEEP SOUTH CONGRESSIONAL DISTRICTS DOMINATED BY PLANTATIONS

State	*No. of Cong. Districts*	*Av. No. Farms 3–99 Acres*	*Av. No. Farms 100–499 Acres*	*Av. No. Farms 500+ Acres*	*Total Farms Average*	*Average Manufacturing Employees (Male and Female)*
Ala.	3	3,944	2,169	561	6,574	1,421
Ga.	5	2,595	2,422	616	5,633	1,614
La.	3	3,195	1,497	461	5,153	1,036
Miss.	2	3,171	1,926	512	5,610	971
S.C.	4	1,784	1,818	369	3,971	966
Totals of Farms and Employees (not averages):						
Totals	7	47,570	34,233	8,648	90,951	21,250

PART B. DEEP SOUTH CONGRESSIONAL DISTRICTS NEAR PLANTATION DOMINANCE

State	*No. of Cong. Districts*	*Av. No. Farms 3–99 Acres*	*Av. No. Farms 100–499 Acres*	*Av. No. Farms 500+ Acres*	*Total Farms Average*	*Average Manufacturing Employees (Male and Female)*
Ala.	1	5,733	2,498	374	8,608	652
Ga.	1	5,312	1,994	216	7,522	813
Miss.	3	5,174	2,217	356	7,745	845
S.C.	2	3,898	1,938	181	6,017	1,679
Totals of Farms and Employees (not averages):						
Totals	7	34,361	15,019	2,019	51,396	7,353

PART C. DEEP SOUTH CONGRESSIONAL DISTRICTS DOMINATED BY YEOMEN

State	*No. of Cong. Districts*	*Av. No. Farms 3–99 Acres*	*Av. No. Farms 100–499 Acres*	*Av. No. Farms 500+ Acres*	*Total Farms Average*	*Average Manufacturing Employees (Male and Female)*
Ala.	3	5,380	1,358	211	6,948	886
Fla.	1	4,676	1,432	288	6,396	2,454
Ga.	2	5,224	1,670	150	7,044	1,644
Texas	2	14,666	3,373	278	18,317	5,774
Totals of Farms and Employees (not averages):						
Totals	8	60,597	15,592	1,775	77,964	19,152

SOURCES: Same as table 9.1.

Table 9.4 Border South Congressional Districts Divided into Plantation, Near-Plantation, and Yeoman Dominance

PART A. BORDER SOUTH CONGRESSIONAL DISTRICTS HAVING PLANTATION DOMINANCE

State	*No. of Cong. Districts*	*Av. No. Farms 3–99 Acres*	*Av. No. Farms 100–499 Acres*	*Av. No. Farms 500+ Acres*	*Total Farms Average*	*Average Manufacturing Employees (Male and Female)*
Ky.	2	3,932	3,197	218	7,360	2,064
Md.	4	2,855	2,806	81	5,742	1,690
N.C.	1	4,489	2,633	276	7,376	1,176
Tenn.	2	3,470	2,323	200	5,993	1,672
Va.	9	2,880	2,766	316	5,962	2,180
Totals of Farms and Employees (not averages):						
Totals	18	56,634	49,790	4,313	110,713	32,029

PART B. BORDER SOUTH CONGRESSIONAL DISTRICTS HAVING NEAR PLANTATION DOMINANCE

State	*No. of Cong. Districts*	*Av. No. Farms 3–99 Acres*	*Av. No. Farms 100–499 Acres*	*Av. No. Farms 500+ Acres*	*Total Farms Average*	*Average Manufacturing Employees (Male and Female)*
Ky.	3	5,422	2,468	122	8,013	1,053
N.C.	4	4,665	2,292	234	7,196	2,246
Tenn.	4	4,884	2,260	112	7,256	1,419
Totals of Farms and Employees (not averages):						
Totals	11	54,462	25,615	1,715	81,848	12,142

PART C. BORDER SOUTH CONGRESSIONAL DISTRICTS HAVING YEOMAN DOMINANCE

State	*No. of Cong. Districts*	*Av. No. Farms 3–99 Acres*	*Av. No. Farms 100–499 Acres*	*Av. No. Farms 500+ Acres*	*Total Farms Average*	*Average Manufacturing Employees (Male and Female)*
Ark.	2	14,289	2,116	129	16,594	940
Ky.	3	7,515	1,897	77	9,468	1,226
N.C.	3	7,535	2,202	77	9,814	1,167
Tenn.	4	6,199	1,802	57	8,058	773
Va.	3	6,888	2,194	91	9,172	2,043
Totals of Farms and Employees (not averages):						
Total	15	119,183	30,319	1,341	150,786	18,277

INDUSTRIAL COUNTIES IN DEEP SOUTH AND BORDER SOUTH

State	*No. of Cong. Districts*	*No. Farms 3–99 Acres*[a]	*No. Farms 100–499 Acres*[a]	*No. Farms 500+ Acres*[a]	*Total Farms*[a]	*Manufacturing Employees*[a] *(Male and Female)*
Ky.	1	1,653	1,714	124	3,486	7,598
La.	1	365	42	42	456	5,635
Md.	2					21,842
Va.	1	1,995	2,252	385	4,633	9,862
Totals	5	3,920	3,010	551	8,575	44,637

[a] Not averages.

districts. In the table, the only categories for the election winners are Democrats or major opposition to the Democrats. There were often minor parties, fusions, groups, and others that do not properly fit under the categories of Democrat and major opposition, but for purposes of detailing the main development of party allegiance in the congressional districts by economic endeavor, these disparate elements, which can detract attention from the obvious pattern, are eliminated.

For the congressional elections between 1842 and 1851, the information about farms and manufacturing employees in the counties was taken from the U.S. Census of 1860, as the published census of 1850 does not provide the number of farms in a county. The assumption was that the essential nature of the counties (agriculture or manufacturing) stayed the same over the decade, which probably overestimates the extent of industrialization in the 1840s. Although New England congressional districts, save Massachusetts, could be figured for the 1840s by the use of counties, I decided to make New England a separate section so the figures from 1842 to 1851 would be consonant with the figures between 1852 and 1860.[35]

Table 9.5 casts an important light on the realignment of the 1850s. By putting the congressional districts in these categories of agricultural, balanced, and manufacturing, one can see a distinct pattern of economic influence in party victors. In the 1840s, the Democratic Party really did draw its strength from the non–New England rural North (or at least had an edge; one may emerge victorious in a congressional election by only a few votes, and the numbers given in table 9.5 tabulate only the victors). However, the table shows no clear division in party preference toward manufacturing; in some cases, the tilt is to the Whig Party. As well, the strength of the Democrats in the New England states deserves notice. The dip in the Democrats' fortunes in 1846–50 might suggest a response to the Wilmot Proviso and the battle over the admission of California, but it is complicated because of the Free Soil revolt in the state of New York, which divided the Democrats and enabled the Whigs to sweep the state, and also because of the Tariff of 1846, which had a decided impact on Pennsylvania. Therefore, no interpretation of that fluctuation is offered, save to emphasize that the pattern of the early 1840s was recovered in 1850–51.

The economic manifestation of the realignment of the 1850s is clear in the seismic change that occurs in the party victors starting in 1854, which was nothing less than a monstrous upheaval in northern party fortunes.

Table 9.5 demonstrates from a rural-industrial perspective that agriculture was the backbone of the Republican Party. In the northern rural congressional districts, the retreat from electing Democratic congressmen was obvious from 1852 to 1854 (the number fell from 48 to 10 in the non–New England North), and only mildly rebounded to 22 in 1856. The Democrats and their major opposition almost reversed themselves between 1852 and 1856; the Democrats had 48 seats in 1852, only 22 in 1856; the opposition in 1852 (Whigs) had 23 districts and grew to 49 (Republicans) in 1856. The districts balanced between farming and manufacturing showed much the same trend. New England remained a section with its own peculiarities, but what was remarkable was the nearly immediate demise of the Democrats in that region. Because of the lack of data on the economic composition of the districts, a firm conclusion cannot be offered, but the New England Democracy's fatal heart attack in 1854 had to occur in the previous Democratic strongholds of New Hampshire, Maine, and Connecticut—all, save Connecticut, having more of an agricultural basis than a manufacturing one.

The non–New England northern congressional *urban* districts were decidedly different. These areas too felt the upheaval of 1854 when the fortunes of the Democrats sank to remarkably low depths, but by 1856 the Democrats rebounded in northern urban-industrial districts and were stronger in that year than they had been in 1852. Some changes came after 1856, but the rural North became and remained the bedrock of the 1850s Republican Party.

Table 9.5 speaks volumes about the nature of the realignment of the 1850s on the axis of farming and manufacturing. The northern Democratic Party foolishly lost its rural supremacy and practically handed it over to the Republican Party, the foolishness of the northern Democrats being their role in passing the Kansas-Nebraska Act of 1854. The reversal of party fortunes in regard to agriculture is the outstanding feature of this realignment. Simply put, the Democrats once had an iron grasp on northern rural congressional districts. That grip was dissolved in 1854, and one can surmise that the acid performing the dissolution was the threat the family-farm economy felt from the potential of being wounded by an aggressive plantation complex, the threat taking substance in the repeal of the Missouri Compromise line. The family-farm districts then became the mainstays of the Republican Party. The final decision of the industrial districts was yet to be made.[36] The conclusion, however, is inescapable: the antislavery political

TABLE 9.5 Party Victors in Congressional Election Contest, 1842–1859, Categories by Economic Activity

District and Party	Congressional Election Years									
	1842–43	*1844–45*	*1846–47*	*1848–49*	*1850–51*	*1852–53*	*1854–55*	*1856–57*	*1858–59*	*1860*
RURAL DISTRICTS										
Democrats	43	45	32	35	42	48	10	22	19	20
Major Opposition	23	21	34	28	24	23	59[a]	49	43	55
Total Districts	66	66	68	70	70	73	73	73	73	75
BALANCED DISTRICTS										
Democrats	18	16	6	5	14	11	4	5	4	7
Major Opposition	8	9	20	19	10	7	9[a]	9	9	12
Total Districts	26	26	26	26	26	19	19	19	19	19
MANUFACTURING DISTRICTS										
Democrats	9	5	7	2	8	17	3	19	6	8
Major Opposition	7	5	7	12	8	6	11[20][b]	3	2	13
Total Districts	16	16	16	17	17	23	23	23	23	23
NEW ENGLAND										
Democrats	15	9	8	11	13	13	1	2	0	0
Major Opposition	12	16	18	12	9	12	20	22	25	17
Total Districts	27	27	27	27	27	26	26	26	26	18

[a] Combines Whigs and Republicans.

[b] Bracketed number is Republicans plus Union fusion and American groups.

NOTE: The total districts minus the number of Democratic and Major Opposition districts is the number of those belonging to other parties, generally Americans, North Americans, Free Soil, Anti-Rent, and variations of these. In 1844–45 there was one unfilled seat from New Hampshire. New state additions were Iowa in 1846, Wisconsin in 1848, and Minnesota in 1860. The total for New England for 1860 excludes results from the congressional elections of 1861 for New Hampshire, Connecticut, and Rhode Island.

surge in the North arose out of the family-farm/village economy, not from industrialization.

The trend of current historians is to emphasize the clash of South and North as a reflection of a rural South versus a market-driven capitalist North, a North that was quickly transforming itself into an industrial juggernaut. The interpretation has been around since the 1920s. But in fact the controversy between the North and South arose out of the differences and values lodged in the North's family-farm economy and the South's plantation economy. Slavery entered not as a humanitarian concern directly—one can say it did so indirectly—but because slavery made the plantation possible, and the plantation by its voracious appetite for land endangered the yeoman farm as well as the many artisanal occupations the family-farm/village economy rested upon. When the plantation system seemed to turn aggressive at the instigation of its congressional leaders, the northern family-farm community reacted to stop incursions into areas it felt reserved for it and to ensure that plantation slavery would never menace the northern family farm. This was the material clash between the sections. It originated within rural America and arose because of the stark differences between family farming and the plantation. Industrialization was irrelevant. The economic nature of the antebellum sectional conflict was not southern agrarianism versus northern industrialization; it was the southern plantation complex versus the northern family farm.

Epilogue

When in 1854 the people in the northern agricultural congressional districts resoundingly transferred their allegiances from the Democratic Party to an anti-Democratic coalition, which by 1858 was the Republican Party, they were not hoisting the banner of war. The path that took the nation to strife after November 1860 reflected a different set of issues than that of slavery's territorial expansion, and decisions by southern and northern political leaders precipitated the conflict. Rather, the northern farm-village complex emphatically declared in 1854 and thereafter that western territories that had been reserved for northern agriculturalists would not be turned over to the plantation complex. Behind that specific message was the general attitude that the days of expansion of the slave-based plantation system had ended and that the holders of slaves needed to acclimate themselves to an economic and social life that had distinct geographic boundaries. As testified in the speeches of their representatives and editors, the northern farm community saw the slave-plantation system as aggressive, and its aggressiveness would not stop with the territory of Kansas but would move to lands wherever slaveholders thought they could wring a profit; the wherever included the possibility of invading the northern states.

Historians have posited numerous reasons for the northern distaste for slavery. They have waxed at length on the demands of industrial capitalists to guide federal economic policy, the anger of northerners over southern control of national politics, the apprehension that slowly the laws of the nation were becoming slaveholding laws that were transforming the laws of northern states, the ideological drive of reformers to implement a free labor/capitalist regime over the entire nation, and the abolitionist insistence that African Americans be treated not as animals but as human beings. All these influences were operating on northern sensibilities. This work, however, stresses how the plantation system warred on the family farm, both in terms of land engrossment and in elimination of skilled trades. The northern farm community understood this feature of the southern plantation regime, and it mattered not what northern rural people thought about the morality of slavery or the powers of the central government. New Englanders and Upland Southerners in the North could vary on a vast number of social, political, and economic questions, but on the subject of the southern plantation they had one response: the plantation warred on their way of life, and it needed to be excluded from expanding beyond its realm. The people of the northern family-farm/village complex operated politically on the injury the plantation could inflict on them.

Southern congressional leaders and secessionists evidently did not understand the northern farm community as they made so little reference to it. Rather, they aimed their rhetorical barbs at abolitionists, antislavery religious leaders, political entrepreneurs, and especially at Massachusetts manufacturers. Somehow they seemed to have been captured by the tariff battles between 1820 and 1846 into thinking that industrialists were their natural enemies. Southern defenders found a ready response to industrialist claims of progress by insisting on the good condition of their slaves while exhibiting the impoverished condition of industrial laborers. As hostility to slavery extension mounted after 1854, southerners even jettisoned their economic analysis of the North in favor of believing northerners suffered an uncontrollable mental obsession with ending slavery and making Africans equal to white Americans, a total misreading of the northern mind. In all their attempts to grapple with the motivations of northern hostility to the expansion of slavery into the territories, they never focused on the northern family farmer and northern rural society—even though they could have seen the telltale signs of small farmers' hostility to the great plantations within their own states. In any event, southern political leaders were not about to allow any group of northerners to legislate an end to slavery's future expansion. The southern leadership's decision to secede stemmed from the leadership's conviction that the South could not trust northern political parties, elections, and congressional legislators to treat slavery as a legitimate national interest. The leadership might have waited to see if the North might reverse the decision of 1860 in later elections and permit southern expansion into land beneath the 36° 30′ line. By 1861, however, patience was not a Cotton South virtue.

A debate still festers in accounts of the postwar years about how the Civil War did or did not change the nation. Certainly the conflict transformed the South, but perhaps not in its land-ownership pattern. Many argue that the North was vastly altered and a "modern" America sprang into existence. However, looking at the northern family-farm/village complex, one concludes that stasis rather than change was its fate. The dimensions of the northern family farm and its associated villages stayed the same for decades after the Civil War, not altering until nearly 1950 (see tables 4.3 and 4.4.) The aggrandizement of northern farms into monster operations would await the 1960s, and the small rural town that served the family farm would not commence its precipitous decline until the large corporate farm appeared. In most ways, the Civil War had little influence on the North.[1]

In the late twentieth century, massive farms of hundreds of thousands of acres came to dominate American agriculture—although the family farm continues its existence, somewhat precariously—but that was not the fate of the British farm. The aristocracy lost its political power in the great constitutional battle of 1909–11 when Lloyd George produced a budget that increased taxes on land and inheritance; the House of Lords started a confrontation that led to its virtual elimination as a force in British legislation. The political power of the great landowners was falling apart. The final blow came after World War I. Possession of land had lost its charms, and a huge land sale occurred in Great Britain between 1919 and 1925; one-fourth of England's land, between six and eight million acres, was sold. Not all the great estates immediately collapsed, but many of the middling ones (three thousand to ten thousand acres) did. In place of the peers and gentry came a rising tide of owner-occupiers.[2] By 1980, owner-occupiers represented 70 percent of farm owners, and they possessed 60 percent of the arable land.[3] Although agriculturalists compose less than 2 percent of the present British population, it is still comforting to know that three and one-half centuries after the English Civil War, the yeoman came at last to be the chief owner of British soil.

APPENDIX A

Calculation of Acreage Owned by Slaveholders and Nonslaveholders

Methods: The Census of 1860 gave the number of slaveholders in each category but not the number of slaves within the categories, so that number had to be estimated. Using the figures from Joseph Karl Menn, *The Large Slaveholders of Louisiana, 1860,* the average was 95 slaves per large slaveholder for cotton planters having 50 and more slaves. These Louisiana large planters held on average 8.84 improved and 17.05 unimproved acres per slave. These acreages were then multiplied by the state totals for number of slaveholders in the holders of more than 20 slaves. The census gave the number of slaves owned by people in each category 1 to 9, so both the number of slaveholders and slaves were directly calculable. For the number of slaves in the categories 15–19 slaves, 20–29 slaves, 30–39 slaves, and 40–49 slaves, I used a guess at a near midpoint for the holding, and multiplied the number of slaveholders in the respective categories by 17, 24, 34, and 44 to get the number of slaves. The number of urban slaves (given in volume 1 of the census for each state; I did not use Missouri) was subtracted from the total number of slaves in the 1–9 slaveholding category to obtain the number of rural slaveholders. The number of nonslaveholders was derived from the employment totals given in the employment section of each state (*Census of 1860* 1: *Population*); I added the categories of farmer, farm laborer, and planter and subtracted from the sum the number of slaveholders in the state (given in *Census of 1860* 2: *Agriculture*, 224, 247).

I. Basic Facts

	Total Farms	*Improved Acres (millions)*	*Unimproved Acres (millions)*
Ala.	55,128	6.386	12.719
Ga.	62,003	8.063	18.588
Miss.	42,840	5.066	10.774
La.	17,328	2.707	6.501
Fla.	6,568	.654	2.266
S.C.	33,171	4.572	11.624
Tex.	42,891	2.651	22.693
Va.	92,605	11.438	19.679
Md.	25,494	3.002	1.833
N.C.	75,203	6.517	17.246
Ky.	90,814	7.694	11.519
Tenn.	82,368	6.795	13.874
Ark.	39,004	1.983	7.590
Mo.	92,792	6.241	13.738

Source: *Census of 1860* 2: *Agriculture.*

II. Adjusting for Urban Slaves

State	Col. 1	Col. 2	Col. 3	Col. 4	Col. 5	Col. 6
	Number of Slaves in 1–9 Category	*Urban Slaves*	*Col. 1 – Col. 2*	*Col. 3 / Col. 1*	*No. of Slaveholders in 1–9 Category*	*Col. 4 × Col. 5*
Ala.	80,365	20,311	60,054	.747	21,793	16,279
Ga.	102,582	35,555	67,027	.653	27,191	17,756
Miss.	73,367	9,721	63,646	.868	19,559	16,977
La.	52,303	27,072	25,231	.482	14,886	7,175
Fla.	12,399	6,872	5,527	.446	3,368	1,502
S.C.	63,590	16,815	46,775	.736	16,199	11,922
Tex.	56,520	11,668	44,852	.794	16,292	12,936
Va.	132,907	36,417	96,490	.726	37,487	27,213
Md.	34,672	4,415	30,257	.873	11,203	9,780
N.C.	90,554	20,811	69,743	.770	24,520	18,880
Ky.	110,375	21,435	88,940	.805	31,819	25,614
Tenn.	101,572	10,522	91,050	.896	28,987	25,972
Ark.	20,434	3,709	16,725	.818	6,827	47,926
Mo.	69,112			say .800	21,380	17,104

III. Estimated Number of Slaves Held by Planters

	No. of Slaves in Category of Holders of 1–9 Slaves	*No. of Slaves in Category of Holders of 10–19 Slaves*	*No. of Slaves in Category of Holders of 20–49 Slaves*	*No. of Slaves in Category of Holders of >50 Slaves*
Ala.	80,365	81,692	132,146	141,157
Ga.	102,582	104,475	140,956	105,185
Miss.	73,367	69,611	126,930	166,356
La.	52,307	44,449	71,586	162,844
Fla.	12,399	13,457	18,162	17,717
S.C.	63,590	71,900	109,914	156,582
Tex.	56,520	47,006	53,589	25,466
Va.	132,907	117,636	143,096	103,800
Md.	34,672	23,341	21,338	8,840
N.C.	90,554	83,021	97,884	47,400
Ky.	110,375	71,152	40,520	3,436
Tenn.	101,572	75,096	3,310	25,725
Ark.	16,725	23,888	30,374	30,640
Mo.	69,112	2,400	502	38

IV: Estimation of Slaveholders and Nonslaveholders

	Number of Slaveholders in the Categories of Holders of:						
	1–9 Slaves[a]	*10–19 Slaves*	*20–49 Slaves*	*>50 Slaves*	*Estimate of Total Rural Slaveholders*	*Total Farmers, Planters, Farm Laborers*	*Total Nonslaveholders*
Ala.	16,279	5,876	4,344	1,687	28,216	67,743	39,527
Ga.	17,756	7,530	5,049	1,314	31,649	90,083	58,434
Miss.	16,977	5,489	4,200	1,675	28,361	57,318	28,957
La.	7,175	3,222	2,349	1,576	14,322	26,952	12,630
Fla.	1,502	976	603	205	3,286	12,831	9,545
S.C.	11,922	5,210	3,646	1,675	22,424	43,970	21,546
Tex.	12,936	3,423	1,827	142	18,222	58,365	40,143
Va.	27,213	8,774	4,917	860	41,764	125,511	83,747
Md.	9,780	1,718	757	112	12,360	41,234	28,874
N.C.	18,880	6,073	3,321	744	29,018	104,438	75,420
Ky.	25,614	5,271	1,485	70	32,350	147,571	115,221
Tenn.	25,972	5,523	2,550	382	34,432	129,903	95,471
Ark.	6,827	1,777	1,018	184	9,967	57,263	47,296
Mo.	17,104	2,400	502	38	20,049	164,389	144,340

[a] Number of urban slaves subtracted from the categories of holders of 1–9 slaves.

SOURCES: Slaveholders in *Census of 1860* 2: *Agriculture,* 247; occupations in *Census of 1860* 1: *Population.*

V. Estimated Acreage by Slaveholder Category

	PLANTER CLASS, SLAVEHOLDERS POSSESSING 20 OR MORE SLAVES		
	Number of Slaves	*Improved Acres No. Slaves × 8.84 Acres*	*Unimproved Acres No. Slaves × 17.05 Acres*
Ala.	264,521	2,338,366	4,510,083
Ga.	246,006	2,179,693	4,194,402
Miss.	260,409	2,302,016	4,439,973
La.	206,096	1,821,889	3,513,937
Fla.	33,947	300,091	578,796
S.C.	243,874	2,155,846	4,158,052
Tex.	75,768	669,789	1,291,844
Va.	199,708	1,765,419	3,405,021
Md.	28,853	255,061	491,944
N.C.	150,384	1,329,395	2,564,047
Ky.	40,130	354,749	684,217
Tenn.	97,610	862,872	1,664,251
Ark.	60,914	536,652	1,037,974
Mo.	15,778	139,478	269,015

	SMALL PLANTERS, SLAVEHOLDERS POSSESSING 10–19 SLAVES		
	Number of Slaveholders	*Improved Acres No. Slaveholders × 200 Acres*	*Unimproved Acres No. Slaveholders × 400 Acres*
Ala.	5,906	1,181,200	2,362,400
Ga.	7,530	1,506,000	3,012,000
Miss.	5,489	1,077,800	2,195,600
La.	3,222	644,400	1,288,800
Fla.	976	195,200	390,400
S.C.	5,210	1,042,000	2,084,000
Tex.	3,423	684,600	1,369,200
Va.	8,774	1,754,800	3,509,600
Md.	1,718	343,600	687,200
N.C.	6,073	1,214,600	2,429,200
Ky.	5,271	1,054,200	2,108,400
Tenn.	5,223	1,044,600	2,089,200
Ark.	1,777	355,400	710,800
Mo.	2,400	80,000	960,000

(*continued*)

	SMALL SLAVEHOLDERS, POSSESSING 1–9 SLAVES		
	Number of Slaveholders	*Improved Acres No. Slaveholders × 100 Acres*	*Unimproved Acres No. Slaveholders × 200 Acres*
Ala.	16,279	1,627,900	3,255,800
Ga.	17,756	1,775,600	3,551,200
Miss.	16,977	1,697,700	3,395,400
La.	7,175	717,500	1,435,000
Fla.	1,502	150,200	300,400
S.C.	11,922	1,192,200	2,384,400
Tex.	12,936	1,293,600	2,587,200
Va.	27,213	2,721,300	5,442,600
Md.	9,780	978,000	1,956,000
N.C.	18,880	1,888,000	3,776,000
Ky.	25,614	2,561,400	5,122,800
Tenn.	25,972	2,597,200	5,194,400
Ark.	6,827	682,700	1,365,400
Mo.	17,104	1,710,400	3,420,800

APPENDIX B

Counties Used for Comparison

The following counties were used to compare number of farms, farm sizes, and manufacturing employees in chapter 8:

Illinois—Bureau, Fulton, Kane, LaSalle, Sangamon, Vermillion, Will
Indiana—Marion, Montgomery, Randolph, Ripley, Shelby
Maine—Aroostook, Franklin, Lincoln, Oxford, Somerset, Waldo
Michigan—Calhoun, Hillsdale, Jackson, Oakland, Washtenaw
New Hampshire—Carroll, Coos, Grafton, Sullivan
New Jersey—Hunterdon, Monmouth, Salem, Somerset, Sussex
New York—Allegheny, Cattaragas, Chautauqua, Cortland, Franklin, Niagara, Ontario, Otsego, Schoharie, Suffolk
Ohio—Brown, Huron, Knox, Lorain, Monroe, Richland, Seneca, Washington
Pennsylvania—Bradford, Butler, Crawford, Indiana, Mercer, Tioga, Venango
Vermont—Addison, Caledonia, Franklin, Orange, Orleans, Washington, Windham
Wisconsin—Dane, Dodge, Fond du Lac, Jefferson, Rock

For the slaveholding counties:

Georgia—Baker, Burke, Crawford, Greene, Houston, Jasper, Monroe
Kentucky—Boyle, Clark, Garrard, Lincoln, Oldham, Scott, Woodford
Mississippi—Amite, Claiborne, Hinds, Lowndes, Noxubes, Yazoo
Missouri—Audrain, Boone, Cass, Howard, Mississippi, New Madrid, Saline
South Carolina— Abbeville, Beaufort, Chester, Colleton, Fairfield, Greenville, Richland, Williamsburg
Virginia—Amelia, Brunswick, Caroline, Goochland, King and Queens, Loudon, Louisa, Nottoway, Prince Edward, Spotsylvania, Sussex

The other industrial counties that had more than 10,000 manufacturing employees include:

Connecticut—New Haven, 18,778
Massachusetts—Bristol, 25,971; Hampden, 11,028; Norfolk, 17,724; Plymouth, 11,402; Suffolk, 19,093
New York—Kings, 12,758; Renssalaer, 13,555

For some purposes I used nonslaveholding southern counties, where the number of free was two or three times the number of slaves (in the case of Missouri, everything was geared toward nonslaveholding); those counties were:

Alabama—Blount, Calhoun, Cherokee, Coffee, DeKalb, Randolph
Georgia—Carroll, Cherokee, Gwinnett, Walker
Kentucky—Brecken, Campbell, Pendleton, Pulaski
North Carolina—Catawba, Randolph, Surry
Virginia—Grayson, Marion, Marshall, Monongalia.

NOTES

Preface

1. Regarding these controversies, some of the more dominant authors and their works are Eric Foner, *Free Soil, Free Labor, Free Men;* John Ashworth, *Slavery, Capitalism, and Politics in the Antebellum Republic;* Amy Dru Stanley, *From Bondage to Contract;* and David Brion Davis, *Problem of Slavery in Western Culture.*

2. For example, Kulikoff, *Agrarian Origins of American Capitalism;* Henretta, *Origins of American Capitalism;* Larson, *Market Revolution in America;* Clark, *Roots of Rural Capitalism.*

3. The earliest exposition of the industrial North versus the plantation South was in the work of Charles A. Beard and Mary Beard, *Rise of American Civilization.* However, it pitted northern small farmers and industrial capitalists against planter capitalists. Eugene Genovese refined this formulation to a capitalist North versus the patriarchal plantation, dropping the "capitalism" from the Beards' description of the planters (Genovese, *Political Economy of Slavery*).

4. California and Oregon were excluded primarily because of their distance from the states where the antagonism raged hottest. California posed other problems because of the extensive mining in the state, most of it of a primitive quality, which confounds typical economic-political analysis. Oregon would have fit well into the study because of its agrarian nature in 1860, but its inclusion might be considered by some an attempt to bias the basic North-South sectionalism in the eastern and midwestern states.

5. Besides those listed in footnote 2, above, one could also add Gray, *Yankee West,* and Rugh, *Our Common Country.*

6. For examples of those trumpeting the industrialism of the North, see Craven, *Civil War in the Making,* chap. 1; Grant, *North over South,* 44–48; Moore, *Social Origins of Dictatorship and Democracy,* 122, 127, 132–33, 141; Morrison, *Slavery and the American West,* 277–78; Levine, *Half Slave and Half Free,* 56–58; Ford, *Origins of Southern Radicalism,* 351–54; and Genovese, *Political Economy of Slavery,* chap. 8.

1. The English Civil War and the Birth of the Free Labor Ideology

1. The debates are laid out in useful detail in Russell, *Causes of the English Civil War,* 218–19; Aston and Philpin, eds., *Brenner Debate*; Finlayson, *Historians, Puritanism, and the English Revolution,* chaps. 1–3; Hill, "Bourgeois Revolution," 109–62. For political blundering in England, see Edwards, *Making of the Modern English State,* 225–26, 258–60, 284–322.

2. Robert C. Allen made this assessment in 1991 (*Enclosure and the Yeoman,* 304–11).

3. Allen, *Enclosure and the Yeomen,* 58–63; Van Bath and Slicher, *Agrarian History of Western Europe,* 35, 42; Smith, *This Realm of England,* 43–78, 85–96, 164–71; Edwards, *Making of the Modern English State,* 5–8, 85–119.

4. Clarkson, *Pre-industrial Economy in England,* 62; Wrightson, *Earthly Necessities,* 141–43; Coleman, *Economy of England,* 43.

5. Numbers from Overton, *Agricultural Revolution in England,* 168; Edwards, *Making of the Modern English State,* 8; Coleman, *Economy of England,* 12. The basic outline of the agricultural change comes from Allen, *Enclosure and the Yeoman,* 19, 21, 47–75.

6. Edwards, *Making of the Modern English State,* 5–6, 8; Coleman, *Economy of England,* 12.

7. Wrightson, *Earthly Necessities,* 140–46; Clarkson, *Pre-industrial Economy in England,* 9–10; Chambers, *Population, Economy and Society in Pre-Industrial England,* 140.

8. Chambers, *Population, Economy, and Society in Pre-Industrial England,* 141; Coward, *Stuart Age,* 15–18; Coleman, *Economy of England,* 71.

9. On the role of the Tudor courts, see Allen, *Enclosure and the Yeoman,* 56, 61, 66, 71–72; Edwards, *Making of the Modern English State,* 85–119; and Woolrych, *Britain in Revolution,* 19–22.

10. Numbers from Smith, *This Realm of England,* 49; Wrightson, *English Society,* 24; Coleman, *Economy of England,* 43–44. On the ordering of ranks, see Edwards, *Making of the Modern English State,* 11.

11. Wrightson, *English Society,* 24; Beckett, *Aristocracy in England,* 33, 39; Heal and Holmes, *Gentry in England and Wales,* 12; Smith, *This Realm of England,* 49. The numbers of the esquires and gentlemen are for 1689 from Gregory King.

12. Smith, *This Realm of England,* 70, 74, 164–71; Adamson, "Aristocracy and Their Mental World," 174–76. On the last castle built, see Hoskins, *Making of the English Landscape,* 126.

13. Beckett, *Aristocracy in England,* quote on 5, 9–11; Stone, *Crisis of the Aristocracy,* 21–22; Wrightson, *English Society,* 17–18, 20–23.

14. On estate management, see Mingay, *English Landed Society in the Eighteenth Century,* 59–60. On capitalist attitudes of the peers and gentry, Stone, *Crisis of the Aristocracy,* 59; Beckett, *Aristocracy in England,* 4–6.

15. Mingay, *English Landed Society in the Eighteenth Century,* 19–24.

16. Philip Edwards has written that, except for the Marxists, most historians do not think class is an operable concept in early modern Europe (*Making of the Modern English State,* 10).

17. Quote from Coleman, *Economy of England,* 7.

18. Stone, *Crisis of the Aristocracy,* 39–40.

19. Peacham in Hill and Dell, *Good Old Cause,* 56.

20. Quote from Thomas Smith, 1565, in Zagorin, *Court and the Country,* 28. See Heal and Holmes, *Gentry in England and Wales,* 31; Stone and Stone, *Open Elite,* 412–15.

21. Wrightson, *Earthly Necessities,* 274–75; Porter, *English Society in the Eighteenth Century,* 15–17; Hecht, *Domestic Servant Class in Eighteenth-Century England,* 72–79.

22. Adamson, "Aristocracy and Their Mental World," 178; Beckett, *Aristocracy in England,* 9; Stone, *Crisis of the Aristocracy,* quote on 8, 212; Hecht, *Domestic Servant Class in Eighteenth-Century England,* 72–74; Russell, *Crisis of Parliaments,* 173; quote from Levy, *Large and Small Holdings,* 91.

23. Coward, *Stuart Age,* 41; Stone, *Crisis of the Aristocracy,* chap. 10; Ashton, *English Civil War,* 22, 37–40; Woolrych, *Britain in Revolution,* 22–24.

24. Beckett, *Aristocracy in England,* 4–10; Porter, *English Society in the Eighteenth Century,* 16.

25. Kalm, *Kalm's Account of His Visit to England,* 54, 81–82, 144.

26. The middle for seventeenth-century England was postulated by Louis B. Wright in *Middle-Class Culture in Elizabethan England.* The interpretation of Christopher Hill (*Change and Continuity in Seventeenth-Century England,* 100) was attacked by Hexter (*Reappraisals in History,* chap. 5).

27. Coward, *Stuart Age,* 37–39. See the discussion for the eighteenth century in Hunt, *Middling Sort,* 2–15, 18–20.

28. Wrightson, *English Society,* 142–43, 162, 173–74; Hill, *Change and Continuity in Seventeenth-Century England,* 183–87, 202–4, 222; Underdown, *Revel, Riot, and Rebellion,* 24–27, 33–34.

29. Quotes, in order, Wright, *Middle-Class Culture in Elizabethan England,* 3; Taylor, *American Colonies,* 161–62; Wright, *Middle-Class Culture in Elizabethan England,* 182, 183.

30. Hill, *Century of Revolution,* 12.

31. Calculated from table 1 in Coleman, *Economy of England,* 6. Christopher Hill said the yeomen were only 5.5 percent of the population; Underdown said 85 percent of the population were beneath the rank of yeoman (Hill, *Century of Revolution,* 13; Underdown, *Revel, Riot, and Rebellion,* 27).

32. Spufford, *Contrasting Communities,* 38, 52–53; Levy, *Large and Small Holdings,* 5; Wrightson, *English Society,* 29–33, 134–35.

33. Levy, *Large and Small Holdings,* 4; Schmidt, *Yeoman in Tudor and Stuart England,* quote on 24; see Campbell, *English Yeoman under Elizabeth and the Early Stuarts.*

34. On Puritan bourgeois values, see Hill, *Change and Continuity in Seventeenth-Century England,* 94–100.

35. Woolrych, *Britain in Revolution,* 41–43; Craig, "The Growth of English Puritanism," 43–44; Edwards, *Making of the Modern English State,* 225–32; Coward, *Stuart Age,* 71–73.

36. See Hunt, *Puritan Moment,* 44–53, 79–84, 312; Como, "Radical Puritanism," 241–44. For historiography on Puritanism, see John Coffey, "Puritan Legacies," 327–45.

37. Weber, *Protestant Ethic and the Spirit of Capitalism,* 79–128, 155–66; Tawney, *Religion and the Rise of Capitalism,* 111–20, 202–44; Hunt, *Puritan Moment,* 124–28; Mackintosh, "New Sweden, Natives, and Nature," 10–11.

38. For example, Tawney, *Religion and the Rise of Capitalism,* 114–15; Hill, *Society and Puritanism in Pre-Revolutionary England,* 259–88; James, *Social Problems and Policy during the Puritan Revolution,* 18–19.

39. Baxter qtd. in Cliffe, *Puritan Gentry,* 9–10; Haller, *Rise of Puritanism,* 12; quote of Edward Billy in Reay, "Quakerism and Society," 150; Como, "Radical Puritanism," 254; Heal and Holmes, *Gentry in England and Wales,* 30–33; for tensions, see Manning, "Puritanism and Democracy," 140; overall, see Cliffe, *Puritan Gentry,* 10–14, 25–26; Wright, *Middle-Class Culture in Elizabethan England,* 183; and James, *Social Problems and Policy during the Puritan Revolution,* 8–9, 13.

40. Hunt, *Puritan Moment,* 136–37; Tawney, *Religion and the Rise of Capitalism,* 114. The idea that Puritans aimed their criticisms at the poor was the theme of Christopher Hill (*Society and Puritanism in Pre-Revolutionary England,* chaps. 4, 6, 7). For criticism of Hill, see Finlayson, *Historians, Puritans, and the English Revolution,* 68–70.

41. Quotes of William Martyn and Sir Simonds D'Ewes in Cliffe, *Puritan Gentry,* 105, 106; Wright, *Middle-Class Culture in Elizabethan England,* 176.

42. Emphasis on labor is explicit in Tawney, *Religion and the Rise of Capitalism,* 115, 241–42; Wright, *Middle-Class Culture in Elizabethan England,* 170–71, 185–86; James, *Social Problems and Policy during the Puritan Revolution,* 17; Weber, *Protestant Ethic and the Spirit of Capitalism,*

158–61; Cliffe, *Puritan Gentry,* 104–5. Quote of Hill, *Society and Puritanism in Pre-Revolutionary England,* 140, application to lower class, 139, 283–87.

43. Quotes from Wright, *Middle-Class Culture in Elizabethan England,* 179; see Hunt, *Puritan Moment,* 127–28; Reay, "Quakerism and Society," 151.

44. For the historiography of the English Civil War, consult Edwards, *Making of the Modern English State,* 258–60, 276–77, 284–85, 310–11; Richardson, *Debate on the English Revolution,* chaps. 6–9; Ashton, *English Civil War,* 72–106; Zagorin, *Rebels and Rulers* 1: 5–13. The Marxist version is best found in Hill, "Bourgeois Revolution," 109–62.

45. Hill, *Intellectual Origins of the English Revolution,* 5.

46. This narrative is a condensation of Smith, *This Realm of England,* 192–225; Woolrych, *Britain in Revolution,* chap. 3; Ashley, *Oliver Cromwell and the Puritan Revolution,* 42–54; Morrill, *Revolt of the Provinces,* 30–37; Edwards, *Making of the Modern English State,* 310–22.

47. Underdown, *Pride's Purge,* 1–3, 175–201; Smith, *This Realm of England,* 247–65.

48. On the Levellers and the Putney debates, see Brailsford, *Levellers and the English Revolution,* 280–83; Sanderson, *But the People's Creatures,* 102–24; Aylmer, *Levellers in the English Revolution,* 28–29, 46; Edwards, *Making of the Modern English State,* 351–58, 361; Manning, *English People and the English Civil War,* 263–78.

49. Ashley, *Oliver Cromwell and the Puritan Revolution,* 72, 78; Aylmer, *Rebellion or Revolution,* quote on 86, 137; on natural rights, see Brailsford, *Levellers and the English Revolution,* 462; Hill, *Century of Revolution,* 153. On connection to the United States, see Kennedy, *English Revolution,* 73.

50. Brailsford, *Levellers and the English Revolution,* 13–15; Woolrych, *Britain in Revolution,* 355–56, 389–90. On the question of suffrage, see Macpherson, *Political Theory of Possessive Individualism,* 156–58; Sanderson, *But the People's Creatures,* 115–17; Kenyon, *Civil Wars of England,* 172–73.

51. Firth, *House of Lords during the Civil War,* 158; James, *Social Problems and Policy during the Puritan Revolution,* 100.

52. There is only one lengthy study on the abolition of the House of Lords: Firth, *House of Lords during the Civil War,* 150–59, 170–71, 205–13.

53. James, *Social Problems and Policy during the Puritan Revolution,* 24–26.

54. Fox qtd. in Reay, "Quakerism and Society," 150–51; John Lilburne and Chelmsford radicals in Manning, *English People and the English Civil War,* 282; pamphleteer in Firth, *House of Lords during the Civil War,* 161; Winstanley, in James, *Social Problems and Policy during the Puritan Revolution,* 27; Lilburne in Manning, "Levellers and Religion," 72. For aristocratic and gentry fears, see Reay, "Quakerism and Society," 162.

55. James, *Social Problems and Policy during the Puritan Revolution* , 26; Manning, *English People and the English Civil War,* 292–95; Brailsford, *Levellers and the English Revolution,* 11, 417, 423, 437–42; Schenck, *Concern for Justice in the Puritan Revolution,* 49–51.

56. Brailsford, *Levellers and the English Revolution,* 444–45; on the Diggers, see Schenck, *Concern for Social Justice in the Puritan Revolution,* 67–71, 105; Como, "Radical Puritanism," 254–55.

57. Information on Harrington from J. C. Davis, "Equality in an Unequal Commonwealth," 229–42; Macpherson, *Political Theory of Possessive Individualism,* 163–86; Morgan, *Inventing the People,* 85–86, 155–56.

58. For Locke as the philosopher of capitalism, see Macpherson, *Political Theory of Possessive Individualism,* 200–257, description of the labor theory of value as a bourgeois idea for appro-

priation, 220–29. Richard Ashcraft describes Locke as a partial radical who came to accept the republican critique of aristocracy (*Revolutionary Politics and Locke's Two Treatises of Government,* 75–84, 165; Locke attacks gentility, 242–63; his critique of landowners, 264–69, his belief in manual labor and laboring classes, 269–73; his anti-aristocracy, 281–83). Aylmer has disputed Ashcraft's ideas of Locke as Leveller ("Locke No Leveller," 304–22).

59. For summaries of the results of the English Revolution and long-term patterns in English life, see Mingay, *Gentry,* 69–77; Hill, *Century of Revolution,* 150–63; Kenyon, *Civil Wars of England,* 245–46; Aylmer, *Rebellion or Revolution,* 195; Zagorin, *Rebels and Rulers* 2: 184–86; and Allen, *Enclosure and the Yeoman,* 304–11.

60. The historiography has been summarized by Taylor in *American Colonies.*

61. Estimates from Kulikoff, *From British Peasant to American Colonial Farmer,* 65; HStatUS76, 1168.

62. Rutman, *Winthrop's Boston,* 80–88; Manegold, *Ten Hills Farm,* 26. On the land-distribution process and acreages given out, see Bidwell and Falconer, *History of Agriculture in the Northern United States,* 37, 49–54; Lockridge, *New England Town,* 9–19; and Russell, *Long Deep Furrow,* 68–77. On village quality of New England, McCusker and Menard, *Economy of British North America,* 92–94.

63. Lockridge, *New England Town,* 19. On hierarchy and status, see Brewer, *By Birth or Consent,* 345; Foster, *Their Solitary Way,* 6–8, 15–26.

64. A few have stressed the rise of a gentry in New England and their control of resources, but they were insignificant compared to the English magnates (Rutman, *Winthrop's Boston,* 87–88; Innes, *Labor in a New Land,* xvii–xxi, 3, 18, 29, 34, 38–50, and rest of book).

65. Hall, *A Reforming People,* 63.

66. See David Galenson, "Settlement and Growth of the Colonies," 136–37.

67. The essential work on the New England farm is Vickers, *Farmers and Fisherman,* 33–82; Henretta, *Evolution of American Society,* 25–35.

68. Foster, *Their Solitary Way,* 104–14; Anderson, *New England's Generation,* 36; Innes, *Creating the Commonwealth,* 76–77, 88; Morgan, "Puritan Ethic and the American Revolution," 3–6.

69. Stone, *Crisis of the Aristocracy,* 39–40. The competency theme in New England is striking among Anderson, *New England's Generation,* White quote on 124, and chap. 4; Vickers, *Farmers and Fishermen,* chap. 1; and Vickers, "The Northern Colonies," 210–11. Episode of Winthrop's manual labor in Innes, *Creating the Commonwealth,* 88.

70. Episode in Firth, *House of Lords during the Civil War,* 60–61; Hall, *A Reforming People,* 37–38; Foster, *Their Solitary Way,* 38–39.

71. See Fischer, *Albion's Seed,* 218–24; Fischer and Kelly, *Bound Away,* 78–103.

72. For a debate over leisure and labor in the southern colonies, see Morgan, "Puritan Ethic and the American Revolution," 3–43; C. Vann Woodward, "The Southern Ethic in a Puritan World," 343–70.

73. The anti-aristocratic tendencies of the New England Puritans can be found in Greene, *Pursuits of Happiness,* 24, and Hall, *A Reforming People,* xi–xii, 95.

74. See comments of David D. Hall in comparing egalitarianism in Massachusetts and England (*A Reforming People,* 4–5, 50).

75. Gladstone, "Kin Beyond the Sea" (1878), in *Representative Essays* (1885), 355, 356, 357. In terms of agricultural practices, Robert C. Allen agrees: "These ideas were put into practice in America rather than England" (*Enclosure and the Yeoman,* 306).

2. The Standard: British Agriculture, 1660–1870

1. De Vries, *Economy of Europe in an Age of Crisis,* 73–82; Zagorin, *Rebels and Rulers* 2: 184–86; Hill, *Century of Revolution,* 150–61. On strict settlement, see John Habakkuk, *Marriage, Debt, and the Estates System,* 1–51.

2. These debates are discussed in Overton, *Agricultural Revolution in England,* 4–8, 194–205; Mingay, "Size of Farms in the Eighteenth Century," 469, 470–71; and Allen, *Enclosure and the Yeoman,* 2–12, 15, 18–21.

3. Wrigley, *Poverty, Progress, and Population,* 45, 51; Allen, "Agriculture during the Industrial Revolution," 96; Parry, "Transport and Trade Routes," 169.

4. Porter, *English Society in the Eighteenth Century,* 38–44; Deane, *First Industrial Revolution,* 54–57; Mingay, *English Landed Society in the Eighteenth Century,* 98–103; Coleman, *Economy of England,* 66, 137–38, 156–72.

5. For industrialization, see Mathias, *First Industrial Nation.* For relations between agriculture and manufacturing, see Jones, *Agriculture and the Industrial Revolution,* 216–21.

6. Overton, *Agricultural Revolution in England,* 134, 138, 140–43, 178–89; Chambers and Mingay, *Agricultural Revolution,* 127–36; Van Vugt, *Britain to America,* 22–26; Hobsbawm and Rude, *Captain Swing,* 16–19, and entirety; Jones, *Agriculture and the Industrial Revolution,* 215–25.

7. Allen, "Agricultural Change in Industrial Revolution," 106–8.

8. Ibid., 103; Mingay, *English Landed Society in the Eighteenth Century,* 12; Overton, *Agricultural Revolution in England,* 88, 121–25.

9. Jones, "Agriculture and Economic Growth in England," 2–5; Allen, *Enclosure and the Yeoman,* 149, 170, 204–6; On wheat yields per acre, see Orwin and Whetham, *History of British Agriculture,* 381; Allen, "Growth of Labor Productivity in Early Modern English Agriculture," 117; Wrigley, *Poverty, Progress, and Population,* 50–51.

10. Example of the older treatment is in Fordham, *Short History of English Rural Life,* 114–18. For Norfolk system of rotation, see Overton, *Agricultural Revolution in England,* 1–4.

11. For example, Allen, *Enclosure and the Yeoman,* 83–95; Thompson, "Landownership and Economic Growth in England in the Eighteenth Century," 42; Overton, *Agricultural Revolution in England,* 63; Wrightson, *Earthly Necessities,* 270–71.

12. Figures from Grigg, *English Agriculture,* 12; GBCensus1851, 72; Mathias, *First Industrial Nation,* 311; Mingay, *English Landed Society in the Eighteenth Century,* 23. On the social changes, Levy, *Large and Small Holdings,* 4–6; Allen, "Growth of Labor Productivity in Early Modern English Agriculture," 120–26; Beckett, "Decline of the Small Landowner in England and Wales," 92; Thompson, "Landownership and Economic Growth in England in the Eighteenth Century," 42–43.

13. Mingay, *English Landed Society in the Eighteenth Century,* 3–6, 10–11, 32–34.

14. Chambers and Mingay, *Agricultural Revolution,* 77–89; Allen, "Agriculture during the Industrial Revolution," 97–99; Overton, *Agricultural Revolution in England,* 154–55. The classic treatment of the enclosure movement is Hammond and Hammond, *Village Labourer* (1911).

15. Malcolmson, *Life and Labour in England,* 24, 137–52; Chambers and Mingay, *Agricultural Revolution,* 96–97; Beckett, "Decline of the Small Landowner in England and Wales," 89–111; Mingay, "Size of Farms in the Eighteenth Century," 471–72, 480–87; Overton, *Agricultural Revolution in England,* 19–22.

16. Levy, *Large and Small Holdings,* 28–31.

17. On the effect of mortgage financing, see Allen, *Enclosure and the Yeoman,* 15, 102–3; on the land lust, Plumb, *England in the Eighteenth Century,* 17–18.

18. Mingay, *Gentry,* 82–93; Thompson, *English Landed Society in the Nineteenth Century,* chap. 5; Grigg, *English Agriculture,* 140–47.

19. Mingay, *English Landed Society in the Eighteenth Century,* 6, 21–22; Cannon, *Aristocratic Century,* 13–14.

20. The ratio of $5 for £1 can be found in Webb, *A Manual for Emigrants* (1849), 35; and Bird, *Englishwoman in America* (1856), 86. Figures are from Porter, *English Society in the Eighteenth Century,* 59, 66; Colman, *European Life and Manners* 1: 108; Badeau, *Aristocracy in England,* 230–31; "Wealth of the British Aristocracy," New York *Saturday Evening Post,* October 4, 1856, 4. A short description of Henry Colman is in Demaree, *American Agricultural Press,* 109.

21. Porter, *English Society in the Eighteenth Century,* 60; Plumb, *Men and Centuries,* 69–76; Plumb, *England in the Eighteenth Century,* 18–19.

22. Kalm, *Kalm's Account of His Visit to England,* 128–29. On forests, hedges, and lumber, see Garnier, *History of the English Landed Interest,* 39–65; Hoskins, *Making of the English Landscape,* 150–57; the 5 percent figure is given by Overton, *Agricultural Revolution in England,* 90.

23. Olmsted, *Walks and Talks of an American Farmer in England* (1859), quote on 58, 128; Kalm, *Kalm's Account of his Visit to England,* quote on 54, 37, 160; Washington Irving, "Rural Life in England," from *Voyage and Other English Essays from the Sketch Book* (1891), 13–31.

24. Colman qtd. in "European Agriculture and Rural Economy," *Cultivator* (Boston), vol. 1 (June 1844): 178; Olmsted, *Walks and Talks of an American Farmer in England,* 162; Irving, "Rural Life in England," 17; beauty descriptions in "The Aristocracy of England," in *Courier and Enquirer,* qtd. in *Albion* 8 (February 10, 1849): 67.

25. Spring, *English Landed Estate in the Nineteenth Century,* 3–4, 14, 53, 97–112; Thompson, *English Landed Society in the Nineteenth Century,* chap. 6; Mingay, "Eighteenth-Century Land Steward," 3–27; Beckett, *Aristocracy in England,* 139–56.

26. Spring, *English Landed Estate in the Nineteenth Century,* appendix 1, p. 183; Colman, *European Life and Manners* 1: 110; compare Thompson, *English Landed Society in the Nineteenth Century,* chap. 6, with Chandler, *Visible Hand.*

27. Colman, *European Agriculture and Rural Economy* 1: 34; Mathias, *First Industrial Nation,* 54; Beckett, *Aristocracy in England,* 186; Chambers and Mingay, *Agricultural Revolution,* 93; Allen, *Enclosure and the Yeoman,* 86–87.

28. Beckett, "The Debate over Farm Sizes in Eighteenth and Nineteenth Century England," 308.

29. Allen, *Enclosure and the Yeoman,* 15, 86, 93, 246; Overton, *Agricultural Revolution in England,* 121–25.

30. Mingay, "Size of Farms in the Eighteenth Century," 471–72; Overton, *Agricultural Revolution in England,* 173; Daunton, *Progress and Poverty,* 39–49.

31. Allen, *Enclosure and the Yeoman,* 17, 114, 118; Beckett, "Agricultural Landownership and Estate Management," 745; B. A. Holderness, "Prices, Productivity, and Output," 179–80.

32. Mingay, *Gentry,* 82; Garnier, *History of English Landed Interest,* 517.

33. Kim, *Landlord and Tenant in Colonial New York,* x. See historiographical discussion in Atack, "Agricultural Ladder Revisited," 1–7.

34. Porter, *English Society in the Eighteenth Century,* 68, 70.

35. Colman, *European Agriculture and Rural Economy* 1: 38, 110; "Life of an English Nobleman," *Farmer and Gardener* 4 (January 9, 1838): 291; Mingay, "Size of Farms in the Eighteenth

Century," 479; Mingay, *English Landed Society in the Eighteenth Century,* 253–54; a description of rents and prices contained in "British Farming Capital," *Genesee Farmer,* qtd. in *Tennessee Farmer* 1 (October 1835): 173.

36. Beckett, *Aristocracy in England,* 183–85, 187–90; Overton, *Agricultural Revolution in England,* 184–85; Heath, *English Peasantry* (1874), 203–6, 213–15.

37. Colman, *European Agriculture and Rural Economy* 1: 38, 39.

38. Mathias, *First Industrial Nation,* quote on 57; Chambers and Mingay, *Agricultural Revolution,* 96–99; Porter, *English Society in the Eighteenth Century,* 84–94; Kussmaul, *Servants in Husbandry in Early Modern England,* number on 3, 6–27.

39. GBCensus1851, 72; Mathias, *First Industrial Nation,* table 7, p. 239.

40. Chambers and Mingay, *Agricultural Revolution,* 103; Mingay, "The Farmer," 810–39; Edward Higgs, "Occupational Censuses and the Agricultural Workforce in Victorian England and Wales," table 1, pp. 709–10; Wrigley, *Poverty, Progress, and Population,* table 4.12, p. 124.

41. Mathias, *First Industrial Nation,* 197; Chambers and Mingay, *Agricultural Revolution,* 134–37; Peter Dewey, "Farm Labour," 839–54; Armstrong, "Labour I," 700–716; Colman, *European Agriculture and Rural Economy* 1: 44–45. Per year earnings from Armstrong, "Labour I," 716.

42. Brodrick, *English Land and English Landlords,* 232; Mingay, *English Landed Society in the Eighteenth Century,* 239–40. On the ladder, see Mingay, "The Farmer," 79–83; Dewey, "Farm Labour," 812–39; Grigg, *English Agriculture,* 143–44. On mobility, Orwin and Whetham, *History of British Agriculture,* 68, 82–84.

43. The best, lengthy description of the English gang system is in Thirsk, *English Peasant Farming,* 217–19, 268–70; quote of Armstrong, "Labour, I," 685–66, and generally, 680–92; Colman, *European Agriculture and Rural Economy* 1: 51, 55–58; Jones, *Agriculture and the Industrial Revolution,* 216. Walking seven miles in Gregg, *Social and Economic History of Britain,* 357; use of wagons to haul workers ten miles or more in Colman, *European Agriculture and Rural Economy* 1: 56.

44. Colman, *European Agriculture and Rural Economy* 1: 66; Orwin and Whetham, *History of British Agriculture,* 68, 79–82; Chambers and Mingay, *Agricultural Revolution,* 101–3, 138–41.

45. "The English Peasantry," *Boston Review* 2 (July 1862): 411; Badeau, *Aristocracy in England,* 234.

46. Gregg, *Social and Economic History of Britain,* 353; Badeau, *Aristocracy in England,* 234; Olmsted, *Walks and Talks of an American Farmer in England,* 238–39, 279; Colman, *European Agriculture and Rural Economy* 1: 41.

47. Badeau, *Aristocracy in England,* 234; Olmsted, *Walks and Talks of an American Farmer in England,* 40; "Agriculture and Rural Economy of Europe," *Cultivator* 4 (August 1847): 251; Brodrick, *English Land and English Landlords,* 238.

48. Levy, *Large and Small Holdings,* 36–37; Hasback, *History of the English Agricultural Labourer,* 69; Jones, *Agriculture and the Industrial Revolution,* quote on 222; Olmsted, *Walks and Talks of an American Farmer in England,* quotes on 241, 250.

49. Quote from Hall, *From Liverpool to St. Louis* (1870), 121–22; Olmsted, *Walks and Talks of an American Farmer in England,* 236–39; Brodrick, *English Land and English Landlords,* 216; Mingay, *English Landed Society in the Eighteenth Century,* 247.

50. Clapham, *Economic History of Modern Britain,* 272–76, 284–85; Orwin and Whetham, *History of British Agriculture,* 90, 207.

51. Report of James Howard, M.P., in "Free Laborers in Europe," *Massachusetts Ploughman and New England Journal of Agriculture* 29 (March 19, 1870): 2; see also [Lewis Cass,] "Republi-

canism vs. Monarchy," *Boston Cultivator* 2 (October 17, 1840): 4; Joseph A. Wright, "Agriculture in Europe: Letter from Governor Wright," *Indiana Farmer* 7 (October 1858): 210; Colman, *Agriculture and Rural Economy of France, Belgium, Holland, and Switzerland* (1848), 38–40.

52. Beckett, *Aristocracy in England,* 449–50; Labaree, *Conservatism in Early American History,* 1–3; Doyle, *Aristocracy and Its Enemies,* 93–94, 281–86; Guttmann, *Conservative Tradition in America,* 6–10.

53. Cannon, *Aristocratic Century,* 10–19, 132–35. On social ideals and practices, Clarkson, *Preindustrial Economy in England,* 36–38; Wrightson, *Earthly Necessities,* 70–72, 276; Stone, *Crisis of the Aristocracy,* 7, 8, 49, 178–91; Thompson, *English Landed Society in the Nineteenth Century,* 65–66, 82–102, 136–45.

54. The description was pulled together from Wrightson, *English Society,* 19; Clark, *English Society,* 76–82, 89–95, 103; Doyle, *Aristocracy and Its Enemies,* 37–49; Turberville, *House of Lords in the Age of Reform,* 18–19, 416–17; Sanderson, *But the People's Creatures,* 1, 8, 40, 53; Cannon, *Aristocratic Century,* 148–54, 170; Adamson, "The Aristocracy and Their Mental World," 175.

55. "The Condition of England," *Massachusetts Ploughman* 4 (April 12, 1845): 1; Habakkuk, "English Landownership," 8–10; Beckett, *Aristocracy in England,* 198–201; Colman, *European Agriculture and Rural Economy* 1: 34, 36–37; Garnier, *History of the English Landed Interest,* 494–97, 499–505. On local government, David Eastwood, *Governing Rural England,* chaps. 1, 2.

56. Porter, *English Society in the Eighteenth Century,* 2; Wrightson, *English Society,* 37. See Jones, *Agriculture and the Industrial Revolution,* 222; Mingay, *English Landed Society in the Eighteenth Century,* 11–12.

3. English Husbandry on Another Continent: North American Farming, 1607–1815

1. The land-labor ratio is well known among historians: Morgan, *American Slavery, American Freedom,* 218; Huston, *Securing the Fruits of Labor,* 50–53.

2. Main, *Social Structure of Revolutionary America,* 8–24; Taylor, *Liberty Men and Great Proprietors,* 13–14, 32–45, 226–29; Roth, *Democratic Dilemma,* 126; Vickers, *Farmers and Fisherman,* chap. 5.

3. Bidwell and Falconer, *History of Agriculture in the Northern United States,* 61; Nash, *Quakers and Politics,* 21–24, 59, 273–74.

4. Using map 32 in Hawke, *Colonial Experience,* 483, I guess the counties in 1850 that composed the southeast corner of Pennsylvania to be Philadelphia, York, Lancaster, Chester, Delaware, Adams, Cumberland, Montgomery, Bucks, Berks, Lebanon, and Lehigh. By the *Census of 1850,* they had 2.9 million improved acres and 0.7 million unimproved. On inequality, see Nash, *Quakers and Politics,* 324–27; Clemens and Simler, "Rural Labor and the Farm Household in Chester County, Pennsylvania," 111–26; Soltow and Keller, "Rural Pennsylvania in 1800," 24–46.

5. Hedrick, *History of Agriculture in the State of New York,* 40–51, 57–60; Bidwell and Falconer, *History of Agriculture in the Northern United States,* 73–74; Kim, *Landlord and Tenant in Colonial New York,* x, 17; Humphrey, "Conflicting Independence," 163; an instructive map of the holdings is map 21 in Hawke, *Colonial Experience,* 225.

6. Mancall, *Valley of Opportunity,* 72–97. The counties are Albany, Columbia, Dutchess, Putnam, Westchester, Suffolk, Queens, New York, Nassau, Kings, Organe, Ulster, Greene, Rensselaer, and Washington; the number does not include the counties containing the Mohawk Valley.

7. Nugent, *Cavaliers and Pioneers* 1: 1634–40; Walsh, *Motives of Honor, Pleasure, and Profit*, 16, 131–37; Menard, "From Servant to Freeholder," 40–64; Carr and Menard, "Immigration and Opportunity," 212–33. The current stress on yeoman farmers rather than plantations is a marked departure from Morgan's interpretation in *American Slavery, American Freedom*.

8. Morgan, *American Slavery, American Freedom*, 106, 158–84; Kulikoff, *Tobacco and Slaves*, 5–7.

9. Fischer, *Albion's Seed*, 218–224; Fischer and Kelly, *Bound Away*, 32–53; Parent, *Foul Means*, 27–30; Morgan, *American Slavery, American Freedom*, 187, 219–21.

10. On the speculation impulse, see Parent, *Foul Means*, 28; Morgan, *American Slavery, American Freedom*, 219; Clemens, *Atlantic Economy and Colonial Maryland's Eastern Shore*, 23–24, 51; Mitchell, *Commercialism and Frontier*, 79–80.

11. Morgan, *American Slavery, American Freedom*, 223; Robinson, *Mother Earth*, 46; see Russell Menard, "Economic and Social Development of the South," 270–72; Clemens, *Atlantic Economy and Colonial Maryland's Eastern Shore*, 76–82.

12. Fischer and Kelly, *Bound Away*, 83–92, 154–55, map p. 32; Voorhis, "Land Grant Policy of Colonial Virginia," 155; Walsh, *Motives of Honor, Pleasure, and Profit*, 369–70.

13. On the arrival and practice of gentility, see Rozbicki, "Curse of Provincialism," 727–28; Bushman, *Refinement of America*, xii–xv; Jan Lewis, *Pursuit of Happiness*, chap. 1; Burnard, *Creole Gentlemen*, 214–28. On entail and primogeniture, Brewer, "Entailing Aristocracy," 307–46, esp. 317–22, 337–42; and Walsh, *Motives of Honor, Pleasure, and Profit*, 343, 402. On aristocratic qualities of the Virginia gentry, see Breen, *Tobacco Culture*, 9, 32–36; Isaac, *Transformation of Virginia*, 21, 26–29, 36, 42, 70–71, 90–93; Fischer, *Albion's Seed*, 383–88.

14. Greene, "Society, Ideology, and Politics," 21; Greene, *Pursuits of Happiness*, 98; Land, "Economic Behavior in a Planting Society," 475–80; Carlton and Coclanis, *South and the Nation*, 6.

15. Gray, *History of Agriculture in the Southern United States* 1: 406. On planter income, I have been guided by Menard, "Economic and Social Development of the South," 274; Clemens, *Atlantic Economy and Colonial Maryland's Eastern Shore*, 161; Clemens, "Operation of an Eighteenth-Century Chesapeake Tobacco Plantation," 525; Russo, "A Model Planter: Edward Lloyd of Maryland," 70; Walsh, *Motives of Honor, Pleasure, and Profit*, 249, 253–54; Burnard, *Creole Gentlemen*, 48. Jackson Turner Main offers these estimates of income during the 1770s: a Connecticut farmer £28 to £54; Massachusetts farmer £8; New York tenant £9 to £11; Pennsylvania farmer £25; for a 289 acre Maryland farm, £27 (*Social Structure of Revolutionary America*, 104–6).

16. In Prince County, Maryland, in 1800, the average size of a large planter farm was 3,661 acres; planters, 1,163 acres; large yeomen, 461 acres; small yeomen (probably tenants), 153 acres; and small holders 8 acres (Sarson, "Yeoman Farmers in a Planter's Republic," tables 10, 11, 12, pp. 90–92). Many argue that the large plantation starts at about 500 acres; the average in South Carolina was 871 acres (Edelson, *Plantation Enterprise in Colonial South Carolina*, 113; Coclanis, *Shadow of a Dream*, 70).

17. Morgan, *American Slavery, American Freedom*, 319–81.

18. A favorable view of New York tenants is in Kim, *Landlord and Tenant in Colonial New York*, x, 138, 170–83, 219, 256–78; number of lessees in Humphrey, "Conflicting Independence," 163.

19. Rent charges from Kim, "A New Look at the Great Landlords of Eighteenth-Century New York," 608, 600–604, 611–14. For conflict between tenants and landlords, see Humphrey, "Conflicting Independence," 159–82; Countryman, *A People in Revolution*, 46–48; Summerhill, *Harvest of Dissent*, 2–3.

20. Lemon, *Best Poor Man's Country,* chaps. 1, 4. For growing inequality, see Nash, *Quakers and Politics,* 324–27; Clemens and Simler, "Rural Labor and the Farm Household in Chester County, Pennsylvania," 111–26. On the rise of nonlandowning farmers, Soltow and Keller, "Rural Pennsylvania in 1800," 25–47. For tenancy as social mobility, Simler, "Tenancy in Colonial Pennsylvania," 550–63. On the expansion into lands inhabited by Native Americans, Preston, "Squatters, Indians, Proprietary Government, and Land in the Susquehanna Valley," 182–98; Moyer, "'Real' Indians, 'White' Indians, and the Contest for the Wyoming Valley," 221–28.

21. Mancall, *Valley of Opportunity,* 76–77, 98–105; Taylor, *William Cooper's Town,* 45–52, 71–94; Summerhill, *Harvest of Dissent,* 12–23. On land companies, see Nettels, *Emergence of a National Economy,* 152–54; Livermore, *Early American Land Companies,* 82–122, 134–62; and Wyckoff, *Developer's Frontier,* 13–16.

22. Lorena Walsh noted that tenants were used when slaves were unavailable and that tenancy was no way for a landlord to get rich (*Motives of Honor, Pleasure, and Profit,* 476–78, 572). On the economics of indentured servitude, see Galenson, *White Servitude in Colonial America.* On the formation of racial views, see Wood, *Origins of American Slavery.* Earnings given in Clemens, *Atlantic Economy and Colonial Maryland's Eastern Shore,* 161. For an overall assessment of North American tenancy, see Stiverson, *Poverty in a Land of Plenty,* 137–38.

23. Wood, *Origins of American Slavery.*

24. The need for cheap labor to make the plantation feasible was noted by Fox-Genovese, *Within the Plantation Household,* 88. Contrary to Gavin Wright, I maintain slavery was fundamentally a low-wage labor system; see Wright, *Slavery and American Economic Development,* 71, and chap. 2, notes 25, 26.

25. Numbers in McCusker and Menard, *Economy of British America,* 136; see also fig. 4 in Fogel, *Without Consent or Contract,* 33; Walsh, *Motives of Honor, Pleasure, and Profit,* 137.

26. Morgan, *Slave Counterpoint,* 176–78, 191; Wright, *Slavery and American Economic Development,* 86–88; Carville Earle, "Beyond the Appalachians," 167, 178–79; Walsh, "Slave Life, Slave Society, and Tobacco Production in the Tidewater Chesapeake," 181–85; Clemens, *Atlantic Economy and Colonial Maryland's Eastern Shore,* 2–24.

27. Stiverson, *Poverty in a Land of Plenty,* 86; Hofstra, *A Separate Place,* 16; Papenfuse, "Planter Behavior and Economic Opportunity in a Staple Economy," 301; Clemens, *Atlantic Economy and Colonial Maryland's Eastern Shore,* 145–47; Tillson, *Gentry and Common Folk,* 12; Kulikoff, "American Revolution, Capitalism, and the Formation of the Yeoman Classes," 81–97.

28. On the colonial struggle between free workers and slaves, see Lewis, *Artisans in the North Carolina Backcountry,* 2–3; Russo, "Fewnesse of Handicraftsmen," 170–79; Russo, *Free Workers in a Plantation Economy,* 96, 121–28, 371–73; Daniels, "Getting His [or Her] Livelyhood," 125–61; Morgan, *Slave Counterpoint,* 204–5, 212–41; Russo, "Self-Sufficiency and Local Exchange," 413, 419–20.

29. Mancall, *Valley of Opportunity,* 89–95; Taylor, *Divided Ground,* 89.

30. Russell, *Long Deep Furrow,* 177–78, 274–75; Taylor, *Liberty Men and Great Proprietors,* 62; Billington, *Westward Expansion,* 245–51; Kim, *Landlord and Tenant in Colonial New York,* 347–83; Mancall, *Valley of Opportunity,* 163, 169–73, 180–94; Horsman, *Frontier in the Formative Years,* 21–69.

31. A superb study of the clash between the tribes and the Virginians in Kentucky is Aron, *How the West Was Lost,* chaps. 1–3; see also Friend, *Kentucke's Frontiers,* chaps. 1–4. On the land grab by Virginians, see Eslinger, *Citizens of Zion,* 63–73; Dunaway, *First American Frontier,* 51–71.

32. Taylor, *Liberty Men and Great Proprietors,* 9. For population pressure generally, see Hors-

man, *Frontier in the Formative Years,* 2–4; Clark, *Social Change in America,* 85–88, 90; Billington, *Westward Expansion,* 156.

33. Henretta, *Evolution of American Society,* 9, 12; Johnson, *Early American Republic,* 74; Klein, *Population History of the United States,* 43–44, 66, 77–78; Lockridge, "Land, Population, and the Evolution of New England Society," 62–79.

34. For soil exhaustion and/or the lack of available land pushing farmers to the West, see Kennedy, *Mr. Jefferson's Lost Cause,* 14, 21–25; Yarbrough, *Economic Aspects of Slavery in Relation to Southern and Southwestern Migration,* 1–3; Craven, *Soil Exhaustion as a Factor in the Agricultural History of Virginia and Maryland,* 11, 56–57. On planter greed, Dunaway, *First American Frontier,* 51–59. On escaping planter aristocracy, Barnhart, *Valley of Democracy,* 4–6.

35. On the Virginian outward push, see especially Fischer and Kelly, *Bound Away,* 81–85, 137–41, 172, and maps, pp. 138–39. On migration because of "land scarcity" and lack of opportunity, see Carr and Menard, "Immigration and Opportunity," 235–38; Kennedy, *Mr. Jefferson's Lost Cause,* 14–15, 21; Parent, *Foul Means,* 37–39; Kulikoff, *Tobacco and Slaves,* 76–77, 92; Morgan, *American Slavery, American Freedom,* 223.

36. On the racial stereotyping of the Indian, see Griffin, *American Leviathan,* 3–16, 166–80.

37. One writer stated that people in New England did not feel the "pinch" for land until 1819: "Remarks on the Agriculture of Massachusetts by One of the Trustees," *Massachusetts Agricultural Repository* 5 (July 1, 1819): 317.

38. Hurt, *Ohio Frontier,* 156–66, 181–83; Clark, *Social Change in America,* 85–93, 142–45; Billington, *Westward Expansion,* 156, 239–46, 248–57; Taylor, *Liberty Men and Great Proprietors,* 1–6.

39. Size of family in Clark, *Social Change in America,* 85. For the typology given to farms by number of acres, see table 2.4, above.

40. Average acreage in Illinois and Indiana farms found in HStatUS76, 461.

41. Lebergott, "Demand for Land," 182. A full accounting of landownership by farm sizes does not exist for Virginia. For the great planters, see Main, "Distribution of Property in Post-Revolutionary Virginia," 241–58; Main, "The One Hundred," 354–84; Robinson, *Mother Earth,* 37–45; Voorhis, "Land Grant Policy of Colonial Virginia," 155–57, 191–203.

42. Aron, *How the West Was Lost,* 89–95; Barnhart, *Valley of Democracy,* 6; Hammond, *Slavery, Freedom, and Expansion in the Early American West,* chap. 3; Cayton, *Frontier Indiana,* 57, 187–92; Guasco, "Southern Yeomen and Slavery in Illinois," 9–17; Friend, *Kentucke's Frontiers,* 214–15.

4. Dimensions of Antebellum Northern Farming

1. De Tocqueville, *Democracy in America* (1840), vol. 2: 222; Chevalier, *Society, Manners and Politics in the United States* (1839), 398; Baxter, *America and the Americans* (1855), 211.

2. Hugh Tremenheere, *Notes on Public Subjects, made during a Tour in the United States and Canada* (1852), 34; Tocqueville, *Democracy in America* 1: 3.

3. GBCensus1851, 72; Mathias, *First Industrial Nation,* table 7, p. 239; for U.S. northern states, occupation totals at the end of each state's statistics in *Census of 1860* 1: *Population.*

4. On British landownership, see chapter 2, herein; on American northern tenancy, see Atack and Bateman, *To Their Own Soil,* 110–11; Bogue, *From Prairie to Corn Belt,* 56–57; Gates, *Landlords and Tenants on the Prairie Frontier,* 12. For tenancy figures from 1900 onward, see HStat76, Series K 109–53, p. 465.

5. *Census of 1860* 2: *Agriculture,* 221. *De Bow's Review* said in 1858 that South Carolina had 1,200 farms greater than 1,000 acres: "Agriculture: Statistics of American Agriculture," *De Bow's Review* 24 (May 1858): 439.

6. Based on computing the compound interest rate of a number of farms in 1850 of 0.873 million and in 1900 of 2.198 million; rural population in 1850, 10.452 million, to 17.628 million in 1900 (HStatUS76, pp. 24–37); number of farms from table 4.4. On the quality of northern farming, see Clark, "Agrarian Context of American Capitalist Development," 13–38.

7. Quote in Hall, *Travels in Canada and the United States* (1818), 35; Bernhard, Duke of Saxe-Weimar Eisenach, *Travels through North America* (1828), vol. 1: 61; Lyell, *Travels in North America* (1845), vol. 1: 70; Huston, *Land and Reform,* chaps. 3, 4, 9, 10; Summerhill, *Harvest of Dissent,* chaps. 3, 4; Brooks, *Frontier Settlement and Market Revolution,* chap. 6.

8. Bidwell and Falconer, *History of Agriculture in the Northern United States,* 449; Barclay, *Agricultural Tour in the United States and Upper Canada* (1842), 36–47.

9. Gates, *Landlords and Tenants on the Prairie Frontier,* 7, 48–57, 114–25, 136–38, 200–207; "A Large Farm," *Cultivator* 11 (September 1863): 292; Russell, *North America* (1857), 113, 114, 124; Ferguson, *America by River and Rail* (1856), 377, 378; Socolofsky, *Landlord William Scully,* xi–xii, 1–12, 22–29, 64–77; Carlson, *Illinois Military Tract,* 40–63.

10. Russell, *North America,* 13–15; Mellen, *Book of the United States* (1839), 340; Lyell, *Second Visit to the United States* (1849), vol. 1: 65.

11. Atack and Bateman, *To Their Own Soil,* table 7.2, pp. 112–13.

12. For example, see summation after December 31 in Oliver B. Bennett Diary, 1859, WesResHS; Ledger Books of Joseph Williams, 1854–63, summary at end of folder 3, container 1, Joseph F. Williams Papers, WesResHS.

13. Conservator, "The Different Advantages of Large and Small Farms Considered," *Farmer's Register* 4 (January 1837): 564; N. Reed, "Large vs. Small Farms," *Cultivator* 8 (March 1860): 81; Allen, "Division of Agricultural Labor," *American Agriculturist* 6 (June 1846): 181–83.

14. A Small Farmer, "Are Large or Small Farms Most Profitable," *Maine Farmer* 18 (August 22, 1850): 1; "Small Farms," *Maine Farmer* 22 (May 18, 1854): 1; JNS, "Large vs. Small Farms," *Cultivator* 9 (February 1861): 60; Thornton, *A Plea for Peasant Proprietors* (1848), 1–40; Morris, *Ten Acres Enough* (1864).

15. Barclay, *Agricultural Tour in the United States and Upper Canada,* viii–xi, 9–10; Kingdom, *America and the British Colonies* (1820), 58; Buel, *Farmer's Companion* (1842), 17, 23, 63–64, 112–22; Donahue, *Great Meadow,* 57, 70; Stoll, *Larding the Lean Earth,* 14–32, 52–58. Sometime in the 1830s and 1840s, however, U.S. husbandmen began to pay more attention to fertilizing by using plaster, gypsum, clover, and grasses to reinvigorate the soil; Americans did not use as much animal manure as did the English. See Russell, *North America,* 24, 29–33; Cunynghame, *Glimpse at the Great Western Republic,* 97.

16. XYZ, "Profit of Farming East and West, The Rock River Country," *Valley Farmer* 3 (February 1851): 57; Morris, *Ten Acres Enough,* 250–53; Gaylord and Tucker, *American Husbandry* (1840), vol. 2: 39–43; "Large Crops vs. Large Farms," *Cultivator* 8 (January 1860): 13.

17. L. A. Hine, "Liberty—Equality—Fraternity," *Ohio Cultivator* 13 (May 15, 1857): 146; J. Gould, "Small Farms," *New England Farmer and Horticultural Register,* 19 (January 13, 1841): 221.

18. Colman, *Agriculture and Rural Economy of France, Belgium, Holland, and Switzerland,* 33–37; quote in "Notes from France," *Michigan Farmer* 2 (February 1, 1853): 36. On the debate over large and small farms, see Allen, *Enclosure and the Yeoman,* 10–12; Friedlander and Oser, *Economic History of Modern Europe,* 205–16; Hale, *Great Illusion,* 36–44.

19. See comment of Gates, *Landlords and Tenants on the Prairie Frontier,* 12.

20. Atack and Bateman, *To Their Own Soil,* 298; Danhof, *Change in Agriculture,* chap. 1. A summary of the debate is in Parkerson, *Agricultural Transition in New York State,* 55–59; and Clark, *Roots of Rural Capitalism,* 62–76.

21. See the summary in Larson, *Market Revolution in America.*

22. Danhof, *Change in Agriculture,* 14–20, 31–33; Fletcher, *Pennsylvania Agriculture and Country Life* 2: 1–5, 30–33, 98–99; Gates, *Farmers Age,* 158–67, 122–23, 224; Hudson, *Making the Corn Belt,* 3–7, 10–13, 63–65; McMurry, *Transforming Rural Life,* 73–90, 123–25; Wilson, *Hill Country of Northern New England,* 34–57.

23. Clark, *Grain Trade in the Old Northwest*; Danhof, *Change in Agriculture,* 31–37; Meinig, *Shaping of America* 2: 423–26; on the regional trade patterns, Gray, *Hidden Civil War,* 15–21; Egnal, *Clash of Extremes,* chap. 4.

24. Parkerson, *Agricultural Transition in New York State,* xi, 7–8. On the general prosperity of farmers, see Atack and Bateman, *To Their Own Soil,* 95, 247–66.

25. Hurt, *American Agriculture,* 147–56; Danhof, *Change in Agriculture,* chaps. 7, 8, 9; Craig, *To Sow One More Acre,* 29–35; McClelland, *Sowing Modernity,* x, 207–20.

26. See Pred, *Urban Growth and City-Systems in the United States.*

27. Cooper, *Some Information Respecting America* (1794), 35, 37; Baily, *Journal of a Tour in Unsettled Parts of North America* (1856), 241–43; Aron, *How the West Was Lost,* chap. 3.

28. Nettels, *Emergence of a National Economy,* 149–55; Horsman, *Frontier in the Formative Years,* 24–25, 27–30, 36, 42; McGrath, *Yazoo*; Gates, *Farmers Age,* 41–49.

29. Pulszky and Pulszky, *Red, White, Black* (1853), vol. 2: 15–17.

30. Horsman, *Frontier in the Formative Years,* 32–33, 73; Pattison, *Beginning of American Rectangular Land Survey System,* 37–57; Martis, "Geographical Dimensions of a New Nation," 143–64. For a negative view of land companies, see Opie, *Law of the Land,* chaps. 1–3.

31. Gates, *Landlords and Tenants on the Prairie Frontier,* 147. For efforts to enact limitations on land acquisitions, see Lause, *Young America,* 3, 41.

32. See Lebergott, *Americans,* 76–81; Hurt, *Ohio Frontier,* 156–59; Taylor, *William Cooper's Town,* 94–99; Atack, Bateman, and Parker, "Northern Agriculture and the Westward Movement," 285–328; Gates, *Landlords and Tenants on the Prairie Frontier,* chaps. 6, 7, 8. On land speculation profits, see Swierenga, "Land Speculator 'Profits' Reconsidered," 1–28. For federal political struggles, see Feller, *Public Lands in Jacksonian Politics.*

33. Lyell, *Travels in North America* 2: 64; Melish, *Travels through the United States of America* (1818), 436–37; Flint, *Letters from America* (1822), 178–79; Godley, *Letters from America* (1844), vol. 1: 141–42; Robert W. Russell, *America Compared with England* (1848), 74. Donald J. Ratcliffe noted that tax laws in Ohio stymied speculators ("Changing Political World of Thomas Worthington," 45–47).

34. One may infer from Robin Einhorn that the large landholdings of planters induced a low land-tax policy (*American Taxation, American Slavery,* 224–29).

35. Consult Clark, "Ohio Country in the Political Economy of Nation Building," 158.

36. On southern land acquisition, see Rothman, *Slave Country,* 41–45, 165–71, 183–88; Dupre, *Transforming the Cotton Frontier,* 20–29, 36, 42–43; Rothman, *Flush Times and Fever Dreams,* 25–29.

37. For the debate, see Danhof, *Change in Agriculture,* 76–81; Danhof, "Farm-Making Costs and the 'Safety Valve,'" 317–59; Ankli, "Farm-Making Costs in the 1850s," 51–70; Judith Klein,

"Farm Costs in the 1850s," 71–74. Jeremy Atack reaffirmed Danhof's estimate ("Farm and Farm-Making Costs Revisited," 663–76).

38. Danhof, *Change in Agriculture,* 77–84; Billington, *Westward Expansion,* quote on 9. Contemporaries, especially those writing guidebooks for immigrants, offered lower dollar amounts for beginning farmers; see table 8.1 in Atack and Bateman, *To Their Own Soil,* 132–33.

39. Marshall, *Farmers and Emigrants Complete Guide* (1857), 26–29; Regan, *Emigrant's Guide to the Western States* (1852), 353; Beaufoy, *Tour through Parts of the United States* (1828), 81.

40. Marshall, *Farmers and Emigrants Complete Guide,* 26–30; Robinson, "To Western Emigrants," 1843, in *Solon Robinson* 1: 356; Oliver, *Eight Months in Illinois* (1843), 239–43; Ankli, "Farm-Making Costs in the 1850s," 64–70.

41. Thomason, *Hints to Emigrants* (1848), 43–44; Northern Pacific Railroad Co., *Guide to the Northern Pacific Railroad Lands in Minnesota* [1872?], 16; Cockrum, *Pioneer History of Indiana* (1907), 189; Oliver, *Eight Months in Illinois,* 236.

42. Chamberlain, *Indiana Gazetteer* (1849), 34; O'Bryan, *Narrative of Travels in the United States* (1836), 124; O'Hanlon, *Irish Emigrant's Guide for the United States* (1851), 106–7; Thomason, *Hints to Emigrants,* 69–70.

43. Regan, *Emigrant's Guide to the Western States,* 350. On saving, see Fidler, *Observations on Professions, Literature, Manners, and Emigration in the United States and Canada* (1833), 210; Stuart, *Three Years in America* (1833), vol. 2: 416–17.

44. Wages of English farmworkers in Heath, *English Peasantry,* 1–2.

45. Gray, *Yankee West,* 100–103; Bogue, *From Prairie to Corn Belt,* 185; Winkle, *Young Eagle,* 10–11.

46. For buying on credit, see Gerhard, *Illinois as It Is,* 445–46; William C. Holgate to James Phelps, December 16, 1846, in Holgate Account Book, OHS; Flint, *Letters from America,* 176; Finch, *Travels in the United States and Canada* (1833), 318. For a more pessimistic view of mortgaging in the 1880s, see Jonathan Levy, "Mortgage Worked the Hardest," 39–67.

47. Baxter, *America and the Americans,* 125; Prentice, *Tour in the United States* (1850), 79; Holmes, *Account of the United States* (1823), 3.

48. Williams, *Americans and Their Forests,* 3–4; Silver, *New Face in the Countryside,* 117–26; Hudson, *Making the Corn Belt,* 58–59.

49. Priest, *Travels in the United States of America* (1802), 14; Amphlett, *Emigrant's Directory to the Western States of North America* (1819), 79; Dalton, *Travels in the United States of America* (1821), 81; Fordham, *Personal Narrative of Travels* (1906), 152.

50. B. O. Williams in Michigan Pioneer and Historical Society, *Pioneer Collections* 2: 2, 464; Fordham, *Personal Narrative of Travels,* 100, 152. On the 1850s and after, Hall, *From Liverpool to St. Louis,* 119; Bird, *Englishwoman in America,* 135.

51. Dalton, *Travels in the United States of America,* 78–81; Williams, *Americans and Their Forests,* 55; Jordan and Kaups, *American Backwoods Frontier,* 96.

52. Wright, *Views of Society and Manners in America* (1821), 179; Playfair, *Recollections of a Visit to the United States and British Provinces of North America* (1856), 94; Martin Primack said settlers could at most clear five acres per year ("Land Clearing under Nineteenth-Century Techniques," 484–97).

53. Williams, *Americans and Their Forests,* 331–32.

54. Ibid., 69; Russell, *Long Deep Furrow,* 184–89; Primack, "Farm Fencing in the Nineteenth Century," 287–91.

55. [Chandler], An Englishman, *A Summary View of America* (1824), 1; Tudor, *Narrative of a Tour in North America* (1834), vol. 1: 231; Houstoun, *Hesperos* (1850), vol. 1: 90. On the Hawthorne hedge, see Gaylord and Tucker, *American Husbandry* 2: 32.

56. On stump sizes and years to rot, see Buel, *Farmer's Companion*, 55; Beaufoy, *Tour through Parts of the United States and Canada*, 80–81; Brown, *America* (1849), 30; Prentice, *Tour in the United States*, 51–52; "Stump Pullers," *American Farmer's Magazine* 12 (October 1858): 592. Quotes from Prentice, *Tour in the United States*, 51–52; Brown, *America*, 16.

57. MacKay, *Life and Liberty in America* (1859), vol. 1: 72; Lyell, *Travels in North America* 2: 83–84; Brown, *America*, 29.

58. Janson, *Stranger in America* (1807), 59; Jobson, *America and American Methodism* (1857), 115; Prentice, *Tour in the United States*, 16; entries of July 13, 20, 1868, vol. 859, B. A. Linville Diary and Record Book, 1867–1869, OHS.

59. On green trees, see Ephraim S. Williams of Flint, Michigan, in Michigan Pioneer and Historical Society, *Pioneer Collections* 10: 139; for sickness, A. D. Johnson, *Illinois and the West* (1838), 140–44, 155; Holmes, *Account of the United States*, 160, 163; Michael Shoemaker, "Sketches of City of Jackson," in Michigan Pioneer and Historical Society, *Pioneer Collections* 2: 420. Mrs. Felton, *Life in America* (1838), 119.

60. Olmstead and Rhode, "Reshaping the Landscape," 663–98; Rasmussen, "Impact of Technological Change on American Agriculture, 1862–1962," 578–91; Danhof, *Change in Agriculture*, 201–46; McClellan, *Sowing Modernity*, 218–20; Atack and Bateman, *To Their Own Soil*, table 7.2, p. 112.

61. Gates, *Farmer's Age*, 172, 249–51; Olmstead and Rhode, "Reshaping the Landscape," 663–98; Rasmussen, "Impact of Technological Change on American Agriculture," 578–91.

62. Entries of December 12, 1868, B. A. Linville Diaries, OHS; December 29, 1846, Benjamin Coales Business Records, UIll; December 11, 1855, S. W. Marshall Account Book, VHS; December 9, 1846, Fielding Beeler Diary, IndHL; January 3, 1859, Oliver Bennett Diary; Nation, *At Home in the Hoosier Hills*, 78–80.

63. Brown, *America*, 32; Dalton, *Travels in the United States of America*, 88; Stoll, *Larding the Lean Earth*, 53.

64. Craig, *To Sow One More Acre*, 30–36; Danhof, *Change in Agriculture*, chaps. 8, 9.

65. Spacing of furrows in Spence, *Settler's Guide in United States* (1862), 181; Gerhard wrote that a man walked about twenty-five miles a day when plowing (*Illinois as It Is*, 317).

66. Robinson, *Facts for Farmers* 2: 974.

67. Levi Countryman, Diary, in Levi N. Countryman Papers, MnnHS; E. B. Drew Diaries, 1859, in James M. Drew and Family Papers, MnnHS.

68. Thomson, *Tradesman's Travels* (1842), 86–88, quote on 88; Houstoun, *Hesperos* 1: 96.

69. Howells, *Recollections of Life in Ohio* (1895), 123; Ferrall, *Ramble of Six Thousand Miles through the United States* (1832), 67; Nation, *At Home in the Hoosier Hills*, 33–34, 107–10; Faragher, *Sugar Creek*, 101; Ulrich, *Age of Homespun*, 3–7, 12–40.

5. The Actors on the Northern Farm and Their Quest for Competency

1. Farnham, *Life in Prairie Land* (1846), 37–38.

2. For background on these matters, see Woloch, *Women and the American Experience* 1: chaps. 3, 4, 5. On farm women, the question of cooperation has been handled by Nancy Grey

Osterud in *Bonds of Community* and "Gender and the Transition to Capitalism in Rural America," 14–29. For women's obscurity, see Jensen, "Now Your See Her, Now You Don't: Historians and Rural Women," 221–32.

3. I am suggesting that the combination of labor-saving machinery with a declining average farm size indicates that fewer man-hours were necessary to conduct cultivation on the average family farm; that is, a reduction in the demand for labor resulted in less need for farm children. Other explanations involve the advance of urbanization; the inability of parents to stop children from migrating west, thereby losing their old-age protection; and the cost of educating children. For a discussion of these theories, see Klein, *Population History of the United States,* 77–81, and the historiographical discussion in Haines, "Population of the United States, 1790–1920," 163–69.

4. Cockrum, *Pioneer History of Indiana,* 189; Regan, *Emigrant's Guide to the Western States,* 355, 357; quote in Griffiths, *Two Years in the New Settlements of Ohio* (1835), 44; Parkerson, *Agricultural Transition in New York State,* 127–37; on the exploitation issue, see discussions in Rugh, *Our Common Country,* 69; and McMurry, *Transforming Rural Life,* 102–3.

5. The making of clothes is nicely portrayed in Cockrum, *Pioneer History of Indiana,* 190, 327–32; Ulrich, *Age of Homespun*; Wermuth, *Rip Van Winkle's Neighbors,* 51–63; Zakim, *Ready-Made Democracy,* 9–10.

6. McMurry, *Transforming Rural Life*; Atack and Bateman, *To Their Own Soil,* 146–60; Jensen, *Loosening the Bonds,* chaps. 5, 6.

7. Thomas, *Rural Affairs* (1858), vol. 1: 25–26; Mellen, *Book of the United States,* 343; Colman, "Female Farm Laborers," *Indiana Farmer and Gardener,* June 28, 1845, 190; Gaylord and Tucker, *American Husbandry* 2: 30; see Fox-Genovese, "Women in Agriculture during the Nineteenth Century," 29–74.

8. This is the conclusion of Joan Cashin, *Family Venture,* 66, 94, 104–5.

9. Trollope, *Domestic Manners of the Americans* (1832), vol. 1: 61–90.

10. Ibid., 105; C. H. Cleaveland, "The Farmer's Wife," *Cultivator* 9 (June 1852): 206; quote in "Wheat Farming in Western New York with Hints for the Farmers of Ohio," *Ohio Cultivator* 1 (August 15, 1845): 122; Thomas, *Rural Affairs* 1: 25–26; "Marrying a Farmer," *Michigan Farmer* 15 (October 1, 1857): 311.

11. Houstoun, *Hesperos* 1: 169; Trollope, *Domestic Manners of the Americans* 1: 106; Thomas, *Rural Affairs* 1: 25–26. An economic assessment of farm women and children is in Craig, *To Sow One Acre More,* chap. 4. Even by 1900, most observers of American farming saw farm women leading depressed lives (Ziegler, "Burdens and the Narrow Life of Farm Women," 78–83).

12. See Craig, *To Sow One Acre More,* 93. David Blanke sees the hard labor of farm women but interprets their embrace of store merchandise as a turn to consumerism (*Sowing the American Dream,* 40–44).

13. Schob, *Hired Hands and Ploughboys,* 265–71; Gates, *Jeffersonian Dream,* 24–36.

14. Bogue, *From Prairie to Corn Belt,* 184–85; Schob, *Hired Hands and Ploughboys,* 7–14, 67–71, 174; Parkerson, *Agricultural Transition in New York State,* 128–30; Bruegel, *Farm, Shop, Landing,* 110–14, 128–30.

15. Griffiths, *Two Years in the New Settlement of Ohio,* 80; *Wiley and Putnam's Immigrant Guide* (1845), 87; for the British comparison, see Steinfeld, *Coercion, Contract, and Free Labor in the Nineteenth Century,* 40–47, 56–58, 86–95.

16. Only a few examples will be cited here: "Hired Man upon the Farm," *Ohio Cultivator* 10 (July 15, 1854): 10; "A Better Class of Labor Needed on the Farm," *American Farmer* 2 (February

1868): 235; "About Hired Men," *American Agriculturist* 18 (June 1859): 169–70; Robinson, *Facts for Farmers* 2: 1006–7; Blake, *Farmer's Every-Day Book* (1850), 260–65.

17. Jabez Brown Papers. One can find similar entries in almost all farm diaries.

18. Entry of July 29, 1873, Curtis Downer Diary, in Downer Family Diaries, ALLib; see also entries of March 17, April 1, May 1, 1873. On reliance of family help, see Parkerson, *Agricultural Transition in New York State*, 128–36; Bruegel, *Farm, Shop, Landing*, 110–12.

19. A Federal Farmer, "Miscellaneous Remarks on Practical Agriculture," *Farmer's Register* 5 (March 1, 1838): 746.

20. May 27, 1876, Curtis Downer Diary; entry April 25, 1856, Levis Burrel Aldrich Diary, Levis Burrel Aldrich Papers, MnnHS.

21. Entry of August 23, 1851, Spencer and Emily Carrier Diaries, WisHS; entry of December 8, 1853 (the 1854 entry put there later), Darius Pierce Account Book, Darius Pierce Papers, UMich; entry of April 30 1874, Curtis Downer Diary, 1874; entry of October 16, 17, 1852, Edward H. Ingraham Diary, Buell Papers, Box 1, UMich.

22. Entry of October 27, 1871, Edwin and Ephraim Warren Diaries, ALLib; on keeping track of pay rates and days worked, entry under James Martin, April 11, 1859, in William Hamilton Ledger/Daybook, OHS; entry under John Bell, February 23, 1860, Hamilton Ledger/Daybook; for sickness, entries of July 23, 29, August 26, 30, 1846 in Lewis Lesher Farm Accounts, OHS; Memorandum of Time Worked by James Desmon[d?] [1836?, p. 3], in Joseph Beeler Account Book, IndHL.

23. Based on comparing entries in the diaries as to where fathers, sons, and hired workers were laboring; for example, entries for March through August 1873, Curtis Downer Diary; entries of August–September 1882, Martin S. Holt Diary, Martin S. Holt Papers, ALLib; entries for April–June 1875, Downer Family Diaries, ALLib.

24. Entry of April 6, 1855, Jabez Brown Diary, Jabez Brown Papers, WisHS; examples of payments in dollars in entries of June 1, 18, 27, 1869, Barber G. Buell Diary, Buell Papers, Box 1, UMich.

25. Entry of June 12, 1858, Levi B. Aldrich Diary; William H. Tebbetts to Daniel Tebbetts, May 21, 1854, William H. Tebbetts Papers, ALLib; entries of November 23, 1845, and July 2, 1847, Benjamin Coales Diary, Coales Business Records; entry of October 17, 1871, in Edwin and Ephraim Warren Diaries.

26. Entries of March 3, November 30, December 4, 1868, and March 23 and December 9, 1869, John C. Bailey Diary, ALLib.

27. Entry of March 30, 1846, in Lewis Lesher Farm Accounts; entries of September 15, 16, 1882, Martin S. Holt Diary.

28. Entry of April 7, 1874, Curtis Downer Diary, 1874; entry of March 18, 1882, Martin S. Holt Diary.

29. On discontent, see William Henry Timlin Diary, WisHS; Moses A. Cleveland, *Autobiography*, version 1, 22–25, 27, UMich; J. C. Holt to Brother and Sister, August 6, 1866, folder 2 1/2, Martin S. Holt Papers, ALLib. Examples of court cases between farmworkers and farmers are in Schmidt, *Free to Work*, 38–44. Schob indicates some dissatisfaction though he more generally emphasizes the good prospects of farmworkers (*Hired Hands and Ploughboys*, 209, 255–61).

30. Kingdom, *America and the British Colonies*, 12; Spence, *Settler's Guide in the United States*, 208; Steevens, *Land of the Dollar* (1898), 159.

31. Estimates based on G. Grenville Benedict to Geo. H. Paul, Feb. 16, 1848, George H. Paul Correspondence, WisHS. Edward S. Abdy reported in 1834 that a clergyman on the Wadsworth

farm in New York received about $500 per year, which Abdy said was enough "to live comfortably" (*Journal of a Residence and a Tour* [1835], vol. 1: chap. 9). For $225 being insufficient for a family of four, see E. D. Seward to Dudley Seward, Feb. 20, 1837, in Wisconsin Territorial Letters, WisHS.

32. For a different view that emphasizes the consumerism of the Great Lakes farmer by at least 1850, see Blanke, *Sowing the American Dream,* chap. 2.

33. Fordham, *Personal Narrative of Travels,* 72; Ferrall, *Ramble of Six Thousand Miles through the United States,* 66; Steevens, *Land of the Dollar,* 159.

34. Calvin Stevens to Mr. Samuel H. Stevens, November 20, 1840, Wisconsin Territorial Letters.

35. Russell, *North America,* 116–17; Mary Carpenter to Cousin Laura, June 1, 1873, Mary E. Carpenter Correspondence, MnnHS. See Gruenwald, *River of Enterprise*; on village life, Cayton, *Ohio,* 239–45; Bruegel, *Farm, Shop, Landing,* 110–14, 128–33. Examples of paying taxes by road work in entry of June 10, 1853, Diary of Daniel Thomas, in Amasa Bigelow and Daniel Thomas Papers, WisHS; entry under Peter Engel, May 1854, Abraham Hawker Daybook and Ledger, OHS; John Abby, Lovira Hart and Emeline Hart to Lyman Abby, February 10, 1839, in Lovira Hart Papers, UMich.

36. Atack and Bateman, *To Their Own Soil,* 201–2, and table 13.8, p. 243. See discussion by Caird, *Prairie Farming in America* (1859), 54, 87–88; Mitchell, *My Farm of Edgewood,* 212–13. For a different interpretation of the decrease in percentage of crops for the family, see Post, "Agrarian Origins of U.S. Capitalism," 428.

37. Impact of corn-shelling machine in William H. Tebbetts to Mother, February 28, 1854, William H. Tebbets Papers; Rasmussen, "Impact of Technological Change on American Agriculture," 579–84.

38. Entry of December 2, 1854, Warren Lane Diary, Lane and Wheller Family Papers, WesResHS; J. P. Williams to Mother and Sister, July 11, 185? in Hepzibah Dumville Papers, ALLib; summation at end of 1859 in Oliver Bennett Diary, 1859, WesResHS; hauling 820 bushels of wheat, January 26, 27, 1869, Linville Diaries, 1866–70, OHS; entry of August 29, 1871, Oliver Parsons Diary, MnnHS; entry of August 14, 1867, John C. Bailey Diary; William H. Tebbetts to Friends [parents], January 1, 1854, Tebbetts Papers.

39. The best work on this theme is Winkle, *Young Eagle,* 7–22.

40. Mitchell, *Ten Years Residence in the United States* (1862), 194; Houstoun, *Hesperos* 1: 78; Thomas Hamilton, *Men and Manners in America* (1834), vol. 1: 259; Lyell, *Second Visit to the United States* 1: 62.

41. Prentice, *Tour in the United States,* 79; Trollope, *North America* (1862), vol. 1: 204.

42. Russell, *North America,* 92–93; Samuel Ayers to S. L. Howe, July 25, 1852, Apollos Warner Family Papers, WesResHS.

43. "Address of Dr. A. C. Stevenson, at the Montgomery County Fair," in *Indiana Farmer* 7 (December 1858): 257; "The Dignity and Importance of Agricultural Pursuits Must Direct the Policy and Manners of the People," *New York Farmer* 1 (November 1828): 262–65.

44. D. A. Ogden, "The Farmer—His Position, Responsibilities, and Duties," *Genesee Farmer* 7 (September 1846): 204.

45. "What Farmers Should Live For," *Southern Aegis,* qtd. in *American Farmer's Magazine* 7 (June 1858): 349; "The Position Farmers Ought to Occupy in Society," in *Genesee Farmer,* qtd. by *Michigan Farmer* 2 (January 1, 1845): 145; "Mr. Van Bergen's Address Before the New York Agricultural Society (Second Part)," *Boston Cultivator* 22 (March 28, 1840): 1; Samuel Row, "Wash-

tenaw County," in Michigan Pioneer and Historical Society, *Historical Collections* 2: 535; "'He's Nothing but a Farmer,'" *Valley Farmer* 6 (September 1854): 363; "Boys! Stick to the Farm!" *Illinois Farmer* 1 (April 1856): 81.

46. J.B., "Ought Farmers to Labor?" *Cultivator* 12 (November 1864): 340; Trollope, *North America* 1: 190; O'Brien, *Minnesota Pioneer Sketches* (1904), 52.

47. Minneapolis *Journal,* March 29, 1924, 1; found in box 1, Levi Countryman Papers.

48. Quotes, in order, August 12, August 13, July 21, 1858; September 30, 1859; July 30, 1858; March 19, 1859; on his determination to quit farming, July 21, 23, 1858; all in Levi Countryman Diary.

49. Entries of July 30, 1858, August 6, 1858, Levi Countryman Diary.

50. For example, Newton Southworth to [Captain Seth Freeman], April 16, 1860, Newton Southworth and Family Papers, MnnHS; "Why Farmers' Sons Leave the Farm," *Massachusetts Ploughman* 31 (December 30, 1871): 1.

51. "The Hired Girl upon the Farm," *Ohio Cultivator* 10 (August 1, 1854): 10; Howells, *Recollections of Life in Ohio,* 123–24; Cockrum, *Pioneer History of Indiana,* 190, 327–32; Miller, *Needle's Eye,* 63.

52. Howells, *Recollections of Life in Ohio,* 97; Donald, *Lincoln,* 30; Parker, "The Magic of Property," 477.

53. Oldmixon, *Transatlantic Wanderings* (1855), vol. 3; Peck, *Forty Years of Pioneer Life* (1864), 172.

54. For connection between capitalism and the American Civil War, see Hobsbawm, *Age of Capital,* 138–43; Larson, *Market Revolution in America,* 131–40; Zakim and Kornblith, "Introduction: An American Revolutionary Tradition," 1–12.

55. Besides the economic historians listed in the preceding notes, see also Rothenberg, *From Market-Places to a Market Economy*; Olmstead and Rhode, *Creating Abundance.*

56. Henretta, "Families and Farms," 3–32; Merrill, "Putting 'Capitalism' in Its Place," 315–26; Clark, *Roots of Rural Capitalism,* chaps. 2, 7, 8; Kulikoff, *Agrarian Origins of American Capitalism,* 34–52, 100–125, 211–22, 264; Fox-Genovese, *Within the Plantation Household,* 56–58.

57. Headlee, *Political Economy of the Family Farm,* 1–32; Post, "Agrarian Origins of U.S. Capitalism," 393–433. A good summary of the troubles of Marxist theory with understanding the family farmer can be found in Mann, *Agrarian Capitalism in Theory and Practice,* chap. 1. Among those finding the northern farmer oriented to the market and to capitalist values are Gray, *Yankee West,* 2–3, 10–14, 111–17, 160; Rugh, *Our Common Country,* xv–xix, 58–59, 130–38; Cayton and Onuf, *Midwest and the Nation,* xvii, 50–62, 84–85, 90; Dunaway, *First American Frontier,* 1–19; Cayton, *Ohio,* 239–61; and Blanke, *Sowing the American Dream,* 14–25.

58. Barron, *Those Who Stayed Behind,* quote on 7, 4–9; Barron, "Listening to the Silent Majority," 3–16; Barron, "Epilogue: Reaping What Has Been Sown," 287–91.

59. Jones, *History of Agriculture in Ohio,* 314–323; Davis, *Frontier Illinois,* 404–25; Parkerson, *Agricultural Transition in New York State,* 18–23, 79–98; McMurry, *Transforming Rural Life,* 43–45. Christopher Clark has recently found the northern family farmer inadequately characterized as capitalist ("Agrarian Context of American Capitalist Development," 35). Slightly different is Nancy Grey Osterud, who saw around Binghamton, New York, the family farm characteristic of kin cooperation and local commerce, only changing after 1880 or so (*Bonds of Community,* 23–39).

60. See the excellent discussion of the definition of capitalism in Bowman, *Masters and Lords,* 83–100. On Max Weber, see Johnson, *Early American Republic,* 56; Weber, *Protestant Ethic and the Spirit of Capitalism,* chap. 5.

61. Osthaus, "Work Ethic of the Plain Folk," 745–82; Hahn, *Roots of Southern Populism,* 52–63; Etcheson, *Emerging Midwest,* 36–38; Cecil-Fronsman, *Common Folk of the Old South,* 50, 88–89, 131.

62. Danhof, "The Farm Enterprise," 129; "Hints to Young Farmers," from *New England Farmer,* in *Prairie Farmer* 6 (July 1846): 207–8; "How Sound Are Your Accounts," *American Agriculturist* 19 (January 1860): 8.

63. On savings, see Mitchell, *My Farm of Edgewood,* 205–13.

64. Williamson and Lindert, *American Inequality,* 161–65, 170–73; Atack and Bateman, *To Their Own Soil,* 246, 260–63, 273.

65. Anderson, *New England's Generation,* chap. 4; Vickers, *Farmers and Fisherman,* 14–23, 60–71. Competency as a theme is also in McMurry, *Transforming Rural Life,* 45–52; Faragher, *Sugar Creek,* 98, 105; Nation, *At Home in the Hoosier Hills,* 113.

66. Quotes in order from Curtiss, *Western Portraiture, and Emigrants' Guide* (1852), xix; William J. Bradford, *Notes on the Northwest* (1846), 147; Thomason, *Hints to Emigrants,* 31.

67. "Small Farms," from *Cincinnati Gazette,* qtd. in *New England Farmer and Horticultural Register* 22 (July 12, 1843): 13.

68. Certain political and social historians attempt to use the yeoman farmer ideal, when not dismissing it as empty rhetoric, to show that early farmers, 1750–1800, had as their goals civic virtue and public good while later northern farmers, post-1830, sought independence by accumulation of profits. See Berthoff, "Peasants and Artisans, Puritans and Republicans," 581, 582. On the yeoman farmer ideal, Fea, *Way of Improvement Leads Home,* 19–23. On criticism of the yeoman ideal as mostly myth, Hofstadter, *Age of Reform,* 23–59; Smith, *Virgin Land,* 123–31; Johnson, *Early American Republic,* 56–70; Appleby, "Commercial Farming and the 'Agrarian Myth' in the Early Republic," 833–49.

69. Jones, *History of Agriculture in Ohio,* 48; entry of October 28, 1854, Diary of Daniel Thomas; entries of September 27, October 7 (quote), 16, 17, 1852, Edward H. Ingraham Diary; entry of November 18, 1846, Fielding Beeler Diary.

70. Today, this concept goes by two other names; in academia it is called "hierarchy," while somewhat oddly in modern American life it is called "leadership." But the concept is fairly straightforward and simple: some give orders, and others obey them.

71. See the difficult time Jeremy Atack and Fred Bateman have in characterizing American farmers, as they clearly did not fit the description of classical economic actors and they straddled the fence on the question of subsistence versus commercial farming (*To Their Own Soil,* 260–74).

6. The Plantation Assault on the Southern Yeoman

1. Hegemony and planter domination are of course the key concepts of Eugene D. Genovese in his earliest work, *Political Economy of Slavery,* but planter dominance was also the primary interpretation of some of the earliest historians of the Old South, such as Dodd's *Cotton Kingdom.* The legacy continues with Scarborough's *Masters of the Big House.*

2. The older interpretation was of poverty-stricken whites, popularized by Frederick Law Olmsted in the 1850s, and then continued by various historians until challenged in the 1940s by Owsley, *Plain Folk of the Old South.* Those still finding yeoman farmers of the South holding onto land while tenancy was minor include Harris, *Plain Folk and Gentry in a Slave Society,* 6;

Hyde, "*Plain Folk* Reconsidered," 803–30; Winters, "'Plain Folk' of the Old South Reexamined," 565–86; Schaefer, "Yeoman Farmers and Economic Democracy," 421–37; and Bond, "Herders, Farmers, and Markets in the Inner Frontier," 81–97.

3. Noe, *Southwest Virginia's Railroad,* chap. 2; McKenzie, *One South or Many*; Bolton, *Poor Whites of the Antebellum South*; Bode and Ginter, *Farm Tenancy and the Census in Antebellum Georgia*; West, *From Yeoman to Redneck in the South Carolina Upcountry,* 8, 20–21, 39–44; Dunaway, *First American Frontier,* 73–82.

4. Curtin, *Rise and Fall of the Plantation Complex,* 86, 99. The view that planters and small farmers had a symbiotic relationship can be found in Hilliard, *Hog Meat and Hoecake,* 25, 191. The interpretation that the plantation pushed out the nonslaveholder and was a danger to the yeoman's existence is not new, it is simply not well documented on a broad basis. See Gray, *History of Agriculture in the Southern United States* 1: 444; Edwards, "Men of Subsistence and Men of Substance," 304–9; Campbell, "Planters and Plain Folk," 48–77; Lowe and Campbell, *Planters and Plain Folk,* 42.

5. Towers, *Urban South and the Coming of the Civil War.*

6. Russell, *North America,* 269; Majewski, *Modernizing a Slave Economy,* 16–51.

7. Koeniger, "Climate and Southern Distinctiveness," 21–44; Arsenault, "End of the Long Hot Summer," 597–628.

8. Menn, *The Large Slaveholders of Louisiana.*

9. Ibid., 1–6.

10. For method and assumptions, see appendix A.

11. For example, McCurry, *Masters of Small Worlds.*

12. The size of farm used in the analysis above generally concurs with the literature. See Hyde, "*Plain Folk* Reconsidered," 819–29; Weaver, *Mississippi Farmers,* 53–55; Sellers, *Slavery in Alabama,* 23, 24, 26, 27–30; Sydnor, *Slavery in Mississippi,* 19; Davis, *Cotton Kingdom in Alabama,* 57.

13. An example of expansion due to land exhaustion is Kennedy, *Mr. Jefferson's Lost Cause,* 12–14. The findings here agree with Foust, *Yeoman Farmer and Westward Expansion of U.S. Cotton Production,* 171–82, 199–201. For how Republicans should have appealed to southern yeomen, see Goodman, "White over White," 446–52.

14. Family size from *Census of 1860* 4: *Mortality,* 351; Clark, *Social Change in America,* 85. On the relation of family size to rural poverty, see Pudup, "Limits of Subsistence," 61–89; Salstrom, *Appalachia's Path to Dependency,* 21–23; McKenzie, "Oh! Ours Is a Deplorable Condition," 217–18.

15. "The Farms and Farming of the Rich Lexington District in Kentucky," in *Boston Cultivator,* qtd. by *Farmer's Register* 9 (January 1841): 56.

16. Weiman, "Farmers and the Market in Antebellum America," 627–47; Weiman, "Economic Emancipation of the Non-slaveholding Class," 71–93; Harris, *Plain Folk and Gentry in a Slave Society,* 30, 79–90.

17. See Freehling, *Road to Disunion: Secessionists at Bay,* 39–54; on population problems in the South, see Huston, "Theory's Failure," 354–81.

18. Phillips, *American Negro Slavery;* Gray, *History of Agriculture in the Southern United States*; Genovese, *Political Economy of Slavery,* 3–39; Fox-Genovese, *Within the Plantation Household,* 53–69; Conrad and Meyer, "Economics of Slavery in the Antebellum South," 136–75; Stampp, *Peculiar Institution*; Fogel and Engerman, *Time on the Cross.* Since 1976, see Wright, *Po-*

litical Economy of the Cotton South; Wright, *Slavery and American Economic Development;* David et al., *Reckoning with Slavery;* Ransom, *Conflict and Compromise.*

19. On the capitalist modernity of the plantation, see Curtin, *Rise and Fall of the Plantation Complex,* 52–53, 83–109; Oakes, *Ruling Race,* 28–34, 147; Carlton and Coclanis, *South and the Nation,* 5–6. For an intermediate view, see Clark, *Social Change in America,* 130–38; Scarborough, *Masters of the Big House.* For those believing planters continued an anti-bourgeois tradition, see Gallay, "Origins of Slaveholders' Paternalism," 369–94; Genovese and Fox-Genovese, *Fatal Self-Deception,* chap. 1; to the extent that the code of honor included paternalism, see Wyatt-Brown, *Southern Honor,* chap. 2. On the food question, Gallman, "Self-Sufficiency in the Cotton Economy of the Antebellum South," 20–23; Hilliard, *Hog Meat and Hoecake,* 2, 192–211; Battalio and Kagel, "Structure of Antebellum Southern Agriculture," 33–34. For a recent assessment of the capitalist nature of the South, but not of the plantation, see Downey, *Planting a Capitalist Class.*

20. Conrad and Meyer, "Economics of Slavery in the Antebellum South," 136–75; Ransom and Sutch, *One Kind of Freedom,* 19–20, 264–69; Wright, "Slavery and the Cotton Boom," 439; Wright, *Political Economy of the Cotton South,* 87; Fogel and Engerman, *Time on the Cross* 1: 47–50, 191–96. The efficiency hypothesis has received recent affirmation in a highly technical paper, Field-Hendrey and Craig, "Relative Efficiency of Free and Slave Agriculture," 255.

21. Mingay, "Size of Farms in the Eighteenth Century," 470–72.

22. "Relative Expense, of Animal Labour, as Performed by Horses or Oxen" (Baltimore) *Farmer and Gardener* 2 (August 18, 1835): 126; G., "Mules in the Southern States," *Cultivator* 3 (October 1855): 306. Although contemporaries figured mules cost less than horses, George Ellenberg said they cost about the same (*Mule South to Tractor South,* 11–19, 24). Prices from Robinson, "To Western Emigrants," 1842, in *Solon Robinson* 1: 348.

23. The basic standard was to choose counties in Illinois (7), Indiana (5), Michigan (5), Ohio (8), Wisconsin (5), Maine (6), New Hampshire (4), Vermont (7), New York (10), New Jersey (5), and Pennsylvania (7) that had at least twice as many farms as people employed in manufacturing, establishing these counties as rural counties. The numbers in parenthesis above is the number of counties chosen; this is not a sample. For southern counties I looked for those whose slave population was at least twice that of the white population: Alabama (5), Georgia (8), Mississippi (6), South Carolina (8), Kentucky (7), Missouri (7), Virginia (11). For the nonslaveholding South, I looked for counties that had two to three times more whites than slaves: Alabama (6), Georgia (4), Kentucky (4), North Carolina (3), and Virginia (4).

24. Taken from Dalton, *Travels in the United States of America,* 231; Marshall, *Farmers and Emigrants Complete Guide,* 16; Robinson, "To Western Emigrants," 1843, in *Solon Robinson* 1: 351. For an example of how often wagons were used for fencing and lumbering, see William H. Tebbetts to Parents, June 17, 1855, Tebbetts Papers.

25. Bacot Family Papers, July 21, 1853, inventory, UNC; Lewis Bird estate, November 18, 1840, in William H. Sanders Account Book, VHS; list of property of Benjamin Sisson, December 1, 1859, in Benjamin Sisson Account Book, UVa; sale of property at South Bend, September 16, 1851, in William S. Kemper Ledger, UVa; inventories in Phillips, *American Negro Slavery,* 231. For the machinery and animals needed for a sixty-acre farm, I took the suggestions of Dalton, *Travels in the United States of North America,* 231; Marshall, *Farmers and Emigrants Complete Guide,* 16, 24; Robinson, "To Western Emigrants," 1842, in *Solon Robinson* 1: 348; accounts in "Farming in Pennsylvania," *Cultivator* 8 (December 1851): 399; Gerhard, *Illinois as It Is,* 316, 446; "Farm Labor," *Michigan Farmer* 16 (March 1, 1856): 61.

26. Entries of minding mules, September 22, 23, 25, 26, 27, 28, 29, 1843, Rosedew Plantation, Ga., in George Kollock Plantation Record, UNC; entries for John and Pleasant plowing, December 15, 16, 17, 1857, William J. Payne Journal, 1857–61, UNC; hauling with a wagon in December 21, 31, 1855, S. W. Marshall Account Book; cradlers in July 17, 18, 19, 1837, John Blackford Diary, VHS; on hog killing, Dec. 1, 1841, Massie Family Papers [p. 40]; December 15, 1829, Nov. 22, 1831, William F. Wickham Diary, VHS. Specialization of labor on the plantation has long been noted by historians; see Phillips, *American Negro Slavery,* 536; Gray, *History of Agriculture in the Southern United States* 1: 471; Taylor, *Negro Slavery in Louisiana,* 67.

27. Entries of March 1, 14, 15, 16, 17, 20, 21, 22, 23; June 15, 16, 17, 19, 20, 21, 22, 23, 24, 26, 27, 1843, in George Kollock Plantation Journal. Entries of December 15, 1832, William F. Wickham Diary. Entry of July 8, 1857, William J. Payne Journal. Blacksmith "Allen" listed as slave in entry of September 22, 1849, Bayside Plantation Records, UNC; the Bayside Plantation Record is from a Louisiana sugar plantation and has much hiring of free labor to set up the machines and buildings (March–July 1846).

28. Martin Bruegel cautions that these rural mechanics lived precarious lives and depended upon their small plots of land for subsistence (*Farm, Shop, Landing,* 146).

29. Entries of July 3, 1829; July 2, 1834; June 26, 1835, Wickham Diary. Use of skilled slave labor is well known.

30. Sellers, *Slavery in Alabama,* 32; Phillips, *American Negro Slavery,* 252, 309–11.

31. Sellers, *Slavery in Alabama,* 32; Phillips, *American Negro Slavery,* 252, 309–11; Weaver, *Mississippi Farmers,* 52–54; Moore, *Emergence of the Cotton Kingdom in the Old Southwest,* 106–7; Bonner, *History of Georgia Agriculture,* 87; Joyner, *Down by the Riverside,* 60, 70–86.

32. Southern wages remain an element of controversy. In a paper presented at the Mid-America Conference on History (September 23, 2011), Paul Paskoff failed to find statistically a reduction of artisan wages as the percentages of slaves increased ("Free Soil Fallacy"). For those arguing for a high-wage South, see Wright, *Slavery and American Economic Development,* 71; Earle, *Geographical Inquiry and American Historical Problems,* chap. 8; Lebergott, *Manpower in American Economic Growth,* 158–59; Margo, "North-South Wage Gap before and after the Civil War," 324–51.

33. Figures from Fogel and Engerman, *Time on the Cross* 1: 151; Ransom and Sutch, *One Kind of Freedom,* table 1.1, p. 3; Lebergott, *Americans,* 219.

34. Fogel, *Without Consent or Contract,* fig. 5, p. 46, and 49–52; Joyner, *Down by the Riverside,* 60.

35. I computed the ratio of northern manufacturing employees to southern manufacturing employees to be 2.018.

36. Entries in December 1857, William J. Payne Journal; Entries of December 11, 20, 21, 18, S. W. Marshall Account Book; Entry May 19, 1844 (?), Massie Family Farm Memoranda and Account Book; Entries of July 28, 29, 1853, on borrowing horse; October 6, 1853, on road work, Nov. 19, 21, 1853, George Sinclair Journal, UVa.; Moore, *Emergence of the Cotton Kingdom in the Old Southwest,* 33–34, 77–78, 102; Wright, *Slavery and American Economic Development,* 115–16. The point about multiple chores needed on the plantation was also made by Robert E. Gallman and Ralph V. Anderson, "Slaves as Fixed Capital," 29–39. It should be mentioned that it is strange to call slaves "fixed capital" because they are not "fixed" in the sense that much capital stock is fixed. Iron rails and five-ton specialized machinery have few alternatives in production; if their function disappears, then they are deadweights. Consider all the machinery lost during deindustrialization of the Midwest after 1973. But slaves were a form of capital not fixed in the same way as industrial capital stock. Slaves had tremendous flexibility in their capacity to work at

various tasks. They deserved the term of something like "flexible fixed capital" or "fixed capital with multiple uses."

37. Phillips, *Life and Labor in the Old South,* 136; Colman, *European Agriculture and Political Economy* 1: 56.

38. Majewski, *Modernizing a Slave Economy,* 25–31; Genovese, *Political Economy of Slavery,* 89–94; Kennedy, *Mr. Jefferson's Lost Cause,* 12–14; "Southern Agriculture," *American Agriculturist* 5 (September 1846): 273.

39. Entries of Feb. 24, 27, 28; March 13, 14, 15, 16, 17, 18, 1843, Kollock Plantation Record. "Plantation and Farm Instruction, Regulation, Record, Inventory and Account Book. For the Use of the Manager on the Estate of ——," by a Southern Planter (Richmond: J. W. Randolph, 1852), 9, in S. W. Marshall Account Book, 1855–57.

40. Cunynghame, *Glimpse of the Great Western Republic,* 97.

41. Entry of February 22, 1858, William J. Payne Journal. This brings up the question of overseers, which will not be pursued here; see Scarborough, *Overseer.* However, management of slaves was the key to successful plantation income; see James Cuthbert, "On the General Management of Plantations," *Southern Agriculturist* 1 (February 1828): 66; D., "A Memorandum of Tasks," *Southern Agriculturist* 7 (June 1834): 297; Conservator, "The Different Advantages of Large and Small Farms Considered," *Farmer's Register* 4 (January 1837): 564; Gray, *History of Agriculture in the Southern United States* 1: 478–79.

42. Slavery injured urban development because it promoted the large plantation, not because of slavery per se; see Fox-Genovese, *Within the Plantation Household,* 70–77.

43. Huston, *Stephen A. Douglas and the Dilemmas of Popular Sovereignty,* 146–48; Nation, *At Home in the Hoosier Hills,* chap. 1, 83–89, 113, 187–89.

44. Mitchell, *Civil War Soldiers,* 99. The pages of the census are from *Census of 1860* 1: *Population,* 374, 271, 467, 452. For explanations of the lack of town development in the South, see for dispersed population due to plantation, Genovese, *Political Economy of Slavery,* 157–79; the nature of southern staples versus northern cereals, Earle, *Geographical Inquiry and American Historical Problems,* chap. 3; the demands of soil types, Majewski, *Modernizing a Slave Economy,* 51–67; and planters demanding only large towns for the sophisticated services the plantation required, Towers, "Southern Path to Modern Cities," 145–65. A good discussion of the various theories is in O'Mara, *Historical Geography of Urban-System Development,* 34–40.

45. See Guasco, "Southern Yeomen and Slavery in Illinois," 7–29; Hammond, *Slavery, Freedom, and Expansion in the Early American West,* chaps. 5, 6, 7. On relations between nonslaveholding poor whites and slaves in the cotton South, see Forret, *Race Relations at the Margins,* 11–12, 223–26.

7. Inspiring Fear: The Plantation, the British Gentry, and Mastery of Others

1. John H. Cocke gave those production numbers in November 1852. A bale equaled four hundred pounds, though it varied from place to place, and a good crop was two bales per acre. The amount of seed cotton, lint cotton, and price per pound for cotton taken from William J. Cooper's "Cotton Crisis in the Antebellum South," 387–89. Entry of November 28, 1852, in Cocke Day Book, November 1852–January 1853, John H. Cocke Papers, UVa.

2. Large purchases and selling of wheat can also be found in accounts for 1858 and 1859 in George B. Tyler and Robert H. Tyler Farm Accounts, UVa.

3. Edmund Ruffin, "Estimates of Farming Expenses and Profits," *Southern Planter* 9 (August 1849): 227–33. Entries (for marriage and European trip) of October 21, 1854, and November 27, 1855; for corn sale, September 8, 1853, Eppes Account Book, VHS. On Richard Eppes, see Kirby, *Poquosin,* 43–55.

4. Edmund Ruffin, "Estimates of Farming Expenses and Profits," *Southern Planter* 9 (August 1849), 227–33; entries of December 9, 1832, November 25, 1833, and May 3, 1844, William Wickham Diary; entries of January 10, 1839, August 19, 1840, September 22, 1840, Sally Spindle with Ro. P. Waring account; November 13, 1839, Gwynfield account with Ro. P. Waring; June 8 and 21, August 22, September 12, 1839, John Saunders account with Ro. P. Waring, Waring-Baylor Family Papers, UVa; 1855, 1856, 1857. Shelter farm corn accounts, George B. Tyler and Robert H. Tyler Farm Account Book; 1837 and 1838 entries, Hubard "Chellowe" Farm Account, UVa.

5. For general descriptions of wheat farming and improvement in Virginia, see Phillips, *Life and Labor in the Old South,* 127, 137–38; Grivno, *Gleanings of Freedom,* 93–100; Schlebecker, "Farmers in the Lower Shenandoah Valley," 462–76. Phillips did not believe that wheat farming by slaves could extend into the North. Kenneth W. Keller doubts the success of the wheat experiment in Virginia ("Wheat Trade in the Upper Potomac, 1800–1860," 21–33). Claudia Bushman finds slavery and wheat worked together well (*In Old Virginia,* 187–88). James Irwin found effects of size in wheat farming ("Exploring the Affinity of Wheat and Slavery in the Virginia Piedmont," 295–303).

6. Quote in Willoughby Newton, "Virginia and Its Farmers," *American Farmer* 9 (May 1854): 343; "Address of the Hon. Willoughby Newton, delivered before the Maryland Agricultural Society," *American Farmer* 6 (December 1850): 189; "Address of Hon. Willoughby Newton, before the Virginia State Agricultural Society," *American Farmer* 7 (June 1852): 413; "A Large Plantation and Great Wheat Crop," *New York Tribune,* qtd. in *Southern Cultivator* 11 (August 1853): 245; "A Virginia Country Residence," from Easton Pennsylvania *Whig,* in *Niles National Register* 19 (October 11, 1845): 92. For the output of wheat in the Tidewater in 1850, see "Virginia Agricultural Statistics," *De Bow's Review* 2 (August 1852): 192–96.

7. Newton, "Virginia and Its Farmers," 343; quote in "Address before Virginia State Agricultural Society," *American Farmer* 7 (June 1852): 413.

8. Entry of June 14, 1843, John H. Cocke Plantation Memorandum, commenced October 1841, Box 188, Farm Journals, John H. Cocke Papers. Economist Heywood Fleisig wrote an impressive article in 1976 stating that if one modeled the plantation and family farm according to a theoretically constructed isoquant-budget analysis and then added a labor constraint on northern farms, the result would be that the family farm would tend toward adoption of machinery while the plantation would not. I have come to doubt the conclusion because the constraints might have been different. The major constraint was the size of farm. As long as the family farm remained stuck at about one hundred acres, a severe limit was introduced on mechanization. The plantation was not so constrained, and thus I would guess that the plantation would be likely to adopt agricultural machinery more quickly than family farms if the machines were fully adapted to the crop (Fleisig, "Slavery, the Supply of Agricultural Labor, and the Industrialization of the South," 572–97).

9. Deane V. Brown to William J. Brown, February 13, 1854, Brown Family Papers, UVa. Entries of September 23, 1848; September 15, 18, 1849; September 3, October 4, 1850, Box 188 Farm Journals, Agricultural Memorandum 1848–1850, John H. Cocke Papers. Entries of January 6, March 19, 1858, Account Book 1806–1855, Conway Family Papers, UVa.

10. Many have doubted the effectiveness of the antebellum southern fertilizer campaign to reinvigorate the soil. See Stoll, *Larding the Lean Earth,* 21–28, 120–58; Majewski, *Modernizing a Slave Economy,* 18–20; Mathew, *Edmund Ruffin and the Crisis of Slavery in the Old South.* However, William J. Cooper found no reason to doubt the successful implementation of the practice ("Cotton Crisis in the Antebellum South," 390).

11. Quote of Muskogee, Alabama, *Herald* in Crandall, *Early History of the Republican Party,* 77.

12. "Agriculture of North Alabama," *American Agriculturist* 3 (January 1844): 25; report of Dr. Russell from *London Times,* "A Southern Plantation," *Genesee Farmer* 22 (September 1861): 273; "Agriculture: Statistics of American Agriculture," *De Bow's Review* 24 (May 1858): 439.

13. For example, A Southerner, "Information Wanted," *Cultivator* 6 (May 1849): 156; J.S., "Slavery," *Southern Planter* 1 (September 1841): 157; Edmund Ruffin, "Slavery and Free Labor Defined and Compared," *Southern Planter* 19 (December 1859): 1.

14. Hall, *Travels in Canada and the United States,* 320; MacKay, *Western World* (1850), vol. 2: 138; Mead, *Travels in North America* (1820), 13–14.

15. Sutcliff, *Travels in Some Parts of North America* (1815), 113–14; MacKay, *Western World* 3: 298–300; Featherstonhaugh, *Excursion through Slave States* (1844), vol. 1: 129.

16. Russell, *North America,* 277–78.

17. Chandler, *A Summary View of America,* 13; Finch, *Englishwoman's Experience in America* (1853), 294–95; Lyell, *Second Visit to the United States* 1: 355.

18. Houstoun, *Hesperos* 2: 157–61, quote on 202. MacKay, *Western World* 1: 330; vol. 2: 125. Barclay, *Agricultural Tour in the United States and Upper Canada,* 92.

19. Lyell, *Second Visit to the United States* 1: 322; Harris, *Remarks Made during a Tour through the United States* (1821), 49. See Genovese, *Roll, Jordan, Roll,* 53–54.

20. This is a common theme. For example, Baxter, *America and the Americans,* 177–78; MacKay, *Western World* 2: 135–38; Lyell, *Travels in North America* 1: 130–31.

21. Spence, *Settler's Guide in the United States,* 323; Howe, *Emigrants' New Guide* (1822), 6–7; quote in O'Hanlon, *Irish Emigrant's Guide for the United States,* 91; quote in William H. Blane, *Excursion through the United States and Canada* (1824), 202.

22. Hall, *Two-Fold Slavery of the United States* (1854), 8–9; MacKay, *Western World* 2: 134–46; Cairnes, *Slave Power* (1862), 45–50; on land exhaustion, see Gurney, *Journey in North America* (1841), 55, 67; on the efficiency and cost of slaves compared to free labor, Stirling, *Letters from the Slave States* (1857), 234, 239–40, Buckingham, *Slave States of America* (1842), vol. 1: 401.

23. Olmsted, *Cotton Kingdom,* 504, and generally 405–63.

24. Russell, *North America,* 136, 140, 141–43, quote on 143, 144–47, 150, 156, 172, 180–81, 249, 265, 293.

25. Quote from Chandler, *Summary View of America,* 250; quote in Sutcliff, *Travels in Some Parts of North America,* 43; quote in Ferguson, *America by River and Rail,* 133; Grattan, *Civilized America* (1859), 181, 223, 230.

26. Stirling, *Letters from the Slave States,* 63; Tocqueville, *Democracy in America* 1: 47; quote in MacKay, *Western World* 1: 207; quote in Chevalier, *Society, Manners, and Politics in the United States,* 383–84; quote in Ferrall, *Ramble of Six Thousand Miles through the United States,* 230.

27. Fogel, *Without Consent or Contract,* 82–84, 133; Overton, *Agricultural Revolution in England,* 125; Fogel and Engerman, *Time on the Cross* 1: 254.

28. Colman, *European Agriculture and Rural Economy* 1: 39–65; Olmsted, *Walks and Talks*

of an American Farmer in England, 233–50, 279, 291, 302; Badeau, *Aristocracy in England,* 173, 234–35; Gregg, *Social and Economic History of Britain,* 353. On deference, see Mingay, *English Landed Society in the Eighteenth Century,* 286–87; Porter, *English Society in the Eighteenth Century,* 65; Stone and Stone, *Open Elite,* 417–18; Clark, *English Society,* 79, 87. On American slave conditions, Blassingame, *Slave Community.*

29. Ozanne, *South as It Is* (1863), 69; "How the South Is Affected by Her Slave Institutions," *De Bow's Review* 11 (October 1851): 349; R. Wheateley, "English Land Tenure: The Aristocracy," *Christian Advocate and Journal* 39 (March 3, 1864): 65.

30. Thornton, *Politics and Power in a Slave Society,* 204–23.

31. On Burke, see Doyle, *Aristocracy and Its Enemies,* 281–86; Guttmann, *Conservative Tradition in America,* 5–11. Filmer wrote *Patriarcha* in the 1650s, but it was not published until 1680.

32. Baugh, "Introduction" to Baugh, ed., *Aristocratic Government and Society in Eighteenth-Century England,* 2, 17–18; Mingay, *English Landed Society in the Eighteenth Century,* 8–11; Crook, *American Democracy in English Politics,* 146–63.

33. Gladstone, "Kin beyond the Sea," 355; Matthew Arnold, "Equality," in *Essays Religious and Mixed,* 277–305.

34. Brown, *America,* iv; Stirling, *Letters from the Slave States,* 83; Chevalier, *Society, Manners, and Politics in the United States,* 62–66, 289, 396–98; Hodgson, *Remarks during a Journey through North America* (1823), 83–85; Doyle, *Aristocracy and Its Enemies,* 338.

35. Pulszky and Pulszky, *White, Red, Black* (1853), vol. 1: 121–22; Chevalier, *Society, Manners, and Politics in the United States,* 399. Jane Censer argues against southerners' believing that labor was disrespectful (*North Carolina Planters and Their Children,* 48).

36. Thompson, *English Landed Society in the Nineteenth Century,* 145.

37. For additional examples on attention to aristocracy in Europe, see "Life of an English Nobleman," *Farmer and Gardener* 4 (January 9, 1838): 291; H. C. Wright, "Old England and New England Laborers on Land Contrasted," *Boston Liberator,* May 26, 1848, [81].

38. There is a wealth of studies here, but for adequate summaries, see Huston, *Panic of 1857 and the Coming of the Civil War,* 79–87; Jenkins, *Pro-Slavery Thought in the Old South*; Genovese, *World the Slaveholders Made,* 119–244; Tise, *Proslavery*; O'Brien, *Conjectures of Order* 2: chaps. 16, 17; Lacy Ford, "Reconfiguring the Old South," 95–122. For extreme proslavery views, see selections in Elliott, ed., *Cotton Is King* (1860); and Faust, ed., *Ideology of Slavery.*

39. Mitchell, *Edmund Ruffin,* 83; Pollard, *Lost Cause* (1867), 50; Abel Upshur, "Domestic Slavery," *Southern Literary Messenger* 5 (October 1839): 685.

40. See Huston, *Panic of 1857 and the Coming of the Civil War,* 87–97; Osthaus, "Work Ethic of the Plain Folk," 745–82; Startup, *Root of All Evil,* 36–44.

41. Entry of July 4, 1856, in David Gavin Diary, UNC. For recent work on southern conservatism and hostility to democracy, see Read, *Majority Rule versus Consensus,* chap. 1; Allitt, *Conservatives,* chap. 2; Tate, *Conservatism and Southern Intellectuals*; Sinha, *Counterrevolution of Slavery*; 2–5, 20–25; O'Brien, *Conjectures of Order* 2: chap. 16. For Herrenvolk democracy, see Fredrickson, *Black Image in the White Mind,* 61–68.

42. Quotes in Hill, *Lord Acton,* 387, and in Himmelfarb, *Lord Acton,* 79, 83, and 79–82. The restricted idea of liberty in England is not the theme of most writers; for a contrast, see Greene, "Introduction: Empire and Liberty," 3–4; but balance it also against Mancke, "Languages of Liberty in British North America," 29–44.

43. See Clark, *English Society,* 103; Mingay, *English Landed Society in the Eighteenth Century,* 11; Murphy, *L. Q. C. Lamar,* 48; Williams, *Rise and Fall of "The Model Republic"* (1863), 148–53. The

rejection of liberty for everyone is explicit in the "Vindication of Slavery" by Chancellor Harper, contained in Pulszky and Pulszky, *White, Red, Black* 2: 319–20. As to the ideas that composed southern politics, the visitors Francis and Theresa Pulszky decided that the South had the same notions "upon which despotism is based in Europe" (*White, Red, Black* 2: 213).

44. Boaz, *Libertarianism,* 2; Morgan, *American Slavery, American Freedom*; Furstenberg, "Beyond Freedom and Slavery," 1295–1330.

45. A writer complimenting John C. Calhoun's *Disquisitions* came to the same conclusion in 1853: one had to have equality and liberty together; if equality were removed, the liberty became aristocracy ("Democracy—What Is It?" *United States Review* 33 [July 1853]: 4).

46. Greenberg, *Masters and Statesmen,* x, 6, 87–88, 130; Ford, *Origins of Southern Radicalism,* 349–54; Cooper, *Liberty and Slavery,* 14, 43–45; Startup, *Root of All Evil,* 56.

47. On slaveholders' fear of nonslaveholders, see Johnson, *Toward a Patriarchal Republic,* xviii–xxi; Tracy, *In the Master's Eye,* 17, 177–86, 197–212; Freehling, *Road to Disunion: Secessionists Triumphant,* xiv, 16, 68–69, 250–57; for a different view, see Walther, *Shattering of the Union,* xix–xxi.

48. Editor, "Our Slaves," *Southern Agriculturist* 2 (December 1829): 575; James J. Butler, "Agricultural Department," *De Bow's Review* 2 (March 1851): 325; Ruffin, *Political Economy of Slavery* (1857), 9; Dabney, *Defense of Virginia,* 278.

49. One assumes that all the contemporary commentary about the haughty attitudes of slaveholders had some basis in fact, but how much credence to give such expressions is difficult to know. The same haughtiness can also be exuded by people occupying high positions in religious organizations and corporations. Some of the variations on the matter of slavery's impact on the masters' general behavior and attitude include James Oakes, who says the master lost some of his mastery because of the need to care for the slave; John Hope Franklin and Loren Schweninger held that runaway slaves made slavemasters be brutal; Drew Gilpin Faust believes the mastery factor overwhelmed all social traits of the slaveholder; Stephanie McCurry finds yeomen and planters coalesced together in defense of the household and patriarchy; Kenneth Greenberg believes the master class had trouble balancing their obvious mastery over slaves with republican freedom; Lacy Ford seems to imply that mastery was not much of a problem between whites as he describes South Carolina as a democratic society; William Freehling infers mastery because of class tensions between nonslaveholding whites and planters. See Oakes, *Slavery and Freedom,* 54, 88; Franklin and Schweninger, *Runaway Slaves,* chap. 10; Faust, *James Henry Hammond and the Old South,* 307; McCurry, *Masters of Small Worlds,* 5–22; Greenberg, *Masters and Statesmen,* vii–xi, 87–88, 94–97; Ford, *Origins of Southern Radicalism,* 349–51; Freehling, *Road to Disunion: Secessionists at Bay,* chap. 3.

50. Huston, *Calculating the Value of the Union,* 43–57. The idea that northerners acted politically on fears of the slave power is the thesis of Holt, *Political Crisis of the 1850s*; Gienapp, *Origins of the Republican Party*; Richards, *Slave Power*; and Fehrenbacher, *Slaveholding Republic.*

8. The Agricultural Basis of the Free Labor Ideology in the United States

1. For an example, Zakim and Kornblith, "Introduction: An American Revolutionary Tradition," 1–12.

2. Genovese, *Political Economy of Slavery,* 4–39.

3. David Montgomery was one of the first to define free labor ideas (*Beyond Equality,* 14–17).

4. Foner, *Free Soil, Free Labor, Free Men,* chaps. 1, 2; see also Richardson, *Greatest Nation of the Earth,* 4–9, 14–20.

5. Davis, *Problem of Slavery in the Age of Revolution,* 303–42, 455–67, 491–96.

6. Consult Glickstein, "Poverty Is Not Slavery," 195–216; Foner, "Abolitionism and the Labor Movement in Antebellum America," 263–67; Stanley, *From Bondage to Contract,* 8–29; Clark, *Social Change in America,* 229–33. For examples of historians who find the capitalist imperialism of the abolitionists too harsh a judgment, given that they fought the people who possessed real political and economic power, consult Stewart, *Holy Warriors,* 33–44, 79–80, 171–76; Robertson, *Parker Pillsbury,* 50–60; Goodman, *Of One Blood*; Volpe, *Forlorn Hope of Freedom,* 61–65; Davis, *Joshua Leavitt,* 139–40. Seymour Drescher has written that a free labor ideology was not voiced in British antislavery ("Free Labor vs. Slave Labor," 56–76; *From Slavery to Freedom,* chaps. 1, 14).

7. For general interpretations of how the free labor ideology and wage labor operated in the South and in northern cities, see Ford, *Origins of Southern Radicalism,* 351–59; Barnes, *Artisan Workers in the Upper South,* 5–10; Towers, *Urban South and the Coming of the Civil War,* 37–71; Gillespie, *Free Labor in an Unfree World,* 24–28, 150–59; Steinfeld, *Coercion, Contract, and Free Labor in the Nineteenth Century*; Blumin, *Emergence of the Middle Class,* 1–12; Gilkeson, *Middle-Class Providence,* 12–17, 50–54. There are numerous places to obtain the Marxist framework; see Cohen, *Karl Marx's Theory of History,* chaps. 2, 3; Elster, *Making Sense of Marx.*

8. John Ashworth, *Slavery, Capitalism, and Politics in the Antebellum Republic* 1: *Commerce and Compromise,* ix–x, 10–12, 83–86, 117–18, 148–67; vol. 2: *The Coming of the Civil War,* 1–5, quote on 5; Glickstein, *Concepts of Free Labor in Antebellum America,* 13–19, 35–58. See the balanced treatment in Laurie, *Artisans into Workers,* 51–57.

9. Rockman, *Scraping By,* 5–6, quote on 241. Larson, *Market Revolution in America,* 131–34; on Civil War genesis, 160–64. For a different interpretation, based on free labor meaning competency, see Tuchinsky, *Horace Greeley's New-York Tribune,* 239–40, 282. For an application of how the free labor ideology was culturally hostile to African Americans and Native Americans, see Oertel, *Bleeding Borders,* 35–38.

10. This interpretation is fully laid out in Larson, *Market Revolution in America,* 154–62. Others do not follow this interpretation, arguing instead that slaveholders acted aggressively; see Fehrenbacher, *Slaveholding Republic*; Richards, *Slave Power*; Huston, *Calculating the Value of the Union,* chaps. 2, 3.

11. Ginger, *Altgeld's America,* 5. An alternative to the presentation of free labor given in this chapter is Phillips, "Antebellum Agricultural Reform, Republican Ideology, and Sectional Tension," 799–822.

12. The adjective "Marxist" is used here in a shorthand way to identify authors who engage in the style of class and property-ownership analysis, but who may or may not be formal Marxists.

13. To put it another way, there is a recognition that productivity enters into the equation—that if the productivity of the worker was high enough, the capitalist might enjoy more exploitation of a worker even if wages rose above subsistence. For example, say a worker's wage was $1 per day, call it a subsistence wage, out of which he/she made $1.50 of product. The exploitation rate is the amount the worker receives compared to the total value he/she made: or, 100% - [($1.00/1.50) × 100%] or 33 percent. But if incentives were added and the wage increased, the wage might become $1.50 per day but the output worth $3.00. Then the exploitation rate would be 100% - [($1.50/ $3.00) × 100%], or 50 percent. The exploitation is more than in the first

case, but the wage has risen and is above subsistence. On the Marxist theory of wages, see Wolff and Resnick, *Economics,* 31–32, 143–54; criticism of the necessity of subsistence wages for Marxist theory and the labor theory of value may be found in Elster, *Making Sense of Marx,* 119–37, 170–203; Roemer, *Analytical Foundations of Marxian Economic Theory,* 2–13, 52–55.

14. In class analysis, the emergence of a wage-earning group was the signal that economic hell was about to be entered. See Bowman, *Masters and Lords,* 83–109; Cohen, *Karl Marx's Theory of History,* 63–77; Ashworth, *Slavery, Capitalism, and Politics in the Antebellum Republic* 1: *Commerce and Compromise,* ix–x, 83–86.

15. Ware, *Labor Movement in the United States,* 37, 227–30.

16. See Huston, *Securing the Fruits of Labor,* chap. 2.

17. Montgomery, *Beyond Equality,* 31. For twentieth-century American adjustment to the corporation, see Galambos, *Public Image of Big Business in America.*

18. Found in *Complete Works of Abraham Lincoln,* ed. Nicolay and Hay, vol. 5: 249–50. Biographers of Lincoln continue to grapple with his celebration of free labor ideals and especially his emphasis on social mobility; see Oakes, "Natural Rights, Citizenship Rights, States' Rights, and Black Rights," 113–15; Foner, *Fiery Trial,* 12–17; Frayssé, *Lincoln, Land, and Labor,* 166–71; Donald, *Lincoln,* 233–35. For a different view, Winger, *Lincoln, Religion, and Romantic Cultural Politics,* 117–22.

19. De Crèvecoeur, *Letters from an American Farmer* (1782), 52–53; in general, 49–55.

20. Cooper, *Some Information Respecting America,* quote on 37 and 50; see 48–50.

21. An American, *American Husbandry* (1775), 1:70–91, quote on 71. On the early formulation of free labor principles, see Huston, *Securing the Fruits of Labor,* 66–69; Eva Sheppard Wolf, "Early Free-Labor Thought and the Contest over Slavery in the Early Republic," 32–48.

22. On the intensity of the backwoods war, reiterated in all frontier histories, see especially Griffin, *American Leviathan.*

23. Regan, *Emigrant's Guide to the Western States,* vi; Robinson, "To Western Emigrants—No. 2," *Albany Cultivator,* December 1840, in *Solon Robinson* 1: 148; Thomson, *Tradesman's Travels,* 77; Ferrall, *Ramble of Six Thousand Miles through the United States,* 66–67.

24. Mrs. Frances S. Osgood, "The Poetry of Labor," in Blake, ed., *Lessons in Modern Farming* (1852), 407–8; Phillippo, *United States and Cuba* (1857), 328.

25. Shirreff, *Tour through North America* (1835), 450; quote in Abdy, *Journal of a Residence and Tour* 1: chap. 9. The numbers Shirreff gives are these: American farm laborers earned $100 to $120 per year, and an ordinary worker obtained $260 per year; the weekly wages of farm laborers in England, Scotland, and Ireland were 10s., 8s., and 3s. 6p. respectively.

26. Regan, *Emigrant's Guide to the Western States,* 350; report of the English Commission on Agricultural Depression in *New England Farmer and Horticultural Register* 59 (October 30, 1880): 1. See Caird, *Prairie Farming in America,* 116.

27. Flower, *Errors of Emigrants* (1841), 52–53; Shirreff, *Tour through North America,* 400. Stanley Lebergott notes that the social mobility theme was present in the United States and absent in Europe (*Manpower in Economic Growth,* 119–20). A considerable literature surrounds wages in antebellum America, almost all of it dealing with urban and industrial labor, not with agricultural labor; see Margo, *Wages and Labor Markets in the United States*; Temin, "Labor Scarcity in America," 251–64; James and Skinner, "Resolution of the Labor-Scarcity Paradox," 513–40.

28. An American, *American Husbandry* 1: 68; Morse, *American Geography,* 316; Prentice, *Tour in the United States,* 84; Wright, *Views of Society and Manners in America,* 186–89, Martineau, *Retrospect of Western Travel* (1838), vol. 1: 87.

29. Bird, *Englishwoman in America,* 39; *Cassell's Emigrants' Handbook* (1852), 10; Henry Colman, "An Address Delivered before the Massachusetts Agricultural Society at the Brighton Cattle Show," *Massachusetts Agricultural Repository and Journal* 7 (January 1, 1822): 1.

30. Bird, *Englishwoman in America,* 3; Hancock, *Emigrant's Five Years in the Free States* (1860), 86–87.

31. Quote in Freeman, *Some Impressions of the United States* (1883), 215; Charles L. Flint, ed., introduction, *American Farmer* 1: 5; Brydges, *Uncle Sam At Home* (1888), 206; Steevens, *Land of the Dollar,* 157; Campbell, *White and Black* (1879), 27, 97–102.

32. This is in contrast to the idea that American freedom was defined by American slavery, as posited in Morgan, *American Slavery, American Freedom*; and Davis, *Inhuman Bondage,* 177.

33. On Canada and the free labor ideology, see Bird, *Englishwoman in America,* 301–2; Hall, *Travels in North America* (1830), vol. 1: 299; Tudor, *Narrative of a Tour in North America* 1: 287–88; Russell, *North America,* 52; MacKenzie, *Sketches of Canada and the United States* (1833), 346–51.

34. In the studies of American life in the nineteenth century, neither the yeoman farmer nor the ideal of mobility has fared well. Early twentieth-century writers on American farmers believed farmers were not really capitalists (Joseph Schafer, John D. Hicks, Paul Wallace Gates). Later scholars like Richard Hofstadter took the "yeoman and the myth" to the whipping post and administered some strong lashes that left permanent scars—the yeoman farmer was really a businessman with little attachment to the soil. The social mobility theme came in for its share of abuse, referred to as the "cult of the self-made man." The latter applied free labor ideals about social mobility to urban, industrial America, exactly where they did not exist. See Gates, *Farmer's Age*; Hicks, *Populist Revolt*; Schafer, *Social History of American Agriculture,* 288–93; Hofstadter, "The Yeoman and the Myth," in *Age of Reform,* 23–45; see also Griswold, *Farming and Democracy,* 178–204. On social mobility, Burns, *Success in America,* chaps. 2–5; Wyllie, *Self-Made Man in America,* 10–23, 154–60. On the truth of social mobility in the rural North see Jeremy Atack, "Agricultural Ladder Revisited," 1–25. On saving, see Anbinder, "Moving beyond 'Rags to Riches,'" 741–70.

35. Allen Guelzo insists that Lincoln abhorred agriculture and the community attached to it. That Lincoln tired of agricultural labor is agreed upon, but that he separated himself from the farm-village economy is dubious; Springfield, Illinois, was no Chicago (*Abraham Lincoln as a Man of Ideas,* 52–54, 59). On Lincoln's dislike of agriculture and farming, see Winkle, *Young Eagle,* 305; Burlingame, *Lincoln* 1: 571–72; Donald, *Lincoln,* 233–34; Boritt, *Lincoln and the Economics of the American Dream,* 185–89.

36. Colman, *European Agriculture and Rural Economy* 1: 41, 43. It should also be noted that agricultural laborers in England, even if they could have made savings, would never have owned their own land because of high land prices.

37. Ibid., 65.

38. Ibid., 72–73.

39. Chambers, *Things as They Are in America* (1854), 236–37.

40. Quote from Lang, *Religion and Education in America* (1840), 40–41 (on urban workers); "The English Peasantry," *Boston Review* 2 (July 1862): 411, quote on 417–18.

41. Bird, *Englishwoman in America,* 444; Blowe, *Geographical, Historical, Commercial, and Agricultural View of the United States* (1820), 73; Lang, *Religion and Education in America,* 40–41; Mitchell, *Ten Years Residence in the United States,* 164–65; MacKenzie, *Sketches of Canada and the United States,* 356–57.

42. Huston, "Political Response to Industrialism," 35–57; Huston, *Securing the Fruits of Labor,* chap. 5.

43. Mill, *Principles of Political Economy* (1871), 253. This sentiment is echoed elsewhere; for example, by Walker, *Wages Question* (1876), 127.

44. Quote in Griffiths, *Two Years in the New Settlements of Ohio,* 56; Shirreff, *Tour through North America,* 348; Cobbett, *Emigrant's Guide* (1830), 47.

45. Bristed, *Resources of the United States,* 53–54.

46. Fearon, *Sketches of America* (1818), quote on 224, see 173, 225; Dalton, *Travels in the United States of America,* 253–54.

47. Trollope, *North America* 1: 203–4. The abolitionists in many ways are the heart of the interpretation that the battle over slavery was a struggle to obtain the dominance of industrial capitalism because they were the idealists justifying wage labor. I would argue that abolitionists were not infatuated with manufacturing, that they believed in high wages and deprecated large companies that did not pay adequate wages—but they did accept the capitalist market. They represented the farm-village outlook of the North. Political economists, like Henry Charles Carey, also were enamored with high wages but saw some of the problems of social mobility in the city. Northern urban workers did not accept the free labor theory for the obvious reason that it was an agrarian outlook that simply did not comport with urban life and business units employing scores of people. Finally, I have found almost no social mobility themes (or high-wage advocacies) among British abolitionists.

48. "Independence of the Farms," *New England Farmer* 5 (February 1853): 84; Ebeneezer Mack, "Progress of Agriculture," in Blake, ed., *Lessons in Modern Farming,* 265; Josiah T. Marshall, "Agricultural: An Address Delivered before an Agricultural Meeting at Plessis, Jefferson County, N.Y.," *New England Farmer and Horticultural Register* 17 (January 16, 1839): 217.

49. Lyell, *Second Visit to United States* 1: 100; Sullivan, *Rambles and Scrambles in North and South America* (1852), 181.

50. Nichols, *Forty Years of American Life* (1864), vol. 2: 134; Haliburton, *Americans at Home* (1854), vol. 1: vi; Sullivan, *Rambles and Scrambles in North and South America,* 181; quote from Stirling, *Letters from the Slave States,* 65.

51. Cronon, *Nature's Metropolis,* chap. 8. My presentation focuses more on the rural-urban divide as defined by self-employment versus working for others rather than differing cultural values. On the egalitarian rural North, see Atack and Bateman, "'Egalitarian Ideal' and the Distribution of Wealth in the Northern Agricultural Community," 129–32.

52. On social mobility and free labor themes in the South, see Huston, *Panic of 1857 and the Coming of the Civil War,* 84–90; Osthaus, "Work Ethic of the Plain Folk," 746–81; Harris, *Plain Folk and Gentry in a Slave Society,* 6, 20–25, 30, 77–85.

53. There is a considerable historiography here, but for a shortcut to it go to Faust, ed., *Ideology of Slavery*; and Cooper and Terrill, *American South* 1: 257–65.

54. Ruffin, *Political Economy of Slavery,* 6–7, 9.

55. Quote in "Good Farming," *Southern Planter* 7 (November 1847): 332; A Planter, "On the Importance of Farmers," *Farmer's Register* 4 (May 1836): 1.

56. William R. Russell, *My Diary North and South* (1862), 1.

57. R. L. Allen, *New American Farm Book* (1869), 508. See especially the idea of free labor as self-mastery in Tuchinsky, *Horace Greeley's New-York Tribune,* 239–40; for the idea of mastery, see Cayton, *Frontier Indiana,* 76–77; and Marx, *Machine in the Garden,* 121–28.

9. Davids versus Goliaths: The Sectional Controversy as a Fight between Agricultural Regions

1. Three works are most useful in laying out the historiography. First, an older work that looks at historical writing up to 1957 is Pressly, *Americans Interpret Their Civil War*; second, the collection of explanations by Stampp, ed., *Causes of the Civil War*; and third, the recent excellent article by Towers, "Partisans, New History, and Modernization: The Historiography of the Civil War's Genesis," 237–64. The use of religion in Civil War causation can be seen in Carwardine, *Evangelicals and Politics in Antebellum America*. Free labor is explained by Foner, *Free Soil, Free Men, Free Labor*. On southern studies, southern white egalitarianism is examined in Thornton, *Politics and Power in a Slave Society*; for southern republicanism, Ford, *Origins of Southern Radicalism*, 347–60; and Morrison, *Slavery and the American West*, 6–7, 58–61, 114–19; the honor-society thesis is found in Wyatt-Brown, *Southern Honor*; on Celtic origins, McWhiney and Jamieson, *Attack and Die*, chap. 12; on fear of free labor, Towers, *Urban South and the Coming of the Civil War*, 1–10; on planter wealth, Huston, *Calculating the Value of the Union*, chap. 2; on fear of the nonslaveholder, Freehling, *Road to Disunion: Secessionists Triumphant*, 265–66; Johnson, *Toward a Patriarchal Republic*, xx–xxii, 85–101. For Genovese, see his *Political Economy of Slavery*.

2. On contingency theory, see Ayers, *What Caused the Civil War*, 117–43; Freehling, *Road To Disunion: Secessionists Triumphant*, 406–26. For the collapse of the second party system, Holt, *Political Crisis of the 1850s*. On expansion and a weak federal government, Potter, *Impending Crisis*; Peter Parish, *American Civil War*. On the slave power, Richards, *Slave Power*; Fehrenbacher, *Slaveholding Republic*. On revisionists, Randall, "Blundering Generation," 3–28; and Milton, *Eve of Conflict*. For the ideas concerning nationalism and abolitionism, Parish, *North and the Nation in Era of the Civil War*; Grant, *North over South*; Stewart, *Abolitionist Politics and the Coming of the Civil War*, 205–11. The notion that the country split because of the lack of European intrigues in North America is in Schoen, *Fragile Fabric of Union*, 195–96, and Brown, *Agents of Manifest Destiny*, 460–62.

3. Beard and Beard, *Rise of American Civilization* 1: 635–42, 678–82; vol. 2: 1–54; Hacker, *Triumph of American Capitalism*, 329–41. Recent restatements include Bensel, *Yankee Leviathan*; Egnal, *Clash of Extremes*.

4. McPherson, "Antebellum Southern Exceptionalism," 230–44.

5. McPherson and Logue, *Ordeal By Fire*, chap. 1; on modernization, Brown, *Modernization*; early anticipations of modernization are in the revisionist writings of Craven, *An Historian and the Civil War*, 28–45; on free labor, Ashworth, *Slavery, Capitalism, and Politics in the Antebellum Republic* 2: *The Coming of the Civil War*, 3–5. On abolitionism, Stewart, *Holy Warriors*, 79–80, 175–76.

6. For a critique of the modernization thesis, see Barnes, Schoen, and Towers, "Introduction: Reimagining the Old South," 3–22.

7. Cohen, *Self-Ownership, Freedom, and Equality*, 12–16; Cohen, *Karl Marx's Theory of History*, 34–37, 63–79.

8. The comments of Steinfeld and Engerman in this connection are interesting, though I have some qualms about confusing the two terms ("Labor—Free or Coerced?" 107–26). On shirking, see Miller, *Managerial Dilemmas*, 1–18, 31–36, 49–50, 58.

9. Fred A. Shannon gave a classic example the progressives' tendency to misconstrue the sectional contest: the Civil War was "a contest for supremacy between Southern agricultural capitalists and Northern industrial capitalists" (*Farmer's Last Frontier*, 76).

10. Parker, "A Sermon on Slavery," 1841, in Parker, *Collected Works* 5: 1–2; "The Mission of Democracy," Washington, D.C., *National Era,* April 13, 1848, 1; speech of McClernand, c. 1870, in Untitled Speech fragments, box 2, folder "Speeches," D70, in John A. McClernand Papers, ALLib. On northern fears of inequality in southern ideas, see Tewell, *Self-Evident Lie.*

11. The relevant works are Beard and Beard, *Rise of American Civilization*; Hacker, *Triumph of American Capitalism*; Egnal, *Clash of Extremes*; Bensel, *Yankee Leviathan*; Richardson, *Greatest Nation of the Earth*; Curry, *Blueprint for Modern America.*

12. On the tariff battle, see Huston, *Securing the Fruits of Labor,* 152–83; on internal improvements, Paskoff, *Troubled Waters,* chaps. 1, 2; on clashes over the public lands, Feller, *Public Lands in Jacksonian Politics.*

13. For examples of those who doubt the importance of federal economic policies, see McPherson, *This Mighty Scourge,* 4–12; Ayers, *What Caused the Civil War,* 132; Towers, "Partisans, New History, and Modernization," 237–64.

14. See Hammond, *Slavery, Freedom, and Expansion in the Early American West,* chaps. 5, 6; Nation, *At Home in the Hoosier Hills,* 188–90.

15. Robert, *Road from Monticello,* quotes on 22 and 64; see 26–27, 69, 75.

16. Quotes of C. C. Clay and Francis Lieber in Phillips, *American Negro Slavery,* 334, 335; *Weekly Wisconsin* (Milwaukee), October 1, 1856.

17. *Weekly Wisconsin* (Milwaukee), October 25, 1848, 3; Whig county convention in *Madison Express,* August 4, 1846. On labor and land reform, see Bronstein, *Land Reform and Working-Class Experience in Britain and the United States.*

18. The full title of the *Indiana Farmer* reads *Indiana Farmer. Devoted to Agriculture, Horticulture, Mechanics and the Useful Arts*; "Young Mechanics," *New England Farmer* 17 (October 31, 1838): 136; "Mechanics' Department: To The Mechanics of Michigan," *Michigan Farmer* 2 (February 15, 1844): 6.

19. Yates, *Speech at Springfield, June 1860,* 7; Unknown, *What Democratic Leaders Think of Slavery* (1865), 3; Robert, *Road from Monticello,* 76–77.

20. Helper, *Impending Crisis of the South* (1857), 41–44, 120; Olmsted, *Cotton Kingdom,* 148–49, 212–13, 257, 371, 378–81, 402–4, 518–29, 557–63; Palmer, *Personal Recollections* (1901), 9.

21. Flower, *History of English Settlement in Edwards County, Illinois* (1882), 209–10; Helper, *Impending Crisis of the South,* 28; *Brownlow's Knoxville Whig,* October 13, 1860, 1.

22. Huston, "American Revolutionaries, the Political Economy of Aristocracy, and the American Concept of the Distribution of Wealth," 1099–1100.

23. This is spelled out completely by Richards in *Slave Power.*

24. For economic issues see Huston, *Panic of 1857 and the Coming of the Civil War*; for an alternative view, Egnal, *Clash of Extremes.*

25. Craven, "Coming of the War Between the States," 304–22. The idea of the revisionists has come again in the form of "deep contingency"; Ayers, *What Caused the Civil War,* 132–43; Freehling, *Road to Disunion: Secessionists Triumphant,* 406–26; Holt, *Fate of Their Country,* 126–27. On southern disavowal of extending slavery into the North, see Cooper, "Critical Signpost on Journey to Secession," 3–16. On the Democrats' position, see Huston, *Stephen A. Douglas and the Dilemmas of Democratic Equality,* 181–93.

26. Schoen, *Fragile Fabric of Union,* 185–96, 213–37; Silbey, *Storm over Texas,* 146–74.

27. Genovese, *Political Economy of Slavery,* 243–74; Kennedy, *Mr. Jefferson's Lost Cause,* 12–25; Majewski, *Modernizing a Slave Economy,* 4–9, 43–46; Barney, *Secessionist Impulse,* 8–16. Others would say that the territorial issue was in one way or another symbolic: southerners wanted

an equality within the Union and would not be dishonored by having their peculiar institution declared unsuitable for the public lands. See Cooper, *Liberty and Slavery,* 252–67; Wyatt-Brown, *Yankee Saints and Southern Sinners,* 183–92, 203; Morrison, *Slavery and the American West,* 57–61, 190, 277–78. Many writers believe that as well political balance in the Senate was the key to southern views on expansion; they needed more slave states to stop the advance of an overwhelming northern majority, a majority hostile to slavery. See Cooper, *Liberty and Slavery,* 219–21; Weingast, "Political Stability and Civil War," 148–93.

28. Huston, *Panic of 1857 and the Coming of the Civil War,* 247–51; Huston, *Calculating the Value of the Union,* 100–103; see Bronstein, *Land Reform and Working-Class Experience*; Lause, *Young America.*

29. In general, McPherson and Logue, *Ordeal by Fire,* 79–83; May, *Southern Dream of a Caribbean Empire.*

30. Huston, *Calculating the Value of the Union,* 153–89; Holt, *Political Crisis of the 1850s,* 87–103.

31. See Huston, *Calculating the Value of the Union,* 190–212.

32. *Boston Semi-Weekly Advertiser,* May 20, 31, 1854, 1. *New York Tribune,* February 2, 1854, 4; May 24, 1854, 4.

33. On Dred Scott, see Huston, *Calculating the Value of the Union,* 217–19; for Kansas, see Etcheson, *Bleeding Kansas,* 3–7.

34. Hey, *English Rural Community,* 143.

35. The counties composing the congressional districts between 1842 and 1851 were considerably different from those between 1852 and 1861. The division of state congressional districts into agricultural, balanced, and manufacturing (industrial) categories is as follows, the numbers that follow the state names being the numbers of the congressional districts.

1842–51:

Agricultural: New Jersey, 3; New York, 14, 15, 17–19, 21, 22, 23, 25, 26, 27, 29–34; Pennsylvania, 12, 15, 16, 18, 20, 22, 23, 24; Illinois, 1–7; Indiana, 1–10; Iowa, 1–2; Michigan, 2, 3; Ohio, 2–11, 13, 15–19, 21; Wisconsin, 2, 3; Vermont, 1–3.

Balanced: New Jersey, 1, 2, 4; New York, 1, 8–11, 16, 20, 24, 28; Pennsylvania, 5–11, 13, 17, 19; Michigan, 1; Ohio, 12, 14, 20.

Industrial: New Jersey, 5; New York, 2–7, 12, 13; Pennsylvania, 1–4, 14, 21; Ohio, 1; Wisconsin, 1.

1852–60:

Agricultural/Rural: New York, 10, 15–19, 21–23, 28, 31, 33; Pennsylvania, 7, 14, 16, 17, 19, 20, 23–25; Illinois, 1–9; Indiana, 1–11; Minnesota, 1, 2; Ohio, 3–13, 17, 18, 20, 21; Wisconsin, 1–3; Michigan, 2–4; Vermont, 1–3.

Balanced: Michigan, 1; Ohio, 19; New Jersey, 1–4; New York, 9, 11, 12, 20, 29, 32; Pennsylvania, 6, 8, 9, 10, 13, 15, 18.

Industrial: Ohio, 1, 2; New Jersey, 5; New York, 1–9, 3, 14; Pennsylvania, 1–4, 5, 11.

36. Again, it must be stressed that what is offered here is a perspective based wholly on northern congressional-district electoral winners using a farming-industrial axis. Voting studies do not reveal this division because they do not use congressional districts as the units of analysis. They attempt to explain group voting or individual voting, and routinely find evangelical, nonevangelical, ethnicity, and age as the primary explanatory variables. To the extent that

they use a rural-urban split, such historians devise variables as a percentage of a county that is farming, manufacturing, urban (if available); average value of farm; average value of farm equipment; and the like. While some associations exist, most scholars emphasize ethnicity and religious persuasion. For works that generally cover the voting characteristics of northerners and detail political realignment, see Kleppner, *Third Electoral System*; Gienapp, *Origins of the Republican Party*; Holt, *Political Crisis of the 1850s*. For Democrats, see Huston, "Illinois Political Realignment of 1844–1860," 506–35.

Epilogue

1. I will not undertake an extensive examination of the topic here but only point out how little change came to northern agriculture due to the Civil War. For an exposition on how the conflict changed the United States, see McPherson, "Civil War and the Transformation of America," 1–8; a discussion of the topic can be found in Huston, *Calculating the Value of the Union*, 233–36.

2. Arnstein, *Britain Yesterday and Today*, 209–13; Alfred Havighurst adds that land taxes after World War II ultimately crushed the landed aristocracy (*Twentieth Century Britain*, 144); Beckett, *Aristocracy in England*, 13, land sales figures on 474–78; Cannadine, *Decline and Fall of the British Aristocracy*, 27–28; Stone and Stone, *Open Elite*, 425–26.

3. Grigg, *English Agriculture*, table 8.3, pp. 104–5.

BIBLIOGRAPHY

Primary Sources

Manuscripts, Ledgers, and Diaries

Adrich, Levi Burrel. Papers, 1853–96. MnnHS.

Bacot Family Papers. UNC.

Bailey, John Campbell. Diaries. Transcripts. From Rock Island Historical Society, Moline, Ill. Transcript housed at ALLib.

Bayside Plantation. Records. UNC.

Beeler, Fielding. Account Book, 1857–79. IndHL.

Beeler, Joseph Fielding. Diary, 1844–49, and Account Book, 1836–49. IndHL.

Bennett, Oliver. Diary. In Oliver Bennett Family Papers. WesResHS.

Bigelow, Amasa, and Daniel Thomas Papers, 1851–77. WisHS.

Blackford, John. Diary, 1836–38. In Blackford Family Papers, VHS.

Brown Family. Papers. UVa.

Brown, Jabez. Papers. WisHS.

Buell, Barber G. Papers. UMich.

Carpenter, Mary E., and Family. Correspondence. MnnHS.

Carrier, Spencer and Emily. Diaries. WisHS.

Clapp, Thomas J. Account Book, 1844–64. In Collections of Account Books, WesResHS.

Cleveland, Moses A. *Autobiography.* 2 vols. UMich2.

Coales, Benjamin. Business Records, 1845–58. UIll.

Cocke, James Powell. Diary, 1836, 1851–57. In the Cocke Family Papers, VHS.

Cocke, John Hartwell. Papers and Farm Journals. Box 188. UVa.

Conway Family. Papers and Ledgers. UVa.

Countryman, Levi N., and Family. Papers. Typescript. MnnHS.

Dibble, Leonard. Papers. MnnHS.

Downer Family. Diaries. ALLib.

Drew, E. B., Diary. In James M. Drew and Family Papers. MnnHS.

Dumville, Hepzibah. Papers. ALLib.

Eppes, Richard. Account Book, 1851–61, 1865–68. In Eppes Family Papers, VHS.

Farmer, William. Farm Journal. In Irvine, Saunders, Davis, and Watts Families, Papers and Farm Journals. UVa.
Gavin, David. Diary, 1855–74. UNC.
Hamilton, William. Ledger/Daybook. OHS.
Harner, William Mathias. Diary, 1847–61. VHS.
Hart, Lovira. Letters, 1837–53. UMich.
Hawker, Abraham. Daybook and Ledger, 1850–69. OHS.
Holgate, William C. Papers. OHS.
Holt, Martin Sidney. Papers. ALLib.
Hubard, Robert Thurston. Journal of Farm Accounts at Chellowe. UVa.
Ingraham, Edward Henry. Diary, 1852. ALLib.
Kemper, William S. Ledger. UVa.
Kirtland and Morse Families. Papers, 1792–1806. WesResHS.
Kollock, George J. Plantation Journals, 1837–61. UNC.
Lane, Warren. Diary. In Land and Wheller Family Papers, 1821–1918. WesResHS.
Lesher, Lewis. Farm Accounts. OHS.
Linville, B. A. Vol. 859, Diary and Record Book, 1867–1869. OHS.
Marshall, S. W. Account Book, 1855–57. VHS.
Massie Family Papers. Account Books, 1813–34, 1828–34. VHS.
McClernand, John A. Papers. ALLib.
Moulton, Hiram. Ledger/Daybook. OHS.
Muzzy, Issac W. Papers. MnnHS.
Parsons, Oliver. Diary and Account Book. MnnHS.
Paul, George H. Correspondence. WisHS.
Payne, William James. Journal, 1857–61. UNC.
Phelps-Lyon. Family Papers. UMich2.
Pierce, Darius. Papers. UMich.
Sanders, William H. Account Book, 1840–76. VHS.
Sinclair, George. Journal. UVa.
Sisson, Benjamin. Account Book. UVa.
Southworth, Newton, and Family. Papers. MnnHS.
Tebbetts, William H. Papers. ALLib.
Thomas, Daniel. Papers. In Daniel Thomas and Amasa Bigelow Papers, 1851–77. WisHS.
Timlin, William Henry. Diary. WisHS.
Tyler, George B., and Robert H. Farm Account Book. UVa.
Waring-Baylor Family. Papers. UVa.
Warner, Apollos. Family Papers, 1796–1925. WesResHS.
Warren, Edwin and Ephraim. Diaries, 1863–1908. ALLib.
Wickham, William Fanning. Diaries, 1828–47. In Wickham Family Papers, VHS.
Wildman, Ezra. Account Book, 1813–42. Contained in Warner Family Papers, WesResHS.

Williams, Joseph F. Papers and Ledgers. WesResHS.
Wisconsin Territorial Letters, 1837–52. WisHS.
Wright, William W. Diary. WisHS.
Young, John H. Daybook and Account Book. OHS.

Periodicals and Farm Journals

Albion.
American Agriculturist. New York.
American Farmer. Baltimore.
American Farmer's Magazine. Baltimore.
Boston Cultivator.
Boston Review.
Christian Advocate and Journal.
Cultivator. Albany, N.Y.
De Bow's Review. New Orleans.
Farmer and Gardener. Baltimore.
Farmer's Cabinet.
Farmer's Register. Shellbanks, Va.
Genesee Farmer. Rochester, N.Y.
Illinois Farmer. Springfield.
Indiana Farmer.
Indiana Farmer and Gardener.
Maine Farmer. Augusta.
Massachusetts Agricultural Repository [*and Journal*].
Massachusetts Ploughman and New England Journal of Agriculture.
Michigan Farmer. Jackson.
New England Farmer [*and Horticultural Register*]. Boston.
New York Farmer.
Ohio Cultivator. Columbus.
Ohio Farmer. Cleveland.
Prairie Farmer. Chicago.
Saturday Evening Post. New York.
Southern Agriculturist, Horticulturist, and Register of Rural Affairs. Charleston, S.C.
Southern Cultivator. Atlanta.
Southern Literary Messenger.
Southern Planter.
Tennessee Farmer.
United States Review.
Valley Farmer. St. Louis, Mo.

Newspapers

Boston Liberator.

Boston Semi-Weekly Advertiser.

Brownlow's Knoxville Whig.

Indianapolis State Journal.

Madison Express.

National Era. Washington, D.C.

New York Tribune.

Niles National Register. Baltimore.

Weekly Wisconsin. Milwaukee.

Government Documents

Great Britain. *British Parliamentary Papers: Slave Trade.* Shannon: Irish University Press, 1968.

Great Britain. Registrar-General. *The Census of Great Britain in 1851.* London: Longman, Brown, Green, and Longmans, 1854.

U.S. Congress. *Congressional Globe,* 35th Cong., 1st Sess.

U.S. Dept. of the Census. *The Seventh Census of the United States.* Vol. 1: *Population in 1850.* Washington, D.C.: Robert Armstrong, 1853.

U.S. Dept. of the Interior. *Eighth Census.* Vol. 1: *Population of the United States in 1860.* Washington, D.C.: Government Printing Office, 1864.

——. *Eighth Census.* Vol. 2: *Agriculture of the United States in 1860.* Washington, D.C.: Government Printing Office, 1864.

——. *Eighth Census.* Vol. 3: *Manufactures of the United States in 1860.* Washington, D.C.: Government Printing Office, 1865.

Election Information

Moore, John L., ed. *Congressional Quarterly's Guide to U.S. Elections.* 2nd ed. Washington, D.C.: Congressional Quarterly, 1985.

Whig Almanac and Politician's Register. (Title varies.)

Published Memoirs, Travel Accounts, and Letters

Abdy, Edward S. *Journal of a Residence and Tour in the United States of North America, from April, 1833, to October, 1834.* 3 vols. London, 1835. Rpt. New York: Negro University Press, 1969.

Allen, R. L., and rev. by Lewis F. Allen. *New American Farm Book.* New York: Orange Judd & Co., 1869.

An American. *American Husbandry. Containing an Account of the Soil, Climate, Production and Agriculture of the British Colonies in North-America and the West-Indies.* 2 vols. London: J. Bew, 1775.

Amphlett, William. *Emigrant's Directory to the Western States of North America.* London: Longman, Hurst, Rees, Orme, & Brown Co., 1819.

Arnold, Mathew. *Essays, Religious and Mixed.* Ed. R. H. Super. Ann Arbor, Mich.: University of Michigan Press, 1972.

Badeau, Adam. *Aristocracy in England.* New York: Harper and Brothers, 1885.

Baily, Francis. *Journal of a Tour in Unsettled Parts of North America in 1796 and 1797.* London: Baily Brothers, 1856.

Barclay, Captain (Barclay-Allardice, Robert). *Agricultural Tour in the United States and Upper Canada with Miscellaneous Notices.* Edinburgh: William Blackwood, 1842.

Baxter, W. E. *America and the Americans.* London: George Routledge & Co., 1855.

[Beaufoy] A British Subject. *Tour through Parts of the United States and Canada.* London: Longman, Rees, Orme, Brown, and Green, 1828.

Bernhard, Duke of Saxe-Weimar Eisenach. *Travels through North America, during the Years 1825 and 1826.* 2 vols. Philadelphia: Carey, Lea and Carey, 1828.

Bird, Isabella. *The Englishwoman in America.* 2nd ed. London: John Murray, 1856.

Birkbeck, Morris. *Letters from Illinois.* Philadelphia: M. Carey & Son, 1818.

——. *Notes on a Journey in America, From the Coast of Virginia to the Territory of Illinois.* 2nd ed. London: James Ridgway, 1818.

Blake, John L. *The Farmer's Every-Day Book; or, Sketches of Social Life in the Country: with the Popular Elements of Practical and Theoretical Agriculture.* Auburn, N.Y.: Derby, Miller & Co., 1850.

——, ed. *Lessons in Modern Farming; or, Agriculture for Schools.* New York: Newman & Ivison, 1852.

[Blane, William Newnham.] An English Gentleman. *An Excursion through the United States and Canada during the Years 1822–23.* London: Baldwin, Cradock, and Joy, 1824.

[Blowe, Daniel.] *A Geographical, Historical, Commercial, and Agricultural View of the United States of America; Forming a Complete Emigrant's Directory through Every Part of the Republic.* London: Edwards & Knobb, 1820.

Bradford, William J. *Notes on the Northwest, or Valley of the Upper Mississippi.* New York: Wiley and Putnam, 1846.

Brown, William. *America: A Four Years' Residence in the United States and Canada.* Leeds: Kemplay and Bolland, 1849.

[Brydges, Harold] Bridge, James Howard. *Uncle Sam at Home.* New York: Henry Holt & Co., 1888.

Buckingham, James Silk. *America Historical, Statistic, and Descriptive.* 3 vols. New York: Harper and Brothers, 1841.

——. *The Slave States of America*. 2 vols. London: Fisher, Son & Co., 1842.

Buel, Jesse. *The Farmer's Companion, or, Essays on the Principles and Practice of American Husbandry*. 3rd ed. Boston: Marsh, Capen, Lyon, and Webb, 1842.

Caird, James. *Prairie Farming in America. With Notes by the Way on Canada and the United States*. New York: D. Appleton & Co., 1859.

Cairnes, J. E. *The Slave Power: Its Character, Career, and Probable Designs*. 1862. Rpt. Columbia: University of South Carolina Press, 2003.

Campbell, George. *White and Black: The Outcome of a Visit to the United States*. London: Chatto and Windus, 1879.

Carlisle, Earl (George William Frederick Howard Carlisle]. *Travels in America: The Poetry of Pope: Two Lectures Delivered to the Leeds Mechanics' Institution and Literary Society, December 5th and 6th, 1850*. New York: G. P. Putnam, 1851.

Cassell's Emigrants' Handbook: Being a Guide to the Various Fields of Emigration in All Parts of the Globe with an Introductory Essay. London: John Cassell, 1852.

Chamberlain, E. *The Indiana Gazetteer, or Topographical Dictionary of the State of Indiana*. 3rd ed. Indianapolis: E. Chamberlain, 1849.

Chambers, William. *Things As They Are in America*. London: William and Robert Chambers, 1854.

[Chandler, Isaac.] An Englishman. *A Summary View of America*. London: T. Cadell, 1824.

Chevalier, Michael. *Society, Manners and Politics in the United States: Being a Series of Letters on North America*. 3rd ed. Boston: Weeks, Jordan & Co., 1839.

Cobbett, William. *The American Gardener*. New York: Orange Judd & Co., 1819.

——. *The Emigrant's Guide: In Ten Letters, Addressed to the Tax-Payers of England*. New ed.: London: Mills, Jowett, and Mills, 1830.

Cockrum, William M. *Pioneer History of Indiana, Including Stories, Incidents and Customs of the Early Settlers*. Oakland City, Ind.: Oakland City Journal, 1907.

Colman, Henry. *The Agriculture and Rural Economy of France, Belgium, Holland, and Switzerland: From Personal Observation*. London: John Petheram, 1848.

——. *European Agriculture and Rural Economy. From Personal Observation*. 2d ed. 2 vols. Boston: Charles C. Little and James Brown, 1850.

——. *European Life and Manners; in Familiar Letters to Friends*. 2 vols. Boston: Charles C. Little and James Brown, 1849.

Combe, George. *Notes on the United States of North America during a Phrenological Visit in 1838–9–40*. 2 vols. Philadelphia: Carey and Hart, 1841.

Conner, John B. *Indiana Agriculture: Agricultural Resources and Development of the State. The Struggles of Pioneer Life Compared with Present Conditions*. Indianapolis: Wm. B. Burford, 1893.

Cooper, Thomas. *Some Information Respecting America*. London: J. Johnson, 1794.

[Cooper, William.] *A Guide in the Wilderness or the History of the First Settlements in the Western Counties of New York*. Dublin: Gilbert & Hodges, 1810.

De Crèvecoeur, J. Hector St. John. *Letters from an American Farmer.* 1782. Rpt. New York: Fox, Duffield & Co., 1904.

Cunynghame, Sir Arthur Augustus Thurlow. *A Glimpse at the Great Western Republic.* London: Richard Bentley, 1851.

Curtiss, Daniel S. *Western Portraiture, and Emigrants' Guide: A Description of Wisconsin, Illinois, and Iowa.* New York: J. H. Colton, 1852.

Dabney, Robert L. *A Defence of Virginia, [and through her, of the South,] in Recent and Pending Contests Against the Sectional Party.* New York: E. J. Hale Y& Son, 1867.

Dalton, William. *Travels in the United States of America, and Part of Upper Canada.* B. Bateman, 1821.

Davis, Stephen. *Notes of a Tour in America in 1832 and 1833.* Edinburgh: Waugh & Innes, 1833.

Dixon, James. *Personal Narrative of a Tour through the United States and Canada...* New York: Lane and Scott, 1849.

Duncan, John M. *Travels through Part of the United States and Canada in 1818 and 1819.* 2 vols. New York: Howe and Spalding, 1823.

Elliott, E. N., ed. *Cotton Is King, and Pro-Slavery Arguments.* Augusta, Ga.: Pritchard, Abbott and Loomis, 1860.

Farnham, Elizabeth W[oodson]. *Life in Prairie Land.* 1846. Rpt. New York: Arno Press, 1972.

Faux, William. *Faux's Memorable Days in America.* London: W. Simplkin and R. Marshall, 1823.

Fearon, Henry Bradshaw. *Sketches of America: A Narrative of a Journey of Five Thousand Miles through the Eastern and Western States of America.* London: Longman, Hurst, Rees, Orme, and Brown, 1818.

Featherstonhaugh, G[eorge]. *Excursion through the Slave States, From Washington on the Potomac to the Frontier of Mexico.* 2 vols. London: John Murray, 1844.

Felton, Mrs. *Life in America: A Narrative of Two Years City and Country Residence in the United States.* London: Simpkin and Marshall, 1838.

Ferguson, William. *America by River and Rail, or, Notes by the Way on the New World and Its People.* London: James Nisbet & Co., 1856.

Ferrall, S[imon] A. *A Ramble of Six Thousand Miles through the United States of America.* London: Effingham Wilson, 1832.

Fidler, Isaac. *Observations on Professions, Literature, Manners, and Emigration in the United States and Canada Made during a Residence There in 1832.* New York: J. & J. Harper, 1833.

Finch, I. [John]. *Travels in the United States of America and Canada.* London: Longman, Rees, Orme, Brown, Green and Longman, 1833.

Finch, Marianne. *An Englishwoman's Experience in America.* London: Richard Bentley, 1853.

Flint, Charles Louis, ed. *The American Farmer: A Complete Agricultural Library, with Useful Facts for the Household.* 2 vols. Hartford, Conn.: Ralph H. Park, 1884.

Flint, James. *Letters from America; 1818–1820 Containing Observations on the Climate and Agriculture of the Western States, the Manners of the People, the Prospects of Emigrants, etc.* Rpt. Cleveland: Arthur H. Clark, 1904.

Flower, George. *The Errors of Emigrants: Pointing Out Many Popular Errors Hitherto Unnoticed.* 1841. Rpt. New York: Arno Press, 1975.

——. *History of the English Settlement in Edwards County, Illinois, Founded in 1817 and 1818, by Morris Birkbeck and George Flower.* Chicago: Fergus Printing Co., 1882.

Fordham, Elias Pym. *Personal Narrative of Travels in Virginia, Maryland, Pennsylvania, Ohio, Indiana, Kentucky.* Ed. Frederic Austin Ogg. Cleveland: Arthur H. Clark, 1906.

Freeman, Edward Augustus. *Some Impressions of the United States.* London: Longman, Green, and Co., 1883.

Gaylord, Willis, and Luther Tucker. *American Husbandry: Being a Series of Essays on Agriculture. Compiled Principally from "The Cultivator" and "The Genesee Farmer."* 2 vols. New York: Harper & Brothers, 1840.

Gerhard, Fred. *Illinois as It Is; Its History, Geography, Statistics.* Chicago: Keen and Lee, 1857.

Gladstone, William Ewart. "Kin beyond the Sea." In George P. Putnam, ed., *Representative Essays Selected from the Series of Prose Masterpieces from the Modern Essayists,* 349–95. New York: G. P. Putnam's Sons, 1885.

Godley, John Robert. *Letters from America.* 2 vols. London: John Murray, 1844.

Grattan, Thomas Colley. *Civilized America.* 2 vols. London: Bradbury & Evans, 1859.

Griffiths, D., Jr. *Two Years in the New Settlements of Ohio.* 1835. Rpt. Ann Arbor: University Microfilms, 1966.

Gurney, Joseph John. *A Journey in North America, Described in Familiar Letters to Amelia Opie.* Norwich: Josiah Fletcher, 1841.

[Haliburton, Thomas C.] "Sam Slick." *The Americans at Home: Or, Byeways, Backwoods, and Prairies.* 3 vols. London: Hurst & Blackett, 1854.

Hall, Basil. *Travels in North America, in the Years 1827 and 1828.* 3 vols. 3rd ed. Edinburgh: Robert Cadell, 1830.

Hall, Edward H[epple]. *Ho! For the West!! The Traveller and Emigrants' Hand-Book to Canada and the North-West of the American Union.* 3rd ed. London: Tweedie, Strand, Kent & Co., 1858.

Hall, Francis. *Travels in Canada and the United States in 1816 and 1817.* London: Longman, Hurst, Rees, Orme, & Brown, 1818.

Hall, Marshall. *The Two-Fold Slavery of the United States; with a Project of Self-Emancipation.* London: Adam, Scott, 1854.

Hall, Newman. *From Liverpool to St. Louis.* London: George Routledge & Sons, 1870.

Hamilton, Thomas. *Men and Manners in America.* 2 vols. 2nd ed. Edinburgh: William Blackwood, 1834.

Hancock, William. *An Emigrant's Five Years in the Free States of America.* London: T. Cautley Newby, 1860.

Harris, William Tell. *Remarks Made during a Tour through the United States of America, in the Years 1817, 1818, and 1819.* London: Sherwood, Neely, & Jones, 1821.

Heath, Francis George. *The English Peasantry.* London: Frederick Warne & Co., 1874.

Helper, Hinton Rowan. *The Impending Crisis of the South: How To Meet It.* New York: Burdick Brothers, 1857.

Hodgson, Adam. *Remarks during a Journey through North America in the Years 1819, 1820, and 1821, in a Series of Letters.* New York: Samuel Whiting, 1823.

Holmes, Isaac. *An Account of the United States of America, Derived from Actual Observation, during a Residence of Four Years in that Republic.* London: Canton Press, 1823.

Houstoun, Mrs. [Matilda Charlotte]. *Hesperos: Or, Travels in the West.* 2 vols. London: John W. Parker, 1850.

Howe, John. *The Emigrants' New Guide: Shewing a Description of the United States, and the British Possessions in Canada.* Leeds: John Howe, 1822.

Howells, William Cooper. *Recollections of Life in Ohio, from 1813 to 1840.* Cincinnati: Robert Clarke, 1895.

Howitt, William. *The Rural Life of England.* 2 vols. London: Longman, Orme, Brown, Green, and Longmans, 1838.

Irving, Washington. *The Voyage and Other English Essays from the Sketch Book.* Boston: Houghton Mifflin and Co., 1891.

Janson, Charles William. *The Stranger in America.* 1807. Rpt. New York: Burt Franklin, 1971.

Jobson, Frederick J. *America and American Methodism.* New York: Virtue, Emmins, & Co., 1857.

Johnson, A. D. *Illinois and the West.* 1838.

Johnson, Charles Britten. *Letters from the British Settlement in Pennsylvania.* Philadelphia: H. Hall, 1819.

Johnson, Cuthbert William. *Farmer's and Planter's Encyclopedia.* Philadelphia: Lippincott, 1841.

Kalm, Peter. *Kalm's Account of His Visit to England on His Way to America in 1748.* Trans. Joseph Lucas. 1748. London: Macmillan & Co., 1892.

———. *Travels Into North America.* Trans. John Reinhold Forster. 2 vols. London: T. Lowndes, 1771.

Kendall, Edward Augustus. *Travels through the Northern Parts of the United States in 1806–08.* 3 vols. New York: I. Riley, 1809.

Kingdom, William, Jun. *America and the British Colonies: An Abstract of All the Most Useful Information Relative to the United States of America, and the British Colonies.* London: G. and W. B. Whittaker, 1820.

Lambert, John. *Travels through Canada, and the United States of North America, in the Years 1806, 1807, & 1808.* 2 vols. London: Baldwin, Cradock, and Joy, 1816.

Lang, John Dunmore. *Religion and Education in America: With Notices of the State and Prospects of American Unitarianism, Property, and African Colonization.* London: Thomas Ward & Co., 1840.

Levinge, R[ichard] G[eorge] A[ugustus]. *Echoes from the Backwoods: or Sketches of Transatlantic Life.* 2 vols. London: Henry Colburn, 1846.

Lewis, Rev. G[eorge]. *Impressions of America and the American Churches.* Edinburgh: W. P. Kennedy, 1845.

Lincoln, Abraham. *Complete Works of Abraham Lincoln.* Ed. John G. Nicolay and John Hay. New York: Francis D. Tandy Co., 1905.

——. *The Political Thought of Abraham Lincoln.* Edited by Richard N. Current. New York: Macmillan, 1967.

Lindsey, Charles. *The Prairies of the Western States: The Advantages and Their Drawbacks.* Toronto: Leader & Patriot, 1860.

Lyell, Charles. *A Second Visit to the United States of North America.* 2 vols. New York: Harper & Bros., 1849.

——. *Travels in North America.* 2 vols. London: 1845.

Macaulay, James. *Across the Ferry: First Impressions of America and its People.* 2nd ed. London: Hodder and Stoughton, 1872.

MacKay, A[lexander]. *The Western World; or, Travels in the United States in 1846–47.* 3 vols. 3rd ed. London: Richard Bentley, 1850.

MacKay, Charles. *Life and Liberty in America: Or, Sketches of a Tour in the United States and Canada, in 1857–8.* 2 vols. London: Smith, Elder & Co., 1859.

MacKenzie, William L[yon]. *Sketches of Canada and the United States.* London: Effingham Wilson, 1833.

Marshall, Josiah T. *The Farmers and Emigrants Complete Guide, or a Hand Book, with Copious Hints, Recipes, and Tables, Designed for the Farmer and Emigrant.* Cincinnati: Applegate & Co., 1857.

Martineau, Harriet. *Retrospect of Western Travel.* 2 vols. London: Saunders & Otley, 1838.

Mattson, Hans. *The Story of an Emigrant.* Saint Paul: D. D. Merrill Co., 1891.

Mead, Whitman. *Travels in North America.* 3 parts. New York: C. S. Van Winkle, 1820.

Mellen, Grenville. *A Book of the United States: Exhibiting Its Geography, Divisions, Constitution and Government.* Hartford: H. F. Sumner, 1839.

Melish, John. *Travels through the United States of America, in the Years 1806 & 1807, and 1808, 1810, & 1811.* Philadelphia: John Melish, 1818.

Michigan Pioneer and Historical Society. *Pioneer Collections.* 40 vols. East Lansing: Michigan Pioneer and Historical Society, 1876–1929.

Mill, John Stuart. *Principles of Political Economy with Some of Their Applications to Social Philosophy.* 1871. Rpt. London: Longmans, Green and Co., 1926.

Mitchell, D[avid] W. *Ten Years Residence in the United States.* London: Smith, Elder & Co., 1862.

[Mitchell, Donald Grant.] *My Farm of Edgewood: A Country Book.* New York: Charles Scribner & Co., 1867.

Moore, George. *Journal of a Voyage Across the Atlantic: With Notes on Canada & the United States; and Return to Great Britain in 1844.* London: private printing, 1845.

Morris, Edmund. *Ten Acres Enough: A Practical Experience, Showing How a Very Small Farm May Be Made To Keep a Very Large Family.* New York: James Miller, 1872.

Morse, Jedediah. *The American Geography: Or, A View of the Present Situation of the United States of America.* 2nd ed. London: John Stockdale, 1792.

Murray, Hugh. *The United States of America; Their History from the Earliest Period.* 3 vols. Edinburgh: Oliver & Boyd, 1844.

Nichols, Thomas Low. *Forty Years of American Life.* 2 vols. London: John Maxwell & Co., 1864.

Northern Pacific Railroad. *Guide to the Northern Pacific Railroad Lands in Minnesota.* New York: Northern Pacific Railroad, [1872?].

O'Brien, Frank G. *Minnesota Pioneer Sketches: From the Personal Recollections and Observations of a Pioneer Resident.* Minneapolis: H. H. S. Roswell, 1904.

O'Bryan, W[illiam]. *A Narrative of Travels in the United States of America.* London: S. Thorne, Shebbear, Hatherleigh, Devon, Gilbert & Co., 1836.

O'Hanlon, Rev. J[ohn]. *The Irish Emigrant's Guide for the United States.* Boston: Patrick Donahoe, 1851.

Oldmixon, [John W.] *Transatlantic Wanderings: Or, A Last Look at the United States.* London: Geo. Routledge, 1855.

Oliver, William. *Eight Months in Illinois.* 1843. Rpt. Ann Arbor: University Microfilms, 1968.

Olmsted, Frederick Law. *The Cotton Kingdom: A Traveller's Observations on Cotton and Slavery in the American Slave States.* Ed. Arthur M. Schlesinger. 1861. Rpt. New York: Alfred A. Knopf, 1953.

———. *Walks and Talks of an American Farmer in England.* 1859. Rpt. Ann Arbor: University of Michigan Press, 1967.

O'Neill, J. W. *The American Farmers' New and Universal Handbook: Containing Complete and Well-Digested Instructions.* 2nd ed. Philadelphia: Charles Desilver, 1869.

Ozanne, T. D. *The South as It Is, or Twenty-One Years Experience in the Southern States of America.* London: Saunders, Otter & Co., 1863.

Palmer, John M. *Personal Recollections of John M. Palmer: The Story of an Earnest Life.* Cincinnati: Robert Clarke Co., 1901.

Parker, Theodore. *The Collected Works of Theodore Parker.* Ed. Francis Power Cobbe et al. 14 vols. London: Trubner & Co., 1864–71.

Patten, Edmund. *A Glimpse at the United States and the Northern States of America, with the Canadas.* London: Effingham Wilson, 1853.

Peck, John Mason. *Forty Years of Pioneer Life: Memoir of John Mason Peck D.D.* Ed. Rufus Babcock. Philadelphia: American Baptist Public Society, 1864.

Phillippo, James M[ursell]. *The United States and Cuba.* London: Petress & Co., 1857.

Pickering, Joseph. *Inquiries of an Emigrant: Being the Narrative of an English Farmer from the Year 1824 to 1830; during which Period he Traversed the United States and Canada.* 3rd ed. London: Effingham Wilson, 1832.

Playfair, Hugo. *Brother Jonathan, the Smartest Nation in all Creation.* 3 vols. London: Saunders & Otley, 1844.

Playfair, Robert. *Recollections of a Visit to the United States and British Provinces of North America, in the Years 1847, 1848, and 1849.* Edinburgh: Thomas Constable & Co., 1856.

Pollard, Edward A. *The Lost Cause: A New Southern History of the War of the Confederates.* 2nd ed. New York: E. B. Treat & Co., 1867.

Prentice, Archibald. *A Tour in the United States with Two Lectures on Emigration, delivered in the Mechanics' Institution Manchester.* 6th ed. London: John Johnston, 1850.

Priest, William. *Travels in the United States of America, Commencing in the Year 1793 and Ending in 1797, with the Author's Journals of His Two Voyages Across the Atlantic.* London: J. Johnson, 1802.

Pulszky, Francis, and Theresa Pulszky. *White, Red, Black: Sketches of American Society in the United States.* 2 vols. New York: Redfield, 1853.

Regan, John. *The Emigrant's Guide to the Western States of America; or, Backwoods and Prairies.* 2nd ed. Edinburgh: Oliver & Boyd, 1852.

Robinson, Solon. *Facts for Farmers: also for the Family Circle.* 2 vols. New York: Johnson and Ward, 1865.

———. *Solon Robinson: Pioneer Agriculturist: Selected Writings.* Ed. Herbert Anthony Kellar. 2 vols. Indianapolis: Indiana Historical Bureau, 1936.

Ruffin, Edmund. *The Political Economy of Slavery: or, The Institution Considered in regard to its Influence on Public Wealth and the General Welfare.* [Richmond, Va.?]: Lemuel Towers, 1857[?].

Russell, Robert. *North America: Its Agriculture and Climate: Containing Observations on the Agriculture and Climate of Canada, the United States, and the Island of Cuba.* Edinburgh: Adam & Charles Black, 1857.

[Russell, Robert W.] *America Compared with England: Respective Social Effects of the American and English Systems of Government and Legislation: and the Mission of Democracy.* London: Effingham Wilson, 1848.

Russell, William Howard. *My Diary North and South.* Ed. Fletcher Pratt. 1862. Rpt. New York: Harper & Bros., 1954.

——. *Pictures of Southern Life, Social, Political, and Military.* New York: James G. Gregg, 1861.

Shirreff, Patrick. *A Tour through North America; Together with a Comprehensive View of the Canadas and United States as Adapted for Agricultural Emigration.* Edinburgh: Oliver & Boyd, 1835.

Spence, Thomas. *The Settler's Guide in the United States and British North American Provinces.* New York: Davies & Kent, 1862.

Steevens, G[eorge] W. *The Land of the Dollar.* New York: Dodd, Mead & Co., 1898.

Stewart, W. M. *Eleven Years' Experience in the Western States of America.* London: Houlston and Sons, 1870.

Stirling, James. *Letters from the Slave States.* London: John W. Parker, 1857.

Stuart, James. *Three Years in North America.* 2 vols. Edinburgh: Robert Cadell, 1833.

Sullivan, Sir Edward [Robert]. *Rambles and Scrambles in North and South America.* London: Richard Bentley, 1852.

Sutcliff, Robert. *Travels in Some Parts of North America in the Years 1804, 1805, & 1860.* York: W. Alexander, 1815.

Thomas, John Jacob, ed. *Rural Affairs: A Practical and Copiously Illustrated Register of Rural Economy and Rural Taste.* Vol. 1. Albany: Luther Tucker & Son, 1858.

Thomason, D. R. *Hints to Emigrants, or to Those Who May Contemplate Emigrating to the United States of America.* Philadelphia: Isaac Ashmead, 1848.

Thomson, William. *A Tradesman's Travels, in the United States and Canada, in the Years 1840, 41, & 42.* Edinburgh: Oliver & Boyd, 1842.

Thornton, William. *A Plea for Peasant Proprietors; with the Outlines of a Plan for their Establishment in Ireland.* London: John Murray, 1848.

De Tocqueville, Alexis. *Democracy in America.* 2 vols. Ed. Phillips Bradley. 1835, 1840. Rpt. New York: Alfred A. Knopf, 1945.

Tremenheere, Hugh Seymour. *Notes on Public Subjects, Made during a Tour in the United States and in Canada.* London: John Murray, 1852.

Trollope, Anthony. *North America.* 3 vols. Leipzig: Bernhard Tauchnitz, 1862.

Trollope, Mrs. [Frances]. *Domestic Manners of the Americans.* 4th ed. London: Whittaker, Treacher & Co., 1832.

Tudor, Henry. *Narrative of a Tour in North America.* 2 vols. London: James Duncan, 1834.

[Unknown.] *General Hints to Emigrants Containing Notices of the Various Fields for Emigration with Practical Hints.* 2nd ed. London: Lockwood & Co., 1875.

[Unknown.] *What Democratic Leaders Think of Slavery.* N.p., n.d. [1865?].

Walker, Francis A. *The Wages Question: A Treatise on Wages and the Wages Class.* New York: Henry Holt and Co., 1876.

Webb, C[harles] H[enry]. *A Manual for Emigrants, Suggested by Circumstances, Proved by Personal Experience and Intended to be Placed in the Hands of All Foreigners, Especially from the "British Isles,"*... New York: William Osborn, 1849.

Wiley & Putnam's Emigrant's Guide: Comprising Advise and Instruction in Every State of the Voyage to America. London: Wiley & Putnam, 1845.

Williams, James. *The Rise and Fall of "The Model Republic."* London: Richard Bentley, 1863.

Wilson, C[harles] H[enry]. *The Wanderer in America, or Truth at Home.* Thirsk: Henry Masterson, 1822.

[Wright, Frances D'Arusmont]. An English Woman. *Views of Society and Manners in America.* London: Longman, Hurst, Rees, Orme, & Brown, 1821.

Yates, Richard. *Speech of Hon. Richard Yates,... Delivered at Springfield, Ill., 7 June 1860.* [Springfield, Ill.?]: n.p., 1860.

Secondary Sources

Adamson, John. "The Aristocracy and Their Mental World." In John Morrill, ed., *The Oxford Illustrated History of Tudor and Stuart Britain,* 173–90. OUP, 1996.

Allen, H. C. *Great Britain and the United States: A History of Anglo-American Relations (1783–1952).* New York: St. Martin's Press, 1955.

Allen, Robert. "Agriculture during the Industrial Revolution." In Roderick Floud and Donald McCloskey, eds., *The Economic History of Britain Since 1700.* Vol. 1: *1700–1860,* 96–122. 2nd ed.: CamUP, 1994.

——. *Enclosure and the Yeoman.* OUP, 1992.

——. "The Growth of Labor Productivity in Early Modern English Agriculture." *EEH* 25 (April 1988): 117–46.

Allitt, Patrick. *The Conservatives: Ideas and Personalities throughout American History.* YUP, 2009.

Anbinder, Tyler. "Moving beyond 'Rags to Riches': New York's Irish Famine Immigrants and Their Surprising Savings Accounts." *JAH* 99 (December 2012): 741–70.

Anderson, Virginia Dejohn. *Creatures of Empire: How Domestic Animals Transformed Early America.* OUP, 2004.

——. *New England's Generation: The Great Migration and the Formation of Society and Culture in the Seventeenth Century.* CamUP, 1991.

Ankli, Robert E. "Farm-Making Costs in the 1850s." In James W. Whitaker, ed., *Farm-*

ing in the Midwest, 1840–1900, 51–70. Washington, D.C.: Agricultural History Society, 1974.

Appleby, Joyce. "Commercial Farming and the 'Agrarian Myth' in the Early Republic." *JAH* 68 (March 1982): 833–49.

Armstrong, W. A. "Labour I: Rural Population Growth, Systems of Employment, and Incomes." In *AgHEW6,* 641–728.

Arnstein, Walter L. *Britain Yesterday and Today: 1830 to the Present.* Lexington, Mass.: D. C. Heath, 1966.

Aron, Stephen. *How the West Was Lost: The Transformation of Kentucky from Daniel Boone to Henry Clay.* JHUP, 1996.

Arsenault, Raymond. "The End of the Long Hot Summer: The Air Conditioner and Southern Culture." *JSH* 50 (November 1984): 597–628.

Ashcraft, Richard. *Revolutionary Politics and Locke's Two Treatises of Government.* PUP, 1986.

Ashley, Maurice. *Oliver Cromwell and the Puritan Revolution.* New York: Collier Books, 1958.

Ashton, Robert. *The English Civil War: Conservatism and Revolution, 1603–1649.* New York: W. W. Norton, 1978.

Ashworth, John. *Slavery, Capitalism, and Politics in the Antebellum Republic.* Vol. 1: *Commerce and Compromise, 1820–1850.* CamUP, 1995.

——. *Slavery, Capitalism, and Politics in the Antebellum Republic.* Vol. 2: *The Coming of the Civil War, 1850–1861.* CamUP, 2007.

Aston, T. H., and C. H. E. Philpin, eds. *The Brenner Debate: Agrarian Class Structure and Economic Development in Pre-Industrial Europe.* CamUP, 1976.

Atack, Jeremy. "The Agricultural Ladder Revisited: A New Look at an Old Question with Some Data for 1860." *AgH* 63 (Winter 1989): 1–25.

——. "Farm and Farm-Making Costs Revisited." *AgH* 56 (October 1982): 663–76.

Atack, Jeremy, and Fred Bateman. "The 'Egalitarian Ideal' and the Distribution of Wealth in the Northern Agricultural Community: A Backward Look." *Review of Economics and Statistics* 63 (February 1981): 129–32.

——. *To Their Own Soil: Agriculture in the Antebellum North.* Ames: Iowa State University Press, 1987.

Atack, Jeremy, Fred Bateman, and William N. Parker. "The Farm, the Farmer, and the Market." In Stanley L. Engerman and Robert E. Gallman, eds., *The Cambridge Economic History of the United States.* Vol. 2: *The Long Nineteenth Century,* 245–84. CamUP, 1996–2000.

——. "Northern Agriculture and the Westward Movement." In Stanley L. Engerman and Robert E. Gallman, eds., *The Cambridge Economic History of the United States.* Vol. 2: *The Long Nineteenth Century,* 285–328. CamUP, 1996–2000.

Ayers, Edward L. *What Caused the Civil War? Reflections on the South and Southern History*. New York: W. W. Norton, 2005.

Aylmer, G. E. "Locke No Leveller." In Ian Gentles, John Morrill, and Blair Worden, eds., *Soldiers, Writers and Statesmen of the English Revolution*, 304–22. CamUP, 1998.

——. *Rebellion or Revolution? England, 1640–1660*. OUP, 1986.

——, ed. *The Levellers in the English Revolution*. London: Thames and Hudson, 1975.

Barnes, L. Diane. *Artisan Workers in the Upper South: Petersburg, Virginia, 1820–1865*. LSUP, 2008.

——, Brian Schoen, and Frank Towers. "Introduction: Reimagining the Old South." In Barnes, Schoen, and Towers, eds., *The Old South's Modern Worlds: Slavery, Region, and Nation in the Age of Progress*, 3–22. OUP, 2011.

Barney, William L. *The Secessionist Impulse: Alabama and Mississippi in 1860*. PrinceUP, 1974.

Barnhart, John D. *Valley of Democracy: The Frontier versus the Plantation in the Ohio Valley, 1775–1818*. UIndP, 1953.

Barron, Hal S. "Epilogue: Reaping What Has Been Sown: The Current State of Rural History." In Koons and Hofstra, eds., *After the Backcountry*, 287–92.

——. "Listening to the Silent Majority: Change and Continuity in the Nineteenth-Century Rural North." In Ferleger, ed., *Agriculture and National Development*, 3–23.

——. *Those Who Stayed Behind: Rural Society in Nineteenth-Century New England*. CamUP, 1984.

Battalio, Raymond C., and John Kagel. "The Structure of Antebellum Southern Agriculture: South Carolina a Case Study." In William N. Parker, ed. *Structure of the Cotton South*, 25–37. Washington, D.-C.: Agricultural History Society, 1970.

Baugh, Daniel A., ed. *Aristocratic Government and Society in Eighteenth-Century England: The Foundations of Stability*. New York: New Viewpoints, 1975.

Beard, Charles A., and Mary Beard. *The Rise of American Civilization*. 2 vols. New York: Macmillan, 1927.

Beckett, J. V. "Agricultural Landownership and Estate Management." In *AgHEW7pt1*, 693–758.

——. *The Aristocracy in England*. Oxford: Basil Blackwell, 1986.

——. "The Debate Over Farm Sizes in Eighteenth and Nineteenth Century England." *AgH* 57 (July 1983): 308–25.

——. "The Decline of the Small Landowner in England and Wales, 1660–1900." In F. M. L. Thompson, ed., *Landowners, Capitalists, and Entrepreneurs: Essays for Sir John Habakkuk*, 89–112. OUP, 1991.

Bensel, Richard Franklin. *Yankee Leviathan: The Origins of Central State Authority in America, 1859–1877*. CamUP, 1990.

Berthoff, Rowland. "Peasants and Artisans, Puritans and Republicans: Personal Liberty and Communal Equality in American History." *JAH* 69 (December 1982): 579–98.

Bidwell, Percy Wells, and John I. Falconer. *History of Agriculture in the Northern United States, 1620–1860.* 1925. Rpt. New York: Peter Smith, 1941.

Billings, Warren M., John E. Selby, and Thad W. Tate. *Colonial Virginia: A History.* White Plains, N.Y.: KTO Press, 1986.

Billington, Ray Allen. *Westward Expansion: A History of the American Frontier.* 4th ed. New York: Macmillan, 1974.

Blanke, David. *Sowing the American Dream: How Consumer Culture Took Root in the Rural Midwest.* Athens: Ohio University Press, 2000.

Blassingame, John W. *The Slave Community: Plantation Life in the Antebellum South.* Rev. ed. OUP, 1979.

Blumin, Stuart M. *The Emergence of the Middle Class: Social Experience in the American City, 1760–1900.* CamUP, 1989.

Boaz, David. *Libertarianism: A Primer.* New York: Free Press, 1997.

Bode, Frederick A., and Donald E. Ginter. *Farm Tenancy and the Census in Antebellum Georgia.* UGaP, 1986.

Boewe, Charles. *Prairie Albion: An English Settlement in Pioneer Illinois.* Carbondale: Southern Illinois University Press, 1962.

Bogue, Allan G. *From Prairie to Corn Belt: Farming on the Illinois and Iowa Prairies in the Nineteenth Century.* UChiP, 1963.

Bolton, Charles C. *Poor Whites of the Antebellum South: Tenants and Laborers in Central North Carolina and Northeast Mississippi.* DukeUP, 1994.

Bond, Bradley G. "Herders, Farmers, and Markets in the Inner Frontier: The Mississippi Piney Woods, 1850–1860." In Samuel Hyde, ed., *Plain Folk of the South Revisited,* 73–99. LSUP, 1997.

Bonner, James C. *A History of Georgia Agriculture, 1732–1860.* UGaP, 1964.

——. "Profile of a Late Ante-Bellum Community." *AHR* 49, no. 4 (July 1944): 663–80.

Boritt, Gabor S. *Lincoln and the Economics of the America Dream.* Memphis: Memphis State University Press, 1978.

Bowman, Shearer Davis. *Masters and Lords: Mid-19th-Century U.S. Planters and Prussian Junkers.* OUP, 1993.

Boydston, Jeanne L. *Home and Work: Housework, Wages, and the Ideology of Labor in the Early Republic.* OUP, 1990.

Brailsford, H. N. *The Levellers and the English Revolution.* Edited by Christopher Hill. Stanford: Stanford University Press, 1961.

Breen, T. H. *Tobacco Culture: The Mentality of the Great Tidewater Planters on the Eve of Revolution.* PrinceUP, 1985.

Brewer, Holly. *By Birth or Consent: Children, Law, and the Anglo-American Revolution in Authority.* UNCP, 2005.

——. "Entailing Aristocracy." *WMQ* 54 (April 1997): 307–46.

Brodrick, George C. *English Land and English Landlords: An Enquiry into the Origin and Character of the English Land System, with Proposals for Its Reform.* London: Cassell, Pitgter, Galpin & Co., 1881.

Bronstein, Jamie L. *Land Reform and Working-Class Experience in Britain and the United States, 1800–1862.* Palo Alto, Calif.: Stanford University Press, 1999.

Brooks, Charles E. *Frontier Settlement and Market Revolution: The Holland Land Purchase.* CorUP, 1996.

Brown, Charles H. *Agents of Manifest Destiny: The Lives and Times of the Filibusters.* UNCP, 1980.

Brown, Richard D. *Modernization: The Transformation of American Life, 1600–1865.* New York: Hill and Wang, 1976.

Brown, Robert E., and B. Katherine Brown. *Virginia, 1705–1786: Democracy or Aristocracy?* East Lansing: Michigan State University Press, 1964.

Bruegel, Martin. *Farm, Shop, Landing: The Rise of a Market Society in the Hudson Valley, 1780–1860.* DukeUP, 2002.

Burlingame, Michael. *Abraham Lincoln: A Life.* 2 vols. JHUP, 2008.

Burnard, Trevor. *Creole Gentlemen: The Maryland Elite, 1691–1776.* New York: Routledge, 2002.

Burns, Rex. *Success in America: The Yeoman Dream and the Industrial Revolution.* UMassP, 1976.

Bushman, Claudia L. *In Old Virginia: Slavery, Farming, and Society in the Journal of John Walker.* JHUP, 2002.

Bushman, Richard L. *The Refinement of America: Persons, Houses, Cities.* New York: Alfred A. Knopf, 1992.

Campbell, Mildred. *The English Yeoman Under Elizabeth and the Early Stuarts.* YUP, 1942.

Campbell, Randolph B. "Planters and Plain Folk: The Social Structure of the Antebellum South." In John B. Boles and Evelyn Thomas Nolen, eds., *Interpreting Southern History: Historiographical Essays in Honor of Sanford W. Higginbotham,* 48–77. LSUP, 1987.

Cannadine, David. *The Decline and Fall of the British Aristocracy.* YUP, 1990.

Cannon, John. *Aristocratic Century: The Peerage of Eighteenth-Century England.* CamUP, 1984.

Carlson, Theodore L. *The Illinois Military Tract: A Study of Land Occupation, Utilization, and Tenure.* UIllP, 1951.

Carlton, David L., and Peter A. Coclanis. *The South and the Nation: Perspectives on Southern Economic Development.* UPVa, 2003.

Carr, Lois Green, and Russell R. Menard. "Immigration and Opportunity: The Freedman in Early Colonial Maryland." In Thad Tate and David L. Ammerman, eds.,

The Chesapeake in the Seventeenth Century: Essays on Anglo-American Society, 206–42. UNCP, 1979.

Carwardine, Richard J. *Evangelicals and Politics in Antebellum America*. YUP, 1993.

——. *Lincoln*. Edinburg Gate: Pearson Education, 2003.

Cashin, Joan E. *A Family Venture: Men and Women on the Southern Frontier*. OUP, 1991.

Cayton, Andrew R. L. *Frontier Indiana*. UIndP, 1996.

——. *Ohio: The History of a People*. Columbus: Ohio State University Press, 2002.

——, and Stuart D. Hobbs, eds., *The Center of a Great Empire: The Ohio Country in the Early American Republic*. Athens: Ohio University Press, 2005.

——, and Peter S. Onuf. *The Midwest and the Nation: Rethinking the History of an American Region*. UIndP, 1990.

Cecil-Fronsman, Bill. *Common Whites: Class and Culture in Antebellum North Carolina*. UPKy, 1992.

Censer, Jane Turner. *North Carolina Planters and Their Children, 1800–1860*. LSUP, 1984.

Chambers, J. D. *Population, Economy, and Society in Pre-Industrial England*. OUP, 1972.

——, and G. E. Mingay. *The Agricultural Revolution, 1750–1880*. New York: Schocken Books, 1966.

Chandler, Alfred D., Jr. *The Visible Hand: The Managerial Revolution in American Business*. Cambridge, Mass.: Harvard, 1977.

Clapham, J. H. *An Economic History of Modern Britain: Free Trade and Steel, 1850–1886*. CamUP, 1932.

Clark, Christopher. "The Agrarian Context of American Capitalist Development." In Zakim and Kornblith, eds., *Capitalism Takes Command*, 13–38.

——. "The Ohio Country in the Political Economy of Nation Building." In Cayton and Hobbs, eds., *Center of a Great Empire*, 146–65.

——. *The Roots of Rural Capitalism: Western Massachusetts, 1780–1860*. CorUP, 1990.

——. *Social Change in America: From the Revolution through the Civil War*. Chicago: Ivan R. Dee, 2006.

Clark, J. C. D. *English Society, 1688–1832: Ideology, Social Structure, and Political Practice during the Ancien Regime*. CamUP, 1985.

Clark, John G. *The Grain Trade in the Old Northwest*. UIllP, 1966.

Clarkson, L. A. *The Pre-industrial Economy in England, 1500–1750*. London: B. T. Batsford, 1971.

Clemens, Paul G. E. *The Atlantic Economy and Colonial Maryland's Eastern Shore: From Tobacco to Grain*. CorUP, 1980.

——. "The Operation of an Eighteenth-Century Chesapeake Tobacco Plantation." *AgH* 49 (July 1975): 517–31.

——, and Lucy Simler. "Rural Labor and the Farm Household in Chester County, Pennsylvania, 1750–1820." In Stephen Innes, ed., *Work and Labor in Early America,* 106–43. UNCP, 1988.

Cliffe, J. T. *The Puritan Gentry: The Great Puritan Families of Early Stuart England.* London: Routledge and Kegan Paul, 1984.

Clinton, Catherine. *The Plantation Mistress: Woman's World in the Old South.* New York: Pantheon Books, 1982.

Coclanis, Peter A. *The Shadow of a Dream: Economic Life and Death in the South Carolina Low Country, 1670–1920.* OUP, 1989.

Coffey, John. "Puritan Legacies." In Coffey and Lim, eds., *Cambridge Companion to Puritanism,* 327–45.

——, and Paul C. H. Lim, eds. *The Cambridge Companion to Puritanism.* CamUP, 2008.

Cohen, G. A. *Karl Marx's Theory of History: A Defence.* PrinceUP, 1978.

——. *Self-Ownership, Freedom, and Equality.* CamUP, 1995.

Coleman, D. C. *The Economy of England, 1450–1750.* OUP, 1977.

Como, David R. "Radical Puritanism, c. 1558–1660." In Coffey and Lim, eds., *Cambridge Companion to Puritanism,* 241–58.

Conrad, Alfred H., and John R. Meyer. "The Economics of Slavery in the Antebellum South." In Hugh G. J. Aitken, ed., *Did Slavery Pay? Readings on the Economics of Black Slavery in the United States,* 136–75. Boston: Houghton Mifflin, 1971.

Cooper, William J., Jr. "The Cotton Crisis in the Antebellum South: Another Look." *AgH* 49 (April 1975): 381–91.

——. "The Critical Signpost on the Journey Toward Secession." *JSH* 77 (February 2011): 3–16.

——. *Liberty and Slavery: Southern Politics to 1860.* New York: Alfred A. Knopf, 1983.

——, and Thomas E. Terrill. *The American South: A History.* 2 vols. 4th ed. Boston: McGraw Hill, 2009.

Countryman, Edward. *A People in Revolution: The American Revolution and Political Society in New York, 1760–1790.* JHUP, 1981.

Coward, Barry. *The Stuart Age: A History of England, 1603–1714.* London: Longman, 1980.

Craig, John. "The Growth of English Puritanism." In Coffey and Lim, eds., *Cambridge Companion to Puritanism,* 34–47.

Craig, Lee A. *To Sow One More Acre: Childbearing and Farm Productivity in the Antebellum North.* JHUP, 1993.

Crandall, Andrew Wallace. *The Early History of the Republican Party, 1854–1856.* Boston: Goreham Press, 1930.

Craven, Avery Odelle. *Civil War in the Making, 1815–1860.* LSUP, 1968.

——. "Coming of the War Between the States: An Interpretation." *JSH* 2 (August 1936): 303–22.

——. *An Historian and the Civil War.* UChiP, 1964.

——. *Soil Exhaustion as a Factor in the Agricultural History of Virginia and Maryland, 1606–1860.* UIllP, 1926.

Cronon, William. *Nature's Metropolis: Chicago and the Great West.* New York: W. W. Norton, 1991.

Crook, David Paul. *American Democracy in English Politics, 1815–1850.* OUP, 1965.

Curry, Leonard P. *Blueprint for Modern America: Nonmilitary Legislation of the First Civil War Congress.* Nashville: Vanderbilt University Press, 1968.

Curtin, Philip D. *The Rise and Fall of the Plantation Complex: Essays in Atlantic History.* 2nd ed. CamUP, 1998.

Danhof, Clarence H. *Change in Agriculture: The Northern United States, 1820–1870.* HUP, 1969.

——. "The Farm Enterprise: The Northern United States, 1820–1860s." *Research in Economic History* 4 (1979): 127–91.

——. "Farm-Making Costs and the 'Safety Value': 1850–60." *Journal of Political Economy* 49 (June 1941): 317–59.

Daniels, Christine. "'Getting His [or Her] Lively-hood': Free Workers in Slave Anglo-America, 1675–1810." *AgH* 71 (Spring 1997): 125–61.

Daunton, M. J. *Progress and Poverty: An Economic and Social History of Britain, 1700–1850.* OUP, 1995.

David, Paul A., Herbert G. Gutman, Richard Sutch, Peter Temin, and Gavin Wright. *Reckoning with Slavery: A Critical Study in the Quantitative History of American Negro Slavery.* OUP, 1976.

Davis, Charles S. *The Cotton Kingdom in Alabama.* Montgomery: Alabama State Department of Archives and History, 1939.

Davis, David Brion. *Inhuman Bondage: The Rise and Fall of Slavery in the New World.* OUP, 2006.

——. *The Problem of Slavery in the Age of Revolution, 1770–1823.* CorUP, 1975.

——. *The Problem of Slavery in Western Culture.* OUP, 1966.

Davis, Hugh. *Joshua Leavitt: Evangelical Abolitionist.* LSUP, 1990.

Davis, J. C. "Equality in an Unequal Commonwealth: James Harrington's Republicanism and the Meaning of Equality." In Ian Gentles, John Morrill, and Blair Worden, eds., *Soldiers, Writers and Statesmen of the English Revolution,* 229–42. CamUP, 1998.

Davis, James E. *Frontier Illinois.* IndUP, 1998.

Dawley, Alan. *Class and Community: The Industrial Revolution in Lynn.* HUP, 1976.

Deane, Phyllis. *The First Industrial Revolution.* 2nd ed. CamUP, 1979.

——, and W. A. Cole. *British Economic Growth, 1688–1959: Trends and Structure.* CamUP, 1962.

Demaree, Albert Lowther. *The American Agricultural Press, 1819–1860.* New York: Columbia University Press, 1941.

Dewey, Peter. "Farm Labour." In *AgHEW7pt1,* 810–62.

Dodd, William E. *The Cotton Kingdom: A Chronicle of the Old South.* YUP, 1919.

Donahue, Brian. *The Great Meadow: Farmers and the Land in Colonial Concord.* YUP, 2004.

Donald, David Herbert. *Lincoln.* New York: Simon and Schuster, 1995.

Dowdy, Clifford. *The Golden Age, A Climate for Greatness: Virginia, 1732–1775.* Boston: Little, Brown, 1970.

Downey, Tom. *Planting a Capitalist Class: Masters, Merchants, and Manufacturers in the Southern Interior, 1790–1860.* LSUP, 2006.

Doyle, William. *Aristocracy and Its Enemies in the Age of Revolution.* OUP, 2009.

Drescher, Seymour. *Capitalism and Antislavery: British Mobilization in Comparative Perspective.* OUP, 1987.

——. "Free Labor vs. Slave Labor: The British and Caribbean Cases." In Stanley L. Engerman, ed., *Terms of Labor: Slavery, Freedom, and Free Labor,* 50–86. Palo Alto, Calif.: Stanford University Press, 1999.

——. *From Slavery to Freedom: Comparative Studies in the Rise and Fall of Atlantic Slavery.* New York: New York University Press, 1999.

Dunaway, Wilma A. *The First American Frontier: Transition to Capitalism in Southern Appalachia, 1700–1860.* UNCP, 1996.

Dupre, Daniel S. *Transforming the Cotton Frontier: Madison County, Alabama, 1800–1840.* UNCP, 1997.

Earle, Carville. "Beyond the Appalachians, 1815–1860." In Thomas E. McIlwraith and Edward K. Muller, eds., *North America: The Historical Geography of a Changing Continent,* 165–87. 2nd ed. Lanham, Md.: Rowman and Littlefield, 2001.

——. *Geographical Inquiry and American Historical Problems.* Palo Alto, Calif.: Stanford University Press, 1992.

Eastwood, David. *Governing Rural England: Tradition and Transformation in Local Development, 1780–1840.* Oxford, U.K.: Clarendon Press, 1994.

Edelson, S. Max. *Plantation Enterprise in Colonial South Carolina.* HUP, 2006.

Edwards, Gary T. "Men of Subsistence and Men of Substance: Agricultural Lifestyles in Antebellum Madison County, Tennessee." *AgH* 73 (Summer 1999): 303–21.

Edwards, Philip. *The Making of the Modern English State, 1460–1660.* Houndmills, U.K.: Palgrave, 2001.

Egnal, Marc. *Clash of Extremes: The Economic Origins of the Civil War.* New York: Hill and Wang, 2009.

Einhorn, Robin. *American Taxation, American Slavery.* UChiP, 2006.

Ellenberg, George B. *Mule South to Tractor South: Mules, Machines, and the Transformation of the Cotton South.* Tuscaloosa: University of Alabama Press, 2007.

Elster, Jon. *An Introduction to Karl Marx.* 1986. Rpt. CamUP, 1999.

——. *Making Sense of Marx.* CamUP, 1985.

Eltis, David. *The Rise of African Slavery in the Americas.* CamUP, 2000.

Eslinger, Ellen. *Citizens of Zion: The Social Origins of Camp Meeting Revivalism.* UTnP, 1999.

——. "The Evolution of Racial Politics in Early Ohio." In Cayton and Hobbs, eds., *Center of a Great Empire,* 81–104.

Etcheson, Nicole. *Bleeding Kansas: Contest Liberty in the Civil War Era.* UPKan, 2004.

——. *The Emerging Midwest: Upland Southerners and the Political Culture of the Old Northwest, 1787–1861.* IndUP, 1996.

Faler, Paul G. *Mechanics and Manufacturers in the Early Industrial Revolution: Lynn, Massachusetts, 1780–1860.* Albany: State University of New York Press, 1981.

Faragher, John Mack. *Sugar Creek: Life on the Illinois Prairie.* YUP, 1986.

Farmer, Charles J. *In the Absence of Towns: Settlement and Country Trade in Southside Virginia, 1730–1800.* Landham, Md.: Rowman and Littlefield, 1993.

Faust, Drew Gilpin. *James Henry Hammond and the Old South: A Design for Mastery.* LSUP, 1982.

——, ed. *The Ideology of Slavery: Proslavery Thought in the Antebellum South, 1830–1860.* LSUP, 1981.

Fea, John. *The Way of Improvement Leads Home: Philip Vickers Fifhian and the Rural Enlightenment in Early America.* UPennP, 2008.

Fehrenbacher, Don E. *The Slaveholding Republic: An Account of the United States Government's Relations to Slavery.* Ed. Ward M. McAfee. OUP, 2001.

Feller, Daniel. *The Public Lands in Jacksonian Politics.* Madison: University of Wisconsin Press, 1984.

Ferleger, Lou, ed., *Agriculture and National Development: Views on the Nineteenth Century.* IaUP, 1990.

Field-Hendrey, Elizabeth B., and Lee A. Craig. "The Relative Efficiency of Free and Slave Agriculture in the Antebellum United States: A Stochastic Production Frontier Approach." In David Eltis, Frank Lewis, and Kenneth L. Sokoloff, eds., *Human Capital and Institutions: A Long Run View,* 236–57. CamUP, 1999.

Finlayson, Michael G. *Historians, Puritanism, and the English Revolution: The Religious Factor in English Politics before and after the Interregnum.* Toronto: University of Toronto Press, 1983.

Firth, Charles Harding. *The House of Lords during the Civil War.* London: Longmans, Green, and Co., 1910.

Fischer, David Hackett. *Albion's Seed: Four British Folkways in America.* OUP, 1989.

——, and James C. Kelly. *Bound Away: Virginia and the Westward Movement.* UPVa, 2000.

Flanders, Ralph Betts. *Plantation Slavery in Georgia.* 1933. Rpt. Cos Cob, Conn.: John E. Edwards, 1967.

Fleisig, Heywood. "Slavery, the Supply of Agricultural Labor, and the Industrialization of the South." *JEH* 36 (September 1976): 572–97.

Fletcher, Stevenson Whitcomb. *Pennsylvania Agriculture and Country Life, 1840–1940.* 2 vols. Harrisburg: Pennsylvania Historical and Museum Commission, 1955.

Fogel, Robert William. *Without Consent or Contract: The Rise and Fall of American Slavery.* New York: Norton, 1989.

——, and Stanley L. Engerman. *Time on the Cross: The Economics of American Negro Slavery.* 2 vols. Boston: Little, Brown and Co., 1974.

Foner, Eric. "Abolitionism and the Labor Movement in Antebellum America." In Christine Bolt and Seymour Drescher, eds., *Anti-Slavery, Religion, and Reform: Essays in Memory of Roger Anstey,* 254–71. Folkestone, U.K.: W. Dawson, 1980.

——*The Fiery Trial: Abraham Lincoln and American Slavery.* New York: W. W. Norton, 2010.

——. *Free Soil, Free Labor, Free Men: The Ideology of the Republican Party before the Civil War.* 2nd ed. OUP, 1995.

Ford, Lacy K., Jr. *Origins of Southern Radicalism: The South Carolina Upcountry, 1800–1860.* OUP, 1988.

——. "Reconfiguring the Old South: 'Solving' the Problem of Slavery, 1787–1838." *JAH* 95 (June 2008): 95–122.

Fordham, Montagu [Edward]. *A Short History of English Rural Life from the Anglo-Saxon Invasion to the Present Time.* London: George Allen and Unwin, 1916.

Forret, Jeff. *Race Relations at the Margins: Slaves and Poor Whites in the Antebellum Southern Countryside.* LSUP, 2006.

Foster, Stephen. "English Puritanism and the Progress of New England Institutions, 1630–1660." In David D. Hall, John M. Murrin, and Thad W. Tate, eds., *Saints and Revolutionaries: Essays in Early American History,* 3–37. New York: W. W. Norton, 1984.

——. *The Long Argument: English Puritanism and the Shaping of New England Culture, 1570–1700.* UNCP, 1991.

——. *Their Solitary Way: The Puritan Social Ethic in the First Century of Settlement in New England.* YUP, 1971.

Foulke, William Dudley. *Life of Oliver P. Morton, including His Important Speeches.* Indianapolis: Bowen-Merrill Co., 1899.

Foust, James D. *The Yeoman Farmer and Westward Expansion of U.S. Cotton Production.* New York: Arno Press, 1975.

Fox-Genovese, Elizabeth. *Within the Plantation Household: Black and White Women of the Old South.* UNCP, 1988.

———. "Women in Agriculture during the Nineteenth Century." In Ferleger, ed., *Agriculture and National Development,* 267–301.

Franklin, John Hope, and Loren Schweninger. *Runaway Slaves: Rebels on the Plantation.* OUP, 1999.

Fraysśe, Olivier. *Lincoln, Land, and Labor, 1809–60.* Trans. Sylvia Neely. UIllP, 1994.

Fredrickson, George M. *The Black Image in the White Mind: The Debate on Afro-American Character and Destiny, 1817–1914.* New York: Harper and Row, 1971.

Freehling, William W. *The Road to Disunion: Secessionists at Bay, 1776–1854.* OUP, 1990.

———. *The Road to Disunion: Secessionists Triumphant.* OUP, 2007.

Friedlaender, Heinrich E., and Jacob Oser. *Economic History of Modern Europe.* New York: Prentice Hall, 1957.

Friend, Craig Thompson. *Kentucke's Frontiers.* IndUP, 2005.

Furstenberg, François. "Beyond Freedom and Slavery: Autonomy, Virtue, and Resistance in Early American Political Discourse." *JAH* 89 (March 2003): 1295–1330.

Galambos, Louis. *Public Image of Big Business in America, 1880–1940.* JHUP, 1975.

Galenson, David W. "The Settlement and Growth of the Colonies: Population, Labor, and Economic Development." In Stanley L. Engerman and Robert E. Gallman, eds., *The Cambridge Economic History of the United States.* Vol. 2: *The Long Nineteenth Century,* 135–207. CamUP, 1996–2000.

———. *White Servitude in Colonial America: An Economic Analysis.* CamUP, 1981.

Gallay, Alan. "The Origins of Slaveholders' Paternalism: George Whitefield, the Bryan Family, and the Great Awakening in the South." *JSH* 53 (August 1987): 369–94.

Gallman, Robert E. "Self-Sufficiency in the Cotton Economy of the Antebellum South." In William N. Parker, ed., *Structure of the Cotton Economy,* 5–37. Washington, D.C.: Agricultural History Society, 1970.

———, and Ralph V. Anderson. "Slavery as Fixed Capital: Slave Labor and Economic Development." *JAH* 64 (June 1977): 24–46.

Garnier, Russell M[ontagu]. *History of the English Landed Interest: Its Customs, Laws and Agriculture (Modern Period).* London: Swan Sonnenschein, 1893.

Gates, Paul Wallace. *The Farmer's Age: Agriculture, 1815–1860.* New York: Holt, Rinehart and Winston, 1960.

———. *The Jeffersonian Dream: Studies in the History of American Land Policy and Development.* Ed. Allan G. Bogue and Margaret Beattie Bogue. Albuquerque: University of New Mexico Press, 1996.

———. *Landlords and Tenants on the Prairie Frontier: Studies in American Land Policy.* CorUP, 1973.

Genovese, Eugene D. *The Political Economy of Slavery: Studies in the Economy and Society of the Slave South.* 1961. Rpt. New York: Pantheon Books, 1965.

——. *Roll, Jordan, Roll: The World the Slaves Made.* New York: Random House, 1972.

——. *The World the Slaveholders Made: Two Essays in Interpretation.* 2nd ed. Middleton, Conn.: Wesleyan University Press, 1988.

——, and Elizabeth Fox-Genovese. *Fatal Self-Deception: Slaveholding Paternalism in the Old South.* CamUP, 2011.

Gienapp, William E. *Origins of the Republican Party, 1852–1856.* OUP, 1987.

Gilkeson, John S., Jr. *Middle-Class Providence, 1820–1940.* PrinceUP, 1986.

Gillespie, Michele. *Free Labor in an Unfree World: White Artisans in Slaveholding Georgia, 1789–1860.* UGaP, 2000.

Ginger, Ray. *Altgeld's America: The Lincoln Ideal versus Changing Realities.* 1958. Rpt. Chicago: Quadrangle Books, 1965.

Gjerde, Jon. *The Minds of the West: Ethnocultural Evolution in the Rural Middle-West, 1830–1917.* UNCP, 1997.

Glickstein, Jonathan A. *American Exceptionalism, American Anxiety: Wages, Competition, and Degraded Labor in the Antebellum United States.* UPVa, 2002.

——. *Concepts of Free Labor in Antebellum America.* YUP, 1991.

——. "'Poverty Is Not Slavery': Abolitionists and the Competitive Labor Market." In Lewis Perry and Michael Fellman, eds., *Antislavery Reconsidered: New Perspectives on the Abolitionists,* 195–218. LSUP, 1979.

Goodman, Paul. *Of One Blood: Abolitionism and the Origins of Racial Equality.* Berkeley: University of California Press, 1998.

——. "White Over White: Planters, Yeomen, and the Coming of the Civil War: A Review Essay." *AgH* 54 (July 1980): 446–52.

Grant, Susan-Mary. *North Over South: Northern Nationalism and American Identity in the Antebellum Era.* UPKan, 2000.

Gray, Lewis Cecil. *History of Agriculture in the Southern United States to 1860.* 2 vols. Washington D.C.: Carnegie Institution of Washington, 1933.

Gray, Susan E. *The Yankee West: Community Life on the Michigan Frontier.* UNCP, 1996.

Gray, Wood. *The Hidden Civil War: The Story of the Copperheads.* New York: Viking Press, 1942.

Greenberg, Kenneth S. *Masters and Statesmen: The Political Culture of American Slavery.* JHUP, 1985.

Greene, Jack P. "Introduction: Empire and Liberty." In Jack P. Greene, ed., *Exclusionary Empire: English Liberty Overseas, 1600–1900,* 1–14. CamUP, 2010.

——. *Pursuits of Happiness: The Social Development of Early Modern British Colonies and the Formation of American Culture.* UNCP, 1988.

——. "Society, Ideology, and Politics: An Analysis of the Political Culture of Mid-Eighteenth-Century Virginia." In Richard M. Jellison, ed., *Society, Freedom, and Conscience: The American Revolution in Virginia, Massachusetts, and New York,* 14–76. New York: W. W. Norton, 1976.

Gregg, Pauline. *A Social and Economic History of Britain, 1760–1970.* 6th ed. London: George G. Harrap, 1971.

Griffin, Patrick. *American Leviathan: Empire, Nation, and Revolutionary Frontier.* New York: Hill and Wang, 2007.

Grigg, David. *English Agriculture: An Historical Perspective.* Oxford: Basil Blackwell, 1989.

Griswold, A. Whitney. *Farming and Democracy.* New York: Harcourt, Brace & Co., 1948.

Grivno, Max. *Gleanings of Freedom: Free and Slave Labor along the Mason-Dixon Line, 1790–1860.* UIllP, 2011.

Gruenwald, Kim. *River of Enterprise: The Commercial Origins of Regional Identity in the Ohio Valley, 1790–1850.* IndUP, 2002.

Guasco, Suzanne C. "Southern Yeomen and Slavery in Illinois." *Civil War History* 47 (March 2001): 7–29.

Guelzo, Allen C. *Abraham Lincoln as a Man of Ideas.* Carbondale: Southern Illinois University Press, 2009.

Guttmann, Allen. *The Conservative Tradition in America.* OUP, 1967.

Habakkuk, John. "English Landownership, 1680–1740." *EHR* 10 (February 1940): 2–17.

——. *Marriage, Debt, and the Estates System: England Landownership 1650–1950.* OUP, 1994.

Hacker, Louis M. *The Triumph of American Capitalism: The Development of Forces in American History to the End of the Nineteenth Century.* New York: Simon and Schuster, 1940.

Hahn, Steven. *The Roots of Southern Populism: Yeoman Farmers and the Transformation of the Georgia Upcountry, 1850–1890.* OUP, 1983.

Haines, Michael R. "The Population of the United States, 1790–1920." In *The Cambridge Economic History of the United States.* Vol. 2: *The Long Nineteenth Century,* ed. Staley L. Engerman and Robert E. Gallman, 143–205. CamUP, 1996–2000.

Hale, Oron J. *The Great Illusion 1900–1914.* New York: Harper & Row, 1971.

Hall, David D. *A Reforming People: Puritanism and the Transformation of Public Life in New England.* New York: Alfred A. Knopf, 2011.

Haller, William. *The Rise of Puritanism: Or, The Way to the New Jerusalem as Set Forth in Pulpit and Press from Thomas Cartwright to John Milburne and John Milton, 1570–1643.* 1938. Rpt. New York: Harper & Brothers, 1957.

Hammond, John Craig. *Slavery, Freedom, and Expansion in the Early American West.* UPVa, 2007.

Hammond, John L., and Barbara Hammond. *The Village Labourer 1760–183: A Study in the Government Before the Reform Bill.* 1911. Rpt. New York: Harper Torchbook, 1970.

Harris, J. William. *Plain Folk and Gentry in a Slave Society: White Liberty and Black Slavery in Augusta's Hinterlands*. Middleton, Conn.: Wesleyan University Press, 1985.

Hasback, Wilhelm. *A History of the English Agricultural Labourer*. Trans. Ruth Kenyon. 2nd ed. London: P. S. King, 1908.

Havighurst, Alfred T. *Twentieth Century Britain*. 2nd ed. New York: Harper and Row, 1962.

Hawke, David. *The Colonial Experience*. Indianapolis: Bobbs Merrill, 1966.

Headlee, Sue. *The Political Economy of the Family Farm: The Agrarian Roots of American Capitalism*. New York: Praeger, 1991.

Heal, Felicity, and Clive Holmes. *The Gentry in England and Wales, 1500–1700*. Palo Alto, Calif.: Stanford University Press, 1994.

Heath, Francis George. *The English Peasantry*. London: Frederick Warne & Co., 1874.

Hecht, J. Jean. *The Domestic Servant Class in Eighteenth-Century England*. London: Routledge & Kegan Paul, 1956.

Hedrick, Ulysses Prentiss. *A History of Agriculture in the State of New York*. 1933. Rpt. New York: Hill and Wang, 1966.

Henretta, James A. *The Evolution of American Society, 1700–1815*. Lexington, Mass.: D. C. Heath & Co., 1973.

——. "Families and Farms: Mentalité in Pre-Industrial America." *WMQ* 35 (January 1978): 3–32.

——. *The Origins of American Capitalism: Collected Essays*. Boston: Northeastern University Press, 1991.

Herndon, G. Melvin. "Elliott L. Story: A Small Farmer's Struggle for Economic Survival in Antebellum Virginia." *AgH* 56 (July 1982): 516–27.

Hexter, J. H. *Reappraisals in History*. London: Longmans, Green & Co., 1967.

Hey, David G. *An English Rural Community: Myddle Under the Tudors and Stuarts*. Bristol, U.K.: Leicester University Press, 1974.

Hicks, John D. *The Populist Revolt: A History of the Farmers' Alliance and the People's Party*. 1931. Rpt. Lincoln: University of Nebraska Press, 1970.

Higgs, Edward. "Occupational Censuses and the Agricultural Workforce in Victorian England and Wales." *EHR* 48 (November 1995): 700–716.

Hill, Christopher. "A Bourgeois Revolution?" In J. G. A. Pocock, ed., *Three British Revolutions: 1641, 1688, 1776*, 109–62. PrinceUP, 1980.

——. *The Century of Revolution*. 1961. Rpt. New York: W. W. Norton, 1980.

——. *Change and Continuity in Seventeenth-Century England*. Rev. ed. YUP, 1991.

——. *Intellectual Origins of the English Revolution*. 1965. Rpt. OUP, 1982.

——. *Society and Puritanism in Pre-Revolutionary England*. 2nd ed. New York: Schocken Books, 1967.

——, and Edmund Dell. *The Good Old Cause: English Revolution of 1640–1660.* Oxford, U.K.: Frank Cass and Co., 1949.

Hill, Roland. *Lord Acton.* YUP, 2000.

Hilliard, Sam Bowers. *Hog Meat and Hoecake: Food Supply in the Old South, 1840–1860.* Carbondale: Southern Illinois University Press, 1972.

Himmelfarb, Gertrude. *Lord Acton: A Study in Conscience and Politics.* UChiP, 1962.

Hobsbawm, E. J. *The Age of Capital, 1848–1875.* New York: Charles Scribner's Sons, 1975.

——. *The Age of Revolution, 1789–1848.* New York: New American Library, 1962.

——, and George Rudé. *Captain Swing.* New York: Random House, 1968.

Hofstadter, Richard. *The Age of Reform: From Bryan to F.D.R.* New York: Random House, 1955.

Hofstra, Warren R. *A Separate Place: The Formation of Clarke County, Virginia.* White Post, Va.: Clarke County Sesquicentennial Committee, 1986.

Holderness, B. A. "Prices, Productivity, and Output." In *AgHEW6,* 84–189.

Holt, Michael F. *The Fate of Their Country: Politicians, Slavery Extension, and the Coming of the Civil War.* New York: Hill and Wang, 2004.

——. *The Political Crisis of the 1850s.* New York: W. W. Norton, 1978.

Horsman, Reginald. *The Frontier in the Formative Years, 1783–1815.* New York: Holt, Rinehart and Winston, 1970.

Hoskins, W. G. *The Making of the English Landscape.* London: Holden & Stoughton, 1955.

Hudson, John C. *Making the Corn Belt: A Geographical History of Middle-Western Agriculture.* IndUP, 1994.

Humphrey, Thomas J. "Conflicting Independence: Land Tenancy and the American Revolution." *JER* 28 (Summer 2008): 159–82.

Hunt, Margaret R. *The Middling Sort: Commerce, Gender, and the Family in England, 1680–1780.* Berkeley: University of California Press, 1996.

Hunt, William. *The Puritan Moment: The Coming of Revolution in an English County.* HUP, 1983.

Hurt, R. Douglas. *American Agriculture: A Brief History.* Rev. ed. West Lafayette, Ind.: Purdue University Press, 2002.

——. *The Ohio Frontier: Crucible of the Old Northwest, 1720–1830.* IndUP, 1996.

Huston, James L. "The American Revolutionaries, the Political Economy of Aristocracy, and the American Concept of the Distribution of Wealth." *AHR* 98, no. 4 (October 1993): 1079–1105.

——. *Calculating the Value of the Union: Slavery, Property Rights, and the Economic Origins of the Civil War.* UNCP, 2003.

——. "The Illinois Political Realignment of 1844–1860: Revisiting the Analysis." *JCWE* 1 (December 2011): 506–35.

——. *The Panic of 1857 and the Coming of the Civil War.* LSUP, 1987.

——. "A Political Response to Industrialism: The Republican Embrace of Protectionist Labor Doctrines." *JAH* 70 (June 1983): 35–57.

——. *Securing the Fruits of Labor: The American Concept of Wealth Distribution, 1765–1900.* LSUP, 1998.

——. *Stephen A. Douglas and the Dilemmas of Democratic Equality.* Lanham, Md.: Rowman and Littlefield, 2007.

——. "Theory's Failure: Malthusian Population Theory and the Projected Demise of Slavery." *CWH* 55 (September 2009), 354–81.

Huston, Reeve. *Land and Reform: Rural Society, Popular Protest, and Party Politics in Antebellum New York.* OUP, 2000.

Hyde, Samuel C., Jr. "*Plain Folk* Reconsidered: Historiographical Ambiguity in Search of Definition." *JSH* 71 (November 2005): 803–30.

Innes, Stephen. *Creating the Commonwealth: The Economic Culture of Puritan New England.* New York: W. W. Norton, 1995.

——. *Labor in a New Land: Economy and Society in Seventeenth-Century Springfield.* PrinceUP, 1983.

Irwin, James R. "Exploring the Affinity of Wheat and Slavery in the Virginia Piedmont." *EEH* 25 (July 1988): 295–322.

Isaac, Rhys. *The Transformation of Virginia, 1740–1790.* 1982. Rpt. New York: W. W. Norton, 1988.

James, John A., and Jonathan S. Skinner. "The Resolution of the Labor-Scarcity Paradox." *JEH* 45 (September 1985): 513–40.

James, Margaret. *Social Problems and Policy during the Puritan Revolution, 1640–1660.* London: George Routledge & Sons, 1930.

Jenkins, William Sumner. *Pro-Slavery Thought in the Old South.* UNCP, 1935.

Jensen, Joan M. *Calling This Place Home: Women on the Wisconsin Frontier, 1850–1925.* Minneapolis: Minnesota Historical Society, 2006.

——. *Loosening the Bonds: Mid-Atlantic Farm Women, 1750–1850.* YUP, 1986.

——. "Now You See Her, Now You Don't: Historians and Rural Women." In Koons and Hofstra, eds., *After the Backcountry,* 221–32.

Johnson, Michael P. *Toward a Patriarchal Republic: The Secession of Georgia.* LSUP, 1977.

Johnson, Paul E. *The Early American Republic, 1789–1829.* OUP, 2007.

Jones, E. L. "Agriculture and Economic Growth in England, 1660–1750: Agricultural Change." *JEH* 25 (March 1965): 1–18.

——. *Agriculture and the Industrial Revolution.* New York: Wiley and Sons, 1974.

Jones, Robert Leslie. *History of Agriculture in Ohio to 1880.* KSUP, 1983.

Jordan, Terry G., and Matti Kaups. *The American Backwoods Frontier: An Ethnic and Ecological Interpretation.* JHUP, 1989.

Jordan, Winthrop D. *White Over Black: American Attitudes toward the Negro, 1550–1812.* UNCP, 1968.

Joyner, Charles. *Down by the Riverside: A South Carolina Slave Community.* 2nd ed. UIllP, 2009.

Keller, Kenneth W. "The Wheat Trade on the Upper Potomac, 1800–1860." In Koons and Hofstra, eds., *After the Backcountry,* 21–33.

Kennedy, D. E. *The English Revolution, 1642–1649.* Houndmills, U.K.: Macmillan Press, 2000.

Kennedy, Roger G. *Mr. Jefferson's Lost Cause: Land, Farmers, Slavery, and the Louisiana Purchase.* OUP, 2003.

Kenyon, J. P. *The Civil Wars of England.* New York: Alfred A. Knopf, 1988.

Kim, Sung Bok. *Landlord and Tenant in Colonial New York: Manorial Society, 1664–1775.* UNCP, 1978.

——. "A New Look at the Great Landlords of Eighteenth-Century New York." *WMQ* 27 (October 1970): 581–614.

Kirby, Jack Temple. *Poquosin: A Study of Rural Landscape and Society.* UNCP, 1995.

Klein, Herbert S. *A Population History of the United States.* CamUP, 2004.

Klein, Judith L. V. "Farm-Making Costs in the 1850s: A Comment." In James W. Whitaker, ed., *Farming in the Midwest, 1840–1900,* 71–74. Washington, D.C.: Agricultural History Society, 1974.

Kleppner, Paul. *The Third Electoral System, 1853–1892: Parties, Voters, and Political Cultures.* UNCP, 1979.

Koeniger, A. Cash. "Climate and Southern Distinctiveness." *JSH* 54 (February 1988): 21–44.

Kolchin, Peter. *American Slavery, 1619–1877.* 1993. Rpt. New York: Hill and Wang, 2003.

Koons, Kenneth E., and Warren R. Hofstra, eds. *After the Backcountry: Rural Life in the Great Valley of Virginia, 1800–1900.* UTnP, 2000.

Kulikoff, Allan. *The Agrarian Origins of American Capitalism.* UPVa, 1992.

——. "The American Revolution, Capitalism, and the Formation of the Yeoman Classes." In Alfred F. Young, ed., *Beyond the American Revolution: Explorations in the History of American Radicalism,* 80–119. DeKalb: Northern Illinois University Press, 1993.

——. *From British Peasants to Colonial American Farmers.* UNCP, 2000.

——. *Tobacco and Slaves: The Development of Southern Cultures in the Chesapeake, 1680–1800.* UNCP, 1986.

Kussmaul, Ann. *Servants in Husbandry in Early Modern England.* CamUP, 1981.

Labaree, Leonard Woods. *Conservatism in Early American History.* New York: New York University Press, 1948.

Land, Aubrey C. "Economic Behavior in a Planting Society: The Eighteenth-Century Chesapeake." *JSH* 33 (November 1967): 467–85.

Larson, John Lauritz. *The Market Revolution in America: Liberty, Ambition, and the Eclipse of the Common Good.* CamUP, 2010.

Laurie, Bruce. *Artisans into Workers: Labor in Nineteenth-Century America.* New York: Noonday Press, 1989.

Lause, Mark. *Young America: Land, Labor, and the Republican Party.* UIllP, 2005.

Lebergott, Stanley. *The Americans: An Economic Record.* New York: W. W. Norton, 1984.

——. "The Demand for Land: The United States, 1820–1860." *JEH* 45 (June 1985): 181–212.

——. *Manpower in Economic Growth: The American Record Since 1800.* New York: McGraw-Hill, 1964.

Lemon, James T. *Best Poor Man's Country: A Geographical Study of Early Southeastern Pennsylvania.* JHUP, 1972.

Levine, Bruce. *Half Slave and Half Free: The Roots of Civil War.* Rev. ed. New York: Hill and Wang, 1992.

Levy, Hermann. *Large and Small Holdings: A Study of English Agricultural Economics.* Trans. Ruth Kenyon. CamUP, 1911.

Levy, Jonathan. "The Mortgage Worked the Hardest: The Fate of Landed Independence in Nineteenth-Century America." In Zakim and Kornblith, eds., *Capitalism Takes Command,* 39–67.

Lewis, Jan. *The Pursuit of Happiness: Family and Values in Jefferson's Virginia.* CamUP, 1983.

Lewis, Johanna Miller. *Artisans in the North Carolina Backcountry.* UPKy, 1995.

Livermore, Shaw. *Early American Land Companies: Their Influence on Corporate Development.* New York: Commonwealth Fund, 1939.

Lockridge, Kenneth A. "Land, Population, and the Evolution of New England Society, 1630–1790." *Past and Present* 39 (April 1968): 62–80.

——. *A New England Town: The First Hundred Years: Dedham, Massachusetts, 1636–1736.* New York: W. W. Norton, 1970.

Lowe, Richard G., and Randolph B. Campbell. *Planters and Plain Folk: Agriculture in Antebellum Texas.* Dallas: Southern Methodist University Press, 1987.

Mackintosh, Michael Dean. "New Sweden, Natives, and Nature." In Pencak and Richter, eds., *Friends and Enemies in Penn's Woods,* 3–17.

Macpherson, C. B. *The Political Theory of Possessive Individualism: Hobbes to Locke.* OUP, 1962.

Main, Gloria L. "The Standard of Living in Southern New England, 1640–1773." *WMQ* 45 (January 1988): 124–34.

Main, Jackson Turner. "The Distribution of Property in Post-Revolutionary Virginia." *MVHR* 41 (June 1941): 241–58.

——. "The One Hundred." *WMQ* 11 (July 1954): 354–88.

——. *The Social Structure of Revolutionary America.* PrinceUP, 1965.

Majewski, John. *A House Dividing: Economic Development in Pennsylvania and Virginia Before the Civil War.* CamUP, 2000.

——. *Modernizing a Slave Economy: The Economic Vision of the Confederate Nation.* UNCP, 2009.

Malcolmson, Robert W. *Life and Labour in England, 1700–1780.* New York: St. Martin's Press, 1981.

Mancall, Peter C. *Valley of Opportunity: Economic Culture along the Upper Susquehanna, 1700–1800.* CorUP, 1991.

Mancke, Elizabeth. "The Language of Liberty in British North America, 1607–1776." In Greene, ed., *Exclusionary Empire,* 25–49.

Manegold, C. S. *Ten Hills Farm: The Forgotten History of Slavery in the North.* PrinceUP, 2009.

Mann, Susan Archer. *Agrarian Capitalism in Theory and Practice.* UNCP, 1990.

Manning, Brian. *The English People and the English Revolution, 1640–1649.* London: Heinemann, 1976.

——. "The Levellers and Religion." In McGregor and Reay, eds., *Radical Religion in the English Revolution,* 65–90.

——. "Puritanism and Democracy, 1640–1642." In Donald Pennington and Keith Thomas, eds., *Puritans and Revolutionaries: Essays in Seventeenth-Century History Presented to Christopher Hill,* 139–60. OUP, 1978.

Margo, Robert A. "The North-South Wage Gap before and after the Civil War." In David Eltis, Frank D. Lewis, and Kenneth L. Sokoloff, eds., *Slavery in the Development of the Americas,* 324–51. CamUP, 2004.

——. *Wages and Labor Markets in the United States, 1820–1860.* UChiP, 2000.

Martis, Kenneth C. "The Geographical Dimensions of a New Nation, 1780s-1820s." In Thomas E. McIlwraith and Edward K. Muller, eds., *North America: The Historical Geography of a Changing Continent,* 143–64. 2nd ed. Lanham, Md.: Rowman and Littlefield, 2001.

——. *The Historical Atlas of United States Congressional Districts, 1789–1983.* New York: Free Press, 1982.

Marx, Leo. *The Machine in the Garden: Technology and the Pastoral Ideal in America.* OUP, 1964.

Mason, Matthew. *Slavery and Politics in the Early American Republic.* UNCP, 2006.

Mathew, William M. *Edmund Ruffin and the Crisis of Slavery in the Old South: The Failure of Agricultural Reform.* UGaP, 1988.

Mathias, Peter. *The First Industrial Nation: An Economic History of Britain, 1700–1914.* 2nd ed. London: Methuen, 1983.

May, Robert F. *The Southern Dream of a Caribbean Empire, 1854–1861.* LSUP, 1973.

McClelland, Peter D. *Sowing Modernity: America's First Agricultural Revolution.* CorUP, 1997.

McCurry, Stephanie. *Masters of Small Worlds: Yeoman Households, Gender Relations, and the Political Culture of the Antebellum South Carolina Low Country.* OUP, 1995.

McCusker, John J., and Russell R. Menard. *The Economy of British America, 1607–1789.* UNCP, 1985.

McGrath, C. Peter. *Yazoo: Law and Politics in the New Republic: The Case of Fletcher v. Peck.* Providence: Brown University Press, 1966.

McGregor, J. F., and B. Reay, eds. *Radical Religion in the English Revolution.* OUP, 1984.

McKenzie, Robert Tracy. "'Oh! Ours Is a Deplorable Condition': The Economic Impact of the Civil War in Upper East Tennessee." In Kenneth W. Noe and Shannon H. Wilson, eds., *The Civil War in Appalachia: Collected Essays,* 199–226. UTnP, 1997.

——. *One South or Many? Plantation Belt and Upcountry in Civil War–Era Tennessee.* CamUP, 1994.

McMurry, Sally. *Transforming Rural Life: Dairying Families and Agricultural Change, 1820–1885.* JHUP, 1995.

McPherson, James M. "Antebellum Southern Exceptionalism: A New Look at an Old Question." *CWH* 29 (September 1983): 230–44.

——. "The Civil War and the Transformation of America." In William J. Cooper Jr. and John M. McCardell Jr., eds., *The Cause of Liberty: How the Civil War Redefined American Ideals,* 1–8. LSUP, 2009.

——. *This Mighty Scourge: Perspectives on the Civil War.* OUP, 2007.

——, and James K. Logue. *Ordeal By Fire: The Civil War and Reconstruction.* 4th ed. Boston: McGraw-Hill Higher Education, 2009.

McWhiney, Grady, and Perry D. Jamieson. *Attack and Die: Civil War Military Tactics and the Southern Heritage.* UAlaP, 1982.

Meinig, D. W. *The Shaping of America: A Geographical Perspective on 500 Years of History.* 4 vols. YUP, 1988–2006.

Menard, Russel R. "Economic and Social Development of the South." In Stanley L. Engerman and Robert E. Gallman, eds., *The Cambridge Economic History of the United States* 2: *The Colonial Era,* 249–95. CamUP, 1996–2000.

——. "From Servant to Freeholder: Status. Mobility and Property Accumulation in Seventeenth Century Maryland." *WMQ* 30 (January 1973): 40–64.

Menn, Joseph Karl. *The Large Slaveholders of Louisiana—1860.* New Orleans: Pelican Publishing, 1964.

Merrill, Michael. "Putting 'Capitalism' in Its Place: A Review of the Recent Literature." *WMQ* 52 (April 1995): 315–26.

Miller, Gary J. *Managerial Dilemmas: The Political Economy of Hierarchy.* CamUP, 1992.

Miller, Marla R. *The Needle's Eye: Women and Work in the Age of Revolution.* UMassP, 2006.

Mills, Dennis R. *Lord and Peasant in Nineteenth Century Britain.* London: Croom Helm, 1980.

Milton, George Fort. *The Eve of Conflict: Stephen A. Douglas and the Needless War.* Boston: Houghton Mifflin, 1934.

Mingay, G. E. "The Eighteenth-Century Land Steward." In E. L. Jones and G. E. Mingay, eds., *Land, Labour, and Population in the Industrial Revolution: Essays Presented to J. D. Chambers,* 3–27. London: Edward Arnold, 1967.

——. *English Landed Society in the Eighteenth Century.* London: Routledge and Kegan Paul, 1963.

——. "The Farmer." *AgHEWVIIpt1,* 759–809.

——. *The Gentry: The Rise and Fall of a Ruling Class.* London: Longman, 1976.

——. "The Size of Farms in the Eighteenth Century." *EHR* 14 (1962): 469–88.

Mitchell, B. R. *European Historical Statistics, 1750–1970.* New York: Columbia University Press, 1975.

——, with Phyllis Deane. *Abstract of British Historical Statistics.* CamUP, 1971.

Mitchell, Betty L. *Edmund Ruffin: A Biography.* IndUP, 1981.

Mitchell, Reid. *Civil War Soldiers: Their Expectations and Their Experiences.* New York: Viking, 1998.

Mitchell, Robert D. *Commercialism and Frontier: Perspectives on the Early Shenandoah Valley.* UPVa, 1977.

Montgomery, David. *Beyond Equality: Labor and the Radical Republicans, 1862–1872.* New York: Random House, 1967.

Moore, Barrington, Jr. *Social Origins of Dictatorship and Democracy.* Boston: Beacon Press, 1966.

Moore, John Hebron. *The Emergence of the Cotton Kingdom in the Old Southwest: Mississippi, 1770–1860.* LSUP, 1988.

Morgan, Edmund S. *American Slavery, American Freedom: The Ordeal of Colonial Virginia.* New York: W. W. Norton, 1975.

——. *Inventing the People: The Rise of Popular Sovereignty in England and America.* New York: W. W. Norton, 1988.

——. "The Puritan Ethic and the American Revolution." *WMQ* 24 (January 1967): 3–43.

Morgan, Marjorie. *Manners, Morals and Class in England, 1774–1858.* New York: St. Martin's Press, 1994.

Morgan, Philip D. *Slave Counterpoint: Black Culture in the Eighteenth-Century Chesapeake and Lowcountry.* UNCP, 1998.

Morrill, J. S. *The Revolt of the Provinces: Conservatives and Radicals in the English Civil War, 1630–1650.* London: George Allen and Unwin, 1976.

Morrison, Michael A. *Slavery and the American West: The Eclipse of Manifest Destiny and the Coming of the Civil War.* UNCP, 1978.

Moyer, Paul. "'Real' Indians, 'White' Indians, and the Contest for the Wyoming Valley." In Pencak and Richter, eds., *Friends and Enemies in Penn's Woods,* 221–37.

Murphy, James B. *L. Q. C. Lamar: Pragmatic Patriot.* LSUP, 1973.

Nash, Gary B. *Quakers and Politics: Pennsylvania, 1681–1726.* PrinceUP, 1968.

Nation, Richard F. *At Home in the Hoosier Hills: Agriculture, Politics, and Religion in Southern Indiana, 1810–1870.* IndUP, 2005.

Nettels, Curtis P. *The Emergence of a National Economy, 1775–1815.* New York: Harper and Row, 1962.

Noe, Kenneth W. *Southwest Virginia's Railroad: Modernization and the Sectional Crisis.* UIllP, 1994.

Nugent, Nell Marion. *Cavaliers and Pioneers: Abstracts of Virginia Land Patents and Grants, 1623–1800.* 5 vols. Richmond, Va.: Dietz Printing Co., 1934.

Oakes, James. "Natural Rights, Citizenship Rights, States Rights, and Black Rights: Another Look at Lincoln and Race." In Eric Foner, ed., *Our Lincoln: New Perspectives on Lincoln and His World,* 109–34. New York: W. W. Norton, 2008.

———. *The Ruling Race: A History of American Slaveholders.* New York: W. W. Norton, 1982.

———. *Slavery and Freedom: An Interpretation of the Old South.* New York: Alfred A. Knopf, 1990.

O'Brien, Michael. *Conjectures of Order: Intellectual Life and the American South, 1810–1860.* 2 vols. UNCP, 2004.

Oertel, Kristen Tegtmeier. *Bleeding Borders: Race, Gender, and Violence in Pre–Civil War Kansas.* LSUP, 2009.

Olmstead, Alan L., and Paul W. Rhode. *Creating Abundance: Biological Innovation and American Agricultural Development.* CamUP, 2008.

———. "Reshaping the Landscape: The Impact and Diffusion of the Tractor in American Agriculture, 1910–1960." *JEH* 56 (September 2001): 663–98.

Olsen, Christopher J. *Political Culture and Secession in Mississippi: Masculinity, Honor, and the Antiparty Tradition, 1830–1860.* OUP, 2000.

O'Mara, James. *An Historical Geography of Urban System Development: Tidewater Virginia in the 18th Century.* York, Canada: Geographical Monographs, No. 13, 1983.

Opie, John. *The Law of the Land: Two Hundred Years of American Farmland Policy.* Lincoln: University of Nebraska Press, 1987.

Orwin, Christabel S., and Edith H. Whetham. *History of British Agriculture, 1846–1914.* London: Archon Books, 1964.

Osterud, Nancy Grey. *Bonds of Community: The Lives of Farm Women in Nineteenth-Century New York.* CorUP, 1991.

———. "Gender and the Transition to Capitalism in Rural America." *AgH* 67 (Spring 1994): 14–29.

Osthaus, Carl R. "The Work Ethic of the Plain Folk: Labor and Religion in the Old South." *JSH* 70 (November 2004): 745–82.

Overton, Mark. *Agricultural Revolution in England: The Transformation of the Agrarian Economy, 1500–1850.* CamUP, 1996.

Owsley, Frank Lawrence. *Plain Folk of the Old South.* LSUP, 1949.

Papenfuse, Edward C., Jr. "Planter Behavior and Economic Opportunity in a Staple Economy." *AgH* 46 (April 1972): 297–312.

Parent, Anthony S., Jr. *Foul Means: The Formation of a Slave Society in Virginia, 1660–1740.* UNCP, 2003.

Parish, Peter J. *The American Civil War.* New York: Holmes and Meier, 1975.

——. *The North and the Nation in the Era of the Civil War.* Ed. Adam I. P. Smith and Susan-Mary Grant. New York: Fordham University Press, 2003.

Parker, William N. "The Magic of Property." *AgH* 54 (October 1980): 477–89.

Parkerson, Donald H. *The Agricultural Transition in New York State: Markets and Migration in Mid-Nineteenth-Century America.* IaUP, 1995.

Parry, J. H. "Transport and Trade Routes." In E. E. Rich and C. H. Wilson, eds., *The Cambridge Economic History of Europe.* Vol. 4: *The Economy of Expanding Europe in the Sixteenth and Seventeenth Centuries,* 155–219. CamUP, 1967.

Paskoff, Paul F. "A Free Soil Fallacy: The Myth of the Geographical Incompatibility of Free and Slave Labor during the 1850s." Paper presented at Mid-America Conference on History, September 23, 2011, Stillwater, Okla.

——. *Troubled Waters: Steamboat Disasters, River Improvements, and American Public Policy, 1821–1860.* LSUP, 2007.

Pattison, William D. *Beginnings of the American Rectangular Land Survey System, 1784–1800.* UChiP, 1957.

Pencak, William A., and Daniel K. Richter, eds. *Friends and Enemies in Penn's Woods: Indians, Colonists, and the Racial Construction of Pennsylvania.* University Park: Pennsylvania State University Press, 2004.

Perry, P. J. "High Farming in Victorian Britain: Prospect and Retrospect." *AgH* 55 (April 1981): 156–66.

Phillips, Sarah T. "Antebellum Agricultural Reform, Republican Ideology, and Sectional Tension." *AgH* 74 (Fall 2000): 799–822.

Phillips, Ulrich Bonnell. *American Negro Slavery: A Survey of the Supply, Employment and Control of Negro Labor as Determined by the Plantation Regime.* 1918. Rpt. New York: Peter Smith, 1952.

——. *Life and Labor in the Old South.* Boston: Little, Brown, and Co., 1929.

Plumb, J. H. *England in the Eighteenth Century.* Middlesex, U.K.: Penguin Books, 1950.

——. *Men and Centuries.* Boston: Houghton Mifflin, 1963.

Porter, J. H. "The Development of Rural Society." In *AgHEW6,* 836–937.

Porter, Roy. *English Society in the Eighteenth Century.* Rev. ed. New York: Penguin Books, 1990.

Post, Charles. "The Agrarian Origins of U.S. Capitalism: The Transformation of the Northern Countryside before the Civil War." *Journal of Peasant Studies* 22 (April 1995): 389–445.

Potter, David M. *The Impending Crisis, 1848–1861.* Ed. Don E. Fehrenbacher. New York: Harper and Row, 1976.

Pred, Allan. *Urban Growth and City-Systems in the United States, 1840–1860.* HUP, 1980.

Pressly, Thomas. *Americans Interpret Their Civil War.* New York: Free Press, 1962.

Preston, David L. "Squatters, Indians, Proprietary Government, and Land in the Susquehanna Valley." In Pencak and Richter, eds., *Friends and Enemies in Penn's Woods,* 180–200.

Primack, Martin L. "Farm Fencing in the Nineteenth Century." *JEH* 29 (June 1969): 287–91.

——. "Land Clearing Under Nineteenth-Century Techniques." *JEH* 22 (December 1962): 484–97.

Prothero, Rowland E. *English Farming Past and Present.* London: Longmans, Green, & Co., 1912.

Pudup, Mary Beth. "The Limits of Subsistence: Agriculture and Industry in Central Appalachia." *AgH* 64 (Winter 1990): 61–89.

Randall, James G. "The Blundering Generation." *MVHR* 27 (June 1940): 3–28.

Ransom, Roger L. *Conflict and Compromise: The Political Economy of Slavery, Emancipation, and the American Civil War.* CamUP, 1989.

——, and Richard Sutch. *One Kind of Freedom: The Economic Consequences of Emancipation.* 2nd ed. CamUP, 2001.

Rasmussen, Wayne D. "The Impact of Technological Change on American Agriculture, 1862–1962." *JEH* 22 (December 1962): 579–84.

Ratcliffe, Donald J. "The Changing Political World of Thomas Worthington." In Cayton and Hobbs, eds., *Center of a Great Empire,* 36–61.

Read, James H. *Majority Rule versus Consensus: The Political Thought of John C. Calhoun.* UPKan, 2009.

Reay, B. "Quakerism and Society." In McGregor and Reay, eds., *Radical Religion in the English Revolution,* 141–64.

——. "Radicalism and Religion in the English Revolution: An Introduction." In McGregor and Reay, eds., *Radical Religion in the English Revolution,* 1–21.

Richards, Leonard L. *The Slave Power: The Free North and Southern Domination, 1780–1860.* LSUP, 2000.

Richardson, Heather Cox. *The Death of Reconstruction: Race, Labor, and Politics in the Post-Civil War North, 1865–1901.* HUP, 2001.

——. *The Greatest Nation of the Earth: Republican Economic Policies during the Civil War.* HUP, 1997.

Richardson, R. C. *The Debate on the English Revolution.* Manchester, U.K.: Manchester University Press, 1998.

Rikoon, J. Sanford. *Threshing in the Midwest, 1820–1940.* IndUP, 1985.

Riley, Glenda. *The Female Frontier: A Comparative View of Women on the Prairie and the Plains.* UPKan, 1988.

Robbins, Roy M. *Our Landed Heritage: The Public Domain, 1776–1936.* 1942. Rpt. New York, Peter Smith, 1950.

Robert, Joseph Clarke. *The Road from Monticello: A Study of the Virginia Slavery Debate of 1832.* DukeUP, 1941.

Robertson, Stacey M. *Parker Pillsbury: Radical Abolitionist, Male Feminist.* CorUP, 2000.

Robinson, W. Stitt. *Mother Earth: Land Grants in Virginia, 1607–1699.* Williamsburg: Virginia 350th Anniversary Celebration Corporation, 1957.

——. *The Southern Colonial Frontier, 1607–1763.* Albuquerque: University of New Mexico Press, 1979.

Rockman, Seth. *Scraping By: Wage Labor, Slavery, and Survival in Early Baltimore.* JHUP, 2009.

Roemer, John E. *Analytical Foundations of Marxian Economic Theory.* CamUP, 1981.

Rohrbough, Malcolm J. *The Trans-Appalachian Frontier: People, Societies, and Institutions, 1775–1850.* OUP, 1978.

Rollison, David. "The Specter of the Commonalty: Class Struggle and the Commonweal in England before the Atlantic World." *WMQ* 63 (April 2006): 221–52.

Rorabaugh, J. J. *The Real Making of the President: Kennedy, Nixon, and the 1960 Election.* UPKan, 2009.

Roth, Randolph A. *The Democratic Dilemma: Religion, Reform, and the Social Order in the Connecticut River Valley of Vermont, 1791–1850.* CamUP, 1987.

Rothenberg, Winifred Barr. "The Emergence of Farm Labor Markets and the Transformation of the Rural Economy: Massachusetts, 1750–1855." *JEH* 48 (September 1988): 537–66.

——. *From Market-Places to a Market Economy: The Transformation of Rural Massachusetts, 1750–1850.* UChiP, 1992.

Rothman, Adam. *Slave Country: American Expansion and the Origins of the Deep South.* HUP, 2005.

Rothman, Joshua D. *Flush Times and Fever Dreams: A Story of Capitalism and Slavery in the Age of Jackson.* UGaP, 2012.

Rozbicki, Michal J. "The Curse of Provincialism: Negative Perceptions of Colonial Plantation Gentry." *JSH* 63 (November 1997): 727–52.

Rugh, Susan Sessions. *Our Common Country: Family Farming, Culture, and Community in the Nineteenth-Century Midwest.* IndUP, 2001.

Russell, Conrad. *The Causes of the English Civil War.* OUP, 1990.

——. *The Crisis of Parliaments: English History, 1509–1660.* OUP, 1971.

Russell, Howard S. *A Long Deep Furrow: Three Centuries of Farming in New England.* Hanover, N.H.: University Press of New England, 1976.

Russo, Jean B. "'The Fewnesse of Handicraftsmen': Artisan Adaptation and Innovation on the Colonial Chesapeake." In Robert Olwell and Alan Tully, eds., *Cultures and Identities in Colonial British America,* 169–96. JHUP, 2006.

——. *Free Workers in a Plantation Economy: Talbot County, Maryland, 1690–1759.* New York: Garland, 1989.

——. "A Model Planter: Edward Lloyd of Maryland." *WMQ* 69 (January 1992): 62–88.

——. "Self-Sufficiency and Local Exchange: Free Craftsmen in the Rural Chesapeake Economy." In Lois Green Carr, Philip D. Morgan, and Jean B. Russo, eds., *Colonial Chesapeake Society,* 389–432. UNCP, 1988

Rutman, Darrett. *Winthrop's Boston: Portrait of a Puritan Town, 1630–1649.* UNCP, 1965.

Salstrom, Paul. *Appalachia's Path to Dependency: Rethinking a Region's Economic History, 1730–1940.* UPKy, 1994.

Sanderson, John. *"But the People's Creatures": The Philosophical Basis of the English Civil War.* Manchester, U.K.: Manchester University Press, 1989.

Sarson, Steven. "Yeoman Farmers in a Planter's Republic: Socioeconomic Condition and Relations in Early National Prince George's County, Maryland." *JER* 29 (Spring 2009): 63–99.

Scarborough, William Kaufman. *Masters of the Big House: Elite Slaveholders of the Mid-Nineteenth-Century South.* LSUP, 2003.

——. *The Overseer: Plantation Management in the Old South.* LSUP, 1966.

Schaefer, Donald. "Yeoman Farmers and Economic Democracy: A Study of Wealth and Economic Mobility in the Western Tobacco Region, 1850 to 1860." *EEH* 15 (October 1978): 421–37.

Schafer, Joseph. *The Social History of American Agriculture.* New York: Macmillan, 1936.

Schenck, W. *The Concern for Justice in the Puritan Revolution.* London: Longmans, Green & Co., 1948.

Schlebecker, John T. "Farmers in the Lower Shenandoah Valley, 1850." *VMHB* 79 (October 1971): 462–76.

Schmidt, Albert J. *The Yeoman in Tudor and Stuart England.* Washington, D.C.: Folger Shakespeare Library, 1961.

Schmidt, James D. *Free to Work: Labor Law, Emancipation, and Reconstruction.* UGaP, 1998.

Schob, David E. *Hired Hands and Plowboys: Farm Labor in the Midwest, 1815–60.* UIllP, 1975.

Schoen, Brian. *The Fragile Fabric of Union: Cotton, Federal Politics and the Global Origins of the Civil War.* JHUP, 2009.

Schultz, Ronald. *The Republic of Labor: Philadelphia Artisans and the Politics of Class, 1720–1830.* OUP, 1993.

Sellers, James Benson. *Slavery in Alabama.* UAlaP, 1950.

Shannon, Fred A. *The Farmer's Last Frontier: Agriculture, 1860–1897.* New York: Harper, 1945.

Silbey, Joel H. *Storm Over Texas: The Annexation Controversy and the Road to War.* OUP, 2005.

Silver, Timothy. *A New Face in the Countryside: Indians, Colonists, and Slaves in South Atlantic Forests, 1500–1800.* CamUP, 1990.

Simler, Lucy. "Tenancy in Colonial Pennsylvania: The Case of Chester County." *WMQ* 43 (October 1986): 542–69.

Sinha, Manisha. *The Counterrevolution of Slavery: Politics and Ideology in Antebellum South Carolina.* UNCP, 2000.

Smith, Henry Nash. *Virgin Land: The American West as Symbol and Myth.* HUP, 1950.

Smith, Lacey Baldwin. *This Realm of England, 1399 to 1688.* Lexington, Mass.: D. C. Heath and Co., 1966.

Socolofsky, Homer E. *Landlord William Scully.* Lawrence: Regents Press of Kansas, 1979.

Soltow, Lee, and Kenneth W. Keller. "Rural Pennsylvania in 1800: A Portrait from the Septennial Census." *Pennsylvania History* 49 (January 1982): 25–47.

Spangler, Jewel L. *Virginians Reborn: Anglican Monopoly, Evangelical Dissent, and the Rise of the Baptists in the Late Eighteenth Century.* Charlottesville: University of Virginia Press, 2008.

Spring, David. *The English Landed Estate in the Nineteenth Century: Its Administration.* JHUP, 1963.

Spufford, Margaret. *Contrasting Communities: English Villagers in the Sixteenth and Seventeenth Centuries.* CamUP, 1974.

Stampp, Kenneth M. *The Peculiar Institution: Slavery in the Ante-Bellum South.* New York: Alfred A. Knopf, 1956.

———, ed. *The Causes of the Civil War.* New York: Touchstone, 1986.

Stanley, Amy Dru. *From Bondage to Contract: Wage Labor, Marriage, and the Market in the Age of Slave Emancipation.* CamUP, 1998.

Startup, Kenneth Moore. *The Root of All Evil: The Protestant Clergy and the Economic Mind of the Old South.* UGaP, 1997.

Steinfeld, Robert J. *Coercion, Contract, and Free Labor in the Nineteenth Century.* CamUP, 2001.

——, and Stanley L. Engerman. "Labor—Free or Coerced? A Historical Reassessment of Differences and Similarities." In Tom Brass and Marcel Van der Linden, eds., *Free and Unfree Labour: The Debate Continues*, 107–26. Bern: Peter Land, 1997.

Stewart, James B. *Abolitionist Politics and the Coming of the Civil War.* UMassP, 2008.

——. *Holy Warriors: The Abolitionists and American Slavery.* 2nd ed. New York: Hill and Wang, 1997.

Stiverson, Gregory A. *Poverty in a Land of Plenty: Tenancy in Eighteenth-Century Maryland.* JHUP, 1977.

Stoll, Steven. *Larding the Lean Earth: Soil and Society in Nineteenth-Century America.* New York: Hill and Wang, 2002.

Stone, Lawrence. *The Causes of the English Revolution, 1529–1642.* New York: Harper and Row, 1972.

——. *The Crisis of the Aristocracy, 1558–1641.* OUP, 1965.

——, and Jeanne C. Fawtier Stone. *An Open Elite? England 1540–1880.* OUP, 1984.

Summerhill, Thomas. *Harvest of Dissent: Agrarianism in Nineteenth-Century New York.* UIllP, 2005.

Swierenga, Robert. "Land Speculator 'Profits' Reconsidered: Central Iowa as a Test Case." *JEH* 26 (March 1966): 1–28.

Sydnor, Charles Sackett. *Slavery in Mississippi.* New York: D. Appleton–Century, 1933.

Tate, Adam L. *Conservatism and Southern Intellectuals, 1789–1861: Liberty, Tradition, and the Good Society.* UMoP, 2005.

Tawney, R. H. *Religion and the Rise of Capitalism: A Historical Study.* 1926. Rpt. Gloucester, Mass.: Peter Smith, 1962.

Taylor, Alan. *American Colonies.* New York: Viking Penguin, 2001.

——. *The Divided Ground: Indians, Settlers, and the Northern Borderland of the American Revolution.* New York: Alfred A. Knopf, 2006.

——. *Liberty Men and Great Proprietors: The Revolutionary Settlement on the Maine Frontier, 1760–1820.* UNCP, 1990.

——. *William Cooper's Town: Power and Persuasion on the Frontier of the Early American Republic.* New York: Alfred A. Knopf, 1995.

Taylor, Joe Gray. *Negro Slavery in Louisiana.* LSUP, 1963.

Temin, Peter. "Labor Scarcity in America." *Journal of Interdisciplinary History* 1 (Winter 1971): 251–64.

Tewell, Jeremy. *A Self-Evident Lie: Southern Slavery and the Threat to American Freedom.* KSUP, 2013.

Thirsk, Joan. *English Peasant Farming: The Agrarian History of Lincolnshire from Tudor to Recent Times.* 1957. Rpt. Oxford, U.K.: Routledge, 2006.

Thompson, F. M. L. *English Landed Society in the Nineteenth Century.* London: Routledge and Kegan Paul, 1963.

——. "Land Ownership and Economic Growth in England in the Eighteenth Century." In E. L. Jones and Woolf, eds., *Agrarian Change and Economic Development,* 41–60. London: Methuen & Co., 1969.

——. "The Social Distribution of Landed Property in England Since the Sixteenth Century." *EHR* 19 (1966): 505–17.

Thornton, J. Mills, III. *Politics and Power in a Slave Society: Alabama, 1800–1860.* LSUP, 1978.

Tillson, Albert H., Jr. *Gentry and Common Folk: Political Culture on a Virginia Frontier, 1740–1789.* UPKy, 1991.

Tise, Larry E. *Proslavery: A History of the Defense of Slavery in America, 1701–1840.* UGaP, 1987.

Towers, Frank. "Partisans, New History, and Modernization: The Historiography of the Civil War's Genesis." *JCWE* 1 (June 2011): 237–64.

——. "The Southern Path to Modern Cities: Urbanization in the Slave States." In L. Diane Barnes, Brian Schoen, and Frank Towers, eds., *The Old South's Modern Worlds: Slavery, Religion, and Nation in the Age of Progress,* 145–64. OUP, 2011.

——. *The Urban South and the Coming of the Civil War.* UPVa, 2004.

Tracy, Susan J. *In the Master's Eye: Representations of Women, Blacks, and Poor Whites in Antebellum Southern Literature.* UMassP, 1995.

Tuchinsky, Adam. *Horace Greeley's New-York Tribune: Civil War–Era Socialism and the Crisis of Free Labor.* CorUP, 2009.

Turberville, A. S. *The House of Lords in the Age of Reform, 1784–1837: With an Epilogue on Aristocracy and the Advent of Democracy, 1837–1867.* 1958. Rpt. Westport, Conn.: Greenwood Press, 1974.

Ulrich, Laurel Thatcher. *The Age of Homespun: Objects and Stories in the Creation of an American Myth.* New York: Alfred A. Knopf, 2001.

Underdown, David. *Pride's Purge: Politics in the Puritan Revolution.* OUP, 1971.

——. *Revel, Riot, and Rebellion: Popular Politics and Culture in England, 1603–1660.* OUP, 1985.

Van Bath, B., and H. Slicher. *The Agrarian History of Western Europe, A.D. 500–1850.* Trans. Olive Ordish. New York: St. Martin's Press, 1963.

Van Vugt, William E. *Britain to America: Mid-Nineteenth-Century Immigrants to the United States.* UIllP, 1999.

Vickers, Daniel. *Farmers and Fishermen: Two Centuries of Work in Essex County, Massachusetts, 1630–1830.* UNCP, 1994.

——. "The Northern Colonies: Economy and Society, 1600–1775." In Stanley L. Engerman and Robert E. Gallman, eds., *Cambridge Economic History of the United States.* Vol. 1: *The Colonial Era,* 209–48. CamUP, 1996–2000.

Volpe, Vernon L. *Forlorn Hope of Freedom: The Liberty Party in the Old Northwest, 1838–1848.* KSUP, 1990.

Voorhis, Manning C. "The Land Grant Policy of Colonia Virginia, 1607–1774." PhD diss., University of Virginia, n.d.

De Vries, Jan. *The Economy of Europe in an Age of Crisis, 1600–1750.* CamUP, 1976.

Walsh, Lorena S. "Land, Landlord, and Leaseholder: Estate Management and Tenant Fortunes in Southern Maryland, 1642–1820. *AgH* 59 (July 1985): 373–96.

——. *Motives of Honor, Pleasure, and Profit: Plantation Management in the Colonial Chesapeake, 1607–1763.* UNCP, 2010.

——. "Plantation Management in the Chesapeake, 1620–1820." *JEH* 49 (June 1989): 393–406.

——. "Slave Life, Slave Society, and Tobacco Production in the Tidewater Chesapeake, 1620–1820." In Ira Berlin and Philip D. Morgan, eds., *Cultivation and Culture: Labor and the Shaping of Slave Life in the Americas,* 170–99. UPVa, 1993.

Walther, Eric H. *The Shattering of the Union: America in the 1850s.* New York: Scholarly Books, 2004.

Ware, Norman J. *The Labor Movement in the United States, 1860–1895: A Study in Democracy.* New York: D. Appleton and Co., 1929.

Weaver, Herbert. *Mississippi Farmers, 1850–1860.* Nashville: Vanderbilt University Press, 1945.

Weber, Max. *The Protestant Ethic and the Spirit of Capitalism.* Trans. Talcott Parsons. New York: Charles Scribner's Sons, 1958.

Weiman, David F. "The Economic Emancipation of the Non-slaveholding Class: Upcountry Farmers in the Georgia Cotton Economy." *JEH* 45 (March 1985): 71–93.

——. "Farmers and the Market in Antebellum America: A View from the Georgia Upcountry." *JEH* 45 (March 1985): 71–93.

Weingast, Barry R. "Political Stability and Civil War: Institutions, Commitment, and American Democracy." In Robert Bates et al., *Analytical Narratives,* 148–93. PrinceUP, 1998

Wentworth, Edward Norris. *America's Sheep Trails: History, Personalities.* Ames: Iowa State College Press, 1948.

Wermuth, Thomas S. *Rip Van Winkle's Neighbors: The Transformation of Rural Society in the Hudson River Valley, 1720–1850.* Albany: State University of New York Press, 2001.

Wertenbacker, Thomas J. *The Planters of Colonial Virginia.* 1922. Rpt. New York: Russell and Russell, 1959.

West, Stephen A. *From Yeoman to Redneck in the South Carolina Upcountry, 1850–1915.* UPVa, 2008.

Williams, Michael. *Americans and Their Forests: A Historical Geography.* CamUP, 1989.

Williamson, Jeffrey G., and Peter H. Lindert. *American Inequality: A Macroeconomic History.* New York: Academic Press, 1980.

Wilson, Harold Fisher. *The Hill Country of Northern New England: Its Social and Economic History, 1790–1830.* New York: Columbia University Press, 1936.

Winger, Stewart. *Lincoln, Religion, and Romantic Cultural Politics.* DeKalb: Northern Illinois University Press, 2003.

Winkle, Kenneth J. *The Young Eagle: The Rise of Abraham Lincoln.* Dallas: Taylor Trade Publishing, 2001.

Winters, Donald L. "'Plain Folk' of the Old South Reexamined: Economic Democracy in Tennessee." *JSH* 53 (November 1987): 565–86.

Wolf, Eva Sheppard. "Early Free-Labor Thought and the Contest Over Slavery in the Early Republic." In John Craig Hammond and Matthew Mason, eds., *Contesting Slavery: The Politics of Bondage and Freedom in the New American Nation,* 32–48. UPVa, 2011.

Wolff, Richard D., and Stephen A. Resnick. *Economics: Marxian versus Neoclassical.* JHUP, 1987.

Woloch, Nancy. *Women and the American Experience: A Concise History.* 2 vols. New York: McGraw-Hill, 1996.

Wood, Betty. *The Origins of American Slavery: Freedom and Bondage in the English Colonies.* New York: Hill and Wang, 1997.

Woodward, C. Vann. "The Southern Ethic in a Puritan World." *WMQ* 25 (July 1968): 343–70.

Woolrych, Austin. *Britain in Revolution, 1625–1660.* OUP, 2002.

Wright, Gavin. *The Political Economy of the Cotton South: Households, Markets, and Wealth in the Nineteenth Century.* New York: W. W. Norton, 1978.

——. *Slavery and American Economic Development.* LSUP, 2006.

——. "Slavery and the Cotton Boom." *EEH* 12 (October 1975): 439–51.

Wright, Louis B. *Middle-Class Culture in Elizabethan England.* 1935. Rpt. CorUP, 1958.

Wrightson, Keith. *Earthly Necessities: Economic Lives in Early Modern Britain.* YUP, 2000.

——. *English Society, 1580–1680.* New Brunswick, N.J.: Rutgers University Press, 1982.

Wrigley, E. A. *Poverty, Progress, and Population.* CamUP, 2004.

Wyatt-Brown, Bertram. *Southern Honor: Ethics and Behavior in the Old South.* OUP, 1982.

——. *Yankee Saints and Southern Sinners.* LSUP, 1985.

Wyckoff, William. *The Developer's Frontier: The Making of the Western New York Landscape.* YUP, 1988.

Wyllie, Irvin G. *The Self-Made Man in America: The Myth of Rags to Riches.* 1954. Rpt. New York: Free Press, 1966.

Yarbrough, W. H. *Economic Aspects of Slavery in Relation to Southern and Southwestern Migration*. Nashville: George Peabody College for Teachers, 1932.

Zagorin, Perez. *The Court and the Country: The Beginning of the English Revolution*. New York: Atheneum, 1970.

——. *Rebels and Rulers, 1500–1600*. 2 vols. CamUP, 1982.

Zakim, Michael. *Ready-Made Democracy: A History of Men's Dress in the American Republic, 1760–1860*. UChiP, 2003.

——, and Gary J. Kornblith, eds., *Capitalism Takes Command: The Social Transformation of Nineteenth-Century America*. UChiP, 2012.

——. "Introduction: An American Revolutionary Tradition." In Zakim and Kornblith, eds., *Capitalism Takes Command*, 1–12.

Ziegler, Edith M. "'The Burdens and the Narrow Life of Farm Women': Women, Gender, and Theodore Roosevelt's Commission on Country Life." *AgH* 86 (Summer 2012): 77–103.

INDEX